Food Allergy
Molecular and Clinical Practice

Food Allergy
Molecular and Clinical Practice

Editor

Andreas L. Lopata

James Cook University
College of Public Health, Medical & Veterinary Sciences
Centre of Biodiscovery and Molecular Development of Therapeutics
Douglas, Queensland, Australia

CRC Press
Taylor & Francis Group
Boca Raton London New York

CRC Press is an imprint of the
Taylor & Francis Group, an **informa** business

A SCIENCE PUBLISHERS BOOK

Cover photograph reproduced by kind courtesy of Dr. Sandip Kamath.

CRC Press
Taylor & Francis Group
6000 Broken Sound Parkway NW, Suite 300
Boca Raton, FL 33487-2742

First issued in paperback 2020

ISBN-13: 978-1-4987-2244-5 (hbk)
ISBN-13: 978-0-367-78199-6 (pbk)

Library of Congress Cataloging-in-Publication Data

Names: Lopata, Andreas Ludwig, editor.
Title: Food allergy : molecular and clinical practice / editor, Andreas Ludwig Lopata.
Other titles: Food allergy (Lopata)
Description: Boca Raton, FL : CRC Press, 2017. | "A Science Publishers book."
| Includes bibliographical references and index.
Identifiers: LCCN 2017005276| ISBN 9781498722445 (hardback : alk. paper) | ISBN 9781498722452 (e-book)
Subjects: | MESH: Food Hypersensitivity--therapy | Allergens--chemistry
Classification: LCC RC596 | NLM WD 310 | DDC 616.97/5--dc23
LC record available at https://lccn.loc.gov/2017005276

Visit the Taylor & Francis Web site at
http://www.taylorandfrancis.com

and the CRC Press Web site at
http://www.crcpress.com

Preface

Allergy-related diseases are today recognized as reaching epidemic proportions, with up to 30% of the general population suffering from clinical symptoms ranging from urticaria, rhinitis and asthma to life-threatening anaphylactic reactions.

The main contributors to the increasing prevalence of allergy seem to be very diverse including increasing immunological predisposition ('atopy'), changing food consumption and well as living conditions. The dramatic increase of allergic diseases is not only seen in the developed world, but increasing evidence indicates that also developing countries are considerably affected. Already over fifty percent of the world population is living in Asia, where not only food consumption, but also food allergies are very different from what is mainly published from Western countries. In the research efforts in the field of food allergy two main questions are often asked: What makes one person allergic to a particular food and not the other? Furthermore, Why are some foods and food proteins more allergenic than others? In addition it is very difficult to predict the severity of clinical reaction and the amount of allergen required to elicit these reactions.

Major food allergens from a small number of sources were identified and purified as early as the 1970s. A boost in the number of newly identified allergens was elicited by the general availability of recombinant DNA technology in the late 1980s. The ever-growing IUIS Allergen Nomenclature Database contains currently over 840 allergens from 252 sources and their isoforms and variants. Currently we know about 290 food allergens from 98 different food sources.

Recent developments into the molecular nature of allergenic proteins enabled us to classify most allergens into few protein families with limited biochemical function. Allergenic proteins can be classified into approximately 130 Pfam protein families, while the most important plant and animal food allergens can be found in 8 protein superfamilies and is discussed in detail in Chapters 1 and 2.

The correct diagnosis of a food allergy can be complex, but includes a convincing clinical history as well as the presence of elevated levels of specific IgE antibody to allergenic proteins in a given food. Therefore, detailed knowledge about the food specific allergenic proteins is central to a specific and sensitive diagnostic approach. The different allergens of peanut, egg, fish, shellfish and food contamination parasites and their diagnostic application are detailed in Chapters 3 to 7.

The food industry is one of the largest employers of workers with about 10% and therefore is the allergic sensitisation to food borne proteins at the workplace not surprising. Workers at increased risk of allergic sensitisation include farmers who grow and harvest crops; factory workers involved in food processing, storage and packing; as well as those involved in food preparation (chefs and waiters) and transport and is detailed in Chapter 8.

Research in food allergies and allergens is much more complex than investigating inhalant allergens since food proteins often undergo extensive modifications during food processing. Furthermore these allergenic proteins are embedded in a complex matrix and may undergo physicochemical changes during digestion and subsequent uptake by the gut mucosal barrier and presentation to the immune system, and have been highlighted in Chapter 9.

Furthermore, food processing results often in water-insoluble proteins, which makes the traditional serological analysis of allergenicity difficult as well as detection and quantification in the food matrix. The approaches and problems of quantifying allergen residues in processed food are detailed in Chapter 10.

To characterize allergens better but also develop better diagnostic and therapeutics, recombinant allergens are increasingly utilized.

Unlike natural allergens or allergen extracts, the production of recombinant proteins is not dependent on biological source material composed of complex mixtures of allergen isoforms. The use of recombinant allergens has revolutionized diagnosis, enabling clinicians to identify disease eliciting allergens as well as cross-reactivity pattern, thereby providing us with the tools necessary for personalized allergy medicine and therapeutics and is detailed in Chapter 11.

Food allergy is a growing problem globally carrying a huge socioeconomic burden for patients, families and the community. Although fatalities are fortunately rare, the fear of death is very real for each patient. Currently, there is no cure for any food allergy available, with management strategies focusing on complete avoidance and utilization of adrenaline as the emergency antidote for anaphylaxis. There is a very strong imperative for safe and effective specific therapeutics for food allergy and one strategy based on T-cell epitopes for peanut allergy is detailed in Chapter 12.

We hope that the joined effort by the authors will not only provide pragmatic information for current food allergy research but also serves as a foundation for significant new research that will advance our current knowledge.

Contents

11. Recombinant Food Allergens for Diagnosis and Therapy   283

Heidi Hofer, Anargyros Roulias, Claudia Asam, Stephanie Eichhorn, Fátima Ferreira, Gabriele Gadermaier and *Michael Wallner*

1

Biomolecular and Clinical Aspects of Food Allergy

Heimo Breiteneder

CONTENTS

Department of Pathophysiology and Allergy Research, Medical University of Vienna, Vienna, Austria.
E-mail: Heimo.Breiteneder@meduniwien.ac.at

1.1 INTRODUCTION

Allergenic proteins are able to elicit Th2-polarized immune responses in predisposed individuals. As compared to the presently known number of protein architectures, allergenic proteins can be classified into a highly limited number of protein families (Radauer et al. 2008a). Version 30.0 of the protein family database Pfam (http://pfam. xfam.org/) describes 16,306 protein families (Finn et al. 2014). The structural database of allergenic proteins (SDAP; http://fermi.utmb. edu/) (Ivanciuc et al. 2003) assigns all allergens to 130 Pfam families. The most important plant and animal food allergens can be found in eight protein superfamilies discussed below. Our understanding why exactly these proteins are able to induce a specific IgE response in certain individuals is still incomplete. Allergenic proteins seem to be able to modulate the communication between innate and adaptive immune cells by interacting with pattern recognition receptors, which results in a Th2 polarization of the adaptive immune response (Karp 2010, Platts-Mills and Woodfolk 2011, Pulendran et al. 2010, Ruiter and Shreffler 2012, Willart and Hammad 2010, Wills-Karp 2010). Recent discoveries have shown that group 2 innate lymphoid cells are able to translate epithelial cell-derived alarmins into downstream adaptive type-2 responses (Scanlon and McKenzie 2015).

The toxin hypothesis of allergy has now gained interest and offers an alternative understanding of why certain proteins are targeted by IgE (Palm et al. 2012, Tsai et al. 2015). This hypothesis offers plausible explanations for allergenic components of insect venoms, proteins that have been altered by environmental toxins or proteins that carry ligands that present a certain danger to a host's cells. Why only few of the individuals who are exposed to the allergen raise an IgE response is most likely rooted in the way the incoming signals are processed. It has been shown that monocyte-derived dendritic cells from birch pollen allergic and non-allergic subjects displayed distinct signal transduction pathways following the contact with the major birch pollen allergen Bet v 1 (Smole et al. 2015). The situation is less clear for food allergens. Certain lipids directly bound as ligands by the allergen or when present in the allergen source seem to play a role in the allergic sensitization process (Bublin et al. 2014). Moreover, plant seed storage proteins of the cupin and prolamin superfamilies have the capacity to damage cells, which might induce danger signals in exposed innate immune cells resulting in allergic sensitization (Candido Ede et al. 2011).

1.2 PROLAMIN SUPERFAMILY

Plant seeds are a major source of dietary proteins. Seed storage proteins such as the prolamins are a source of amino acids for use during germination and seedling growth. The prolamin superfamily comprises several families of proteins with limited sequence homology. The prolamins which gave the superfamily its name are the major seed storage proteins in most cereal seeds. They possess two or more unrelated structural domains, one of which contains repeated sequences. Parts of the non-repetitive domain of one group of the sulfur-rich prolamins are homologous with sequences present in a large group of low molecular seed proteins including the 2S albumins, the non-specific lipid proteins (nsLTPs) and the cereal inhibitors of α-amylase and trypsin (Kreis et al. 1985). They all share a conserved cysteine skeleton, which contains eight cysteine residues. The prolamin superfamily seems to be of a much more recent origin than the cupin seed storage proteins. The 2.2S spore storage protein matteucin of the ostrich fern is related to the 2S albumins of angiosperms whose common ancestors lived more than 300 million years ago (Rodin and Rask 1990). nsLTPs are abundant in liverworts, mosses and land plants but have not been found in any algae indicating that they have evolved only after plants had conquered land (Edstam et al. 2011).

1.2.1 Prolamins

The prolamins which are characterized by high levels of glutamine and proline residues are restricted to the grasses including major cereals such as wheat, barley and rye (Shewry et al. 1995). The prolamin seed storage proteins of wheat are the major components of gluten, which determines the quality of the flour for bread making. The complex mixture of cereal storage proteins, the gluten, consists of roughly equal amounts of gliadins and glutenins (Tatham and Shewry 2008). Gliadins are monomeric proteins, which interact by noncovalent forces. Based on their electrophoretic mobility they are divided into the fast moving α/β-gliadins, the intermediate γ-gliadins, and the slowly moving ω-gliadins. The glutenins are polymers of individual proteins that are linked by interchain disulfide bridges. Glutenins can be classified into high molecular weight (HMW) and low molecular weight (LMW) groups. The sulfur-rich prolamins are quantitatively the major prolamin group in wheat, barley and rye, and they include polymeric and monomeric proteins (Shewry and Tatham 1990). Wheat-dependent exercise-induced anaphylaxis (WDEIA) is associated with ω_5-gliadins (Tatham and Shewry 2008) while both gliadins and glutenins appear to be implicated in baker's asthma (Quirce and Diaz-Perales 2013).

1.2.2 Bifunctional Inhibitors

Plants have evolved a certain degree of resistance to insect pests that feed on plant tissues. Six types of proteinaceous α-amylase inhibitors are found in higher plants (Svensson et al. 2004). The bifunctional inhibitors impede digestion by acting on insect gut α-amylases and proteinases such as trypsin (Franco et al. 2002). A large family of these inhibitors, also referred to as CM proteins for their presence in chloroform/methanol extracts, is found in cereals seeds (Svensson et al. 2004). Several of these proteins are α-amylase/trypsin inhibitors while others inhibit only α-amylase or trypsin. These inhibitors consist of 120 to 160 amino acids, have a high α-helical content, and possess ten cysteine residues which form five disulfide bonds

(Oda et al. 1997). Tri a 28 (syn. 0.19 α-amylase inhibitor form wheat) acts as a homodimer (Oda et al. 1997) whereas the wheat inhibitor 0.28 and the corresponding barley inhibitor BMAI-1 (Hor v 15) are monomers (Sanchez-Monge et al. 1992). Current immunological and clinical data point to the α-amylase/trypsin inhibitor family as the main culprit of Baker's asthma (Salcedo et al. 2011).

1.2.3 2S Albumins

2S albumins are a water-soluble storage protein group widely present in mono- and dicotyledonous seeds (Candido Ede et al. 2011). They are encoded by a multigene family, which results in the presence of several isoforms in individual plants. They are synthesized as a single large precursor, which is then processed to give rise to two subunits that are held together by disulfide bonds. Typically, the 2S albumins comprise four α-helices and four to five disulfide bonds (Moreno and Clemente 2008). Although the major function of 2S albumins is the storage of amino acids, antifungal and antibacterial properties of several 2S albumins and thus their role in plant defense against pathogens were described (Candido Ede et al. 2011). A novel antimicrobial protein, SiAMP2, of the 2S albumin family was identified in sesame seeds and its inhibition of the growth of the human pathogenic bacterium *Klebsiella* was described (Maria-Neto et al. 2011). The 2S albumins of *Brassica napus* were able to significantly damage the fungal plasma lemma and to cause its permeabilization (Barciszewski et al. 2000). The number of 2S albumins that are described as food allergens is still increasing (Moreno and Clemente 2008). Many of the highly important seed, tree nut and legume allergens belong to the 2S albumins. Among them are Ara h 2, Ara h 6, and Ara h 7 from peanut (Burks et al. 1992, Kleber-Janke et al. 1999), Jug r 1 from walnut (Teuber et al. 1998), Ses i 1 and Ses i 2 from sesame seeds (Beyer et al. 2002a, Pastorello et al. 2001), Ber e 1 from Brazil nut (Pastorello et al. 1998), and Ana o 1 from cashew (Robotham et al. 2005). Ber e 1 serves as a model protein for studies of intrinsic allergenicity of food proteins (Alcocer et al. 2012).

1.2.4 Nonspecific Lipid Transfer Proteins (nsLTPs)

The nsLTPs are a family of allergens of high importance. They are divided into the 9 kDa nsLTP1 and the 7 kDa nsLTP2 subfamilies (Kader 1996). NsLTP1 are primarily found in aerial organs while nsLTP2 are expressed in roots. Both nsLTP1 and nsLTP2 are found in seeds. Members of both subfamilies are compact cysteine-rich proteins, which are made up of four or five α-helices that are held together by four conserved disulfide bridges. The α-helices enclose a hydrophobic cavity that enables them to transfer various lipid ligands between lipid bilayers *in vitro* (Lascombe et al. 2008). NsLTPs are involved in key cellular processes such as stabilization of membranes, cell wall organization and signal transduction but they also play important roles in resistance to biotic and abiotic stress, plant growth and development (Liu et al. 2015). Besides their various biologic roles in plants, nsLTPs are a large group of heat- and proteolysis-resistant allergens (Egger et al. 2010). The type 1 nsLTPs are able to elicit severe type 1 reactions to fresh fruits such as peach in predisposed individuals in Southern Europe and the Mediterranean region. NsLTPs are regarded as panallergens due to their presence in a variety of plant tissues including seeds, fruits and vegetative tissues (Salcedo et al. 2007). In addition, nsLTPs1 were described as inhalant allergens in pollen of many flowering plants including *Parietaria judaica* (Duro et al. 1996), olive tree (Tejera et al. 1999), and mugwort (Gadermaier et al. 2009).

Plant food nsLTPs1 have been identified in fruits such as peach (Pastorello et al. 1999), apple (Zuidmeer et al. 2005), and grapes (Pastorello et al. 2003), in vegetables such as asparagus (Diaz-Perales et al. 2002), corn (Pastorello et al. 2000), and celery (Gadermaier et al. 2011), and in various nuts including hazelnut (Offermann et al. 2015). Cross-reactivities between nsLTPs1 from closely related plants are frequently observed but decreases with evolutionary distance. The kiwi fruit nsLTP1 does not cross-react with the peach nsLTP1 (Bernardi et al. 2011). Similarly, the nsLTP1s from olive pollen and *Parietaria judaica* pollen neither cross-react with each other nor with other plant food nsLTP1s such as the one from peach (Tordesillas et al. 2011). In contrast, sensitization to the nsLTP1 from peach is

Table 1.1 Selected allergens of the prolamin superfamily.

Protein family	Allergen source	Allergen designation
Prolamin	Wheat (*Triticum aestivum*)	Tri a 19: ω-5-glaidin
		Tri a 20: γ-gliadin
		Tri a 21: α/β-gliadin
		Tri a 26: high molecular weight glutenin
		Tri a 36: low molecular weight glutenin
Bifunctional inhibitor	Wheat (*Triticum aestivum*)	Tri a 15: monomeric α-amylase inhibitor
		Tri a 28: dimeric α-amylase inhibitor 0.19
		Tri a 29: tetrameric α-amylase inhibitor CM1/CM2
		Tri a 30: tetrameric α-amylase inhibitor CM3
	Rye (*Secale cereale*)	Sec c 38: dimeric α-amylase/trypsin inhibitor
2S albumin	Brazil nut (*Bertholletia excelsa*)	Ber e 1
	Cashew nut (*Anacardium occidentale*)	Ana o 3
	Hazelnut (*Corylus avellana*)	Cor a 14
	Peanut (*Arachis hypogaea*)	Ara h 2, Ara h 6, Ara h 7
	Sesame (*Sesamum indicum*)	Ses i 1, Ses i 2
	Walnut (*Juglans regia*)	Jug r 1
Non-specific lipid transfer protein type 1	Apple (*Malus domestica*)	Mal d 3
	Celeriac (*Apium graveolens*)	Api g 2
	Cherry (*Prunus avium*)	Pru av 3
	Corn (*Zea mays*)	Zea m 14
	Grape (*Vitis vinifera*)	Vit v 1
	Hazelnut (*Corylus avellana*)	Cor a 8
	Peach (*Prunus persica*)	Pru p 3
Non-specific lipid transfer protein type 2	Celeriac (*Apium graveolens*)	Api g 6
	Tomato (*Solanum lyopersicum*)	Sola l 6

frequently present with a sensitization to the mugwort nsLTP1 in the Mediterranean region. A primary sensitization to the peach nsLTP1 can lead to a respiratory allergy based on the cross-reactivity of peach and mugwort nsLTPs (Sanchez-Lopez et al. 2014). The first allergenic type 2 nsLTP, detected as a heat-resistant protein in celeriac, showed only a very limited cross-reactivity to the tape 1 nsLTP from celeriac (Vejvar et al. 2013). Recently, a type 2 nsLTP was identified as an allergen present in tomato seeds (Giangrieco et al. 2015).

1.3 CUPIN SUPERFAMILY

At present, the cupin superfamily contains 57 families. The members of this superfamily possess one or more conserved cupin domain, a characteristic β-barrel (Latin *cupa* = barrel) that evolved in a prokaryotic organism and was then passed on into the plant kingdom (Khuri et al. 2001). The cupin domain is used for a large number of biological functions and is found in fungal spherulins that are produced upon spore formation, in proteins that bind saccharose, or in germins whose function depends on the binding of manganese ions by the cupin domain (Dunwell et al. 2000). Cupins are highly thermostable, a trait that has most likely evolved in thermophilic archaea and that can still be found in today's plant food allergens. The cupin domain was duplicated in flowering plants giving rise to the so-called bicupin seed storage proteins (Dunwell and Gane 1998), the 7S and 11S globulins which are described as major allergens of peanut, tree nuts and various seeds (Mills et al. 2002, Radauer and Breiteneder 2007, Willison et al. 2014). The cupin seed storage proteins are primarily an energy source and provide amino acids during seed germination. In addition, they are also involved in the defense of many plant species against fungi and insects (Candido Ede et al. 2011).

1.3.1 Vicilins (7S globulins)

The 7S globulin seed storage proteins are trimeric proteins that are also referred to as vicilins, as they are primarily found in the Viciae group of legumes. The monomers of these proteins are products of a multigene family that are proteolytically processed during their maturation and glycosylated by varying degrees dependent on the plant species (Marcus et al. 1999). Many major plant food allergens are vicilins, including Ara h 1 from peanut (Burks et al. 1991), Gly m 5 from soybean (Ogawa et al. 1995), Ana o 1 from cashew (Wang et al. 2002), Jug r 2 from walnut (Teuber et al. 1999), Len c 1 from lentil (Lopez-Torrejon et al. 2003), Ses i 3 from sesame (Beyer et al. 2002a), and Cor a 11 from hazelnut (Lauer et al. 2004).

1.3.2 Legumins (11S globulins)

The 11S globulins are the seed storage proteins of many mono- and dicotyledonous plants. They are also referred to as legumins as they were primarily studied in legume seeds. Legumins are hexameric proteins that consist of two associated viclin-like trimers (Dunwell et al. 2000). The monomers, like in their vicilin counterparts, are the products of multigene families. In contrast to the vicilin monomers, the legumin monomer is proteolytically cleaved into an acidic and a basic chain that are held together by a disulfide bond. Legumins are only rarely glycosylated. Various allergens of

Table 1.2 Selected allergens of the cupin superfamily.

Protein family	Allergen source	Allergen designation
Vicilin (7S globulins)	Cashew nut (*Anacardium occidentale*)	Ana o 1
	Hazelnut (*Corylus avellana*)	Cor a 11
	Peanut (*Arachis hypogaea*)	Ara h 1
	Sesame (*Sesamum indicum*)	Ses i 3
	Soybean (*Glycine max*)	Gly m 5
	Walnut (*Juglans regia*)	Jug r 2
Legumin (11S globulins)	Brazil nut (*Bertholletia excelsa*)	Ber e 2
	Cashew nut (*Anacardium occidentale*)	Ana o 2
	Hazelnut (*Corylus avellana*)	Cor a 9
	Peanut (*Arachis hypogaea*)	Ara h 3
	Sesame (*Sesamum indicum*)	Ses i 6, Ses i 7
	Soybean (*Glycine max*)	Gly m 6
	Walnut (*Juglans regia*)	Jug r 4

legume seeds, tree nuts, and seeds belong to the legumin protein family. They include Ara h 3 from peanut (Rabjohn et al. 1999), Gly m 6 from soybean (Beardslee et al. 2000), Ana o 2 from cashew nut (Wang et al. 2003), Cor a 9 from hazelnut (Beyer et al. 2002b), and Ses i 6 and Ses i 7 from sesame seeds (Beyer et al. 2007).

1.4 EF-HAND SUPERFAMILY

The EF-hand motif is the most common calcium-binding motif found in proteins where two α-helices connected by a loop form a calcium-binding structure (Lewit-Bentley and Rety 2000). Proteins that contain EF-hand motifs have functions as diverse as calcium buffering in the cytosol, signal transduction between cellular compartments or muscle contraction. EF-hand motifs are found in certain pollen allergens, the polcalcins, as well as in the major fish allergens, the parvalbumins. Plant EF-hand and animal EF-hand proteins do not cross-react with each other.

1.4.1 Parvalbumins

Parvalbumins are present in high concentration in the white muscle of many fish species and are highly cross-reactive major allergens (Lee et al. 2011). Parvalbumins possess three characteristic EF-hand motifs (Ikura 1996) of which only two are able to bind calcium ions (Declercq et al. 1991). Parvalbumins play an important role in relaxing muscle fibers by binding free intracellular calcium ions (Pauls et al. 1996). Binding of the calcium ligand is necessary for the correct conformation of parvalbumin. Loss of the ligand leads to a change in conformation, which results in the loss of the ability to bind IgE (Bugajska-Schretter et al. 1998, Bugajska-Schretter et al. 2000). Calcium-bound parvalbumin displays a high stability to denaturation by heat or degradation by proteolysis (Elsayed and Aas 1971, Filimonov et al. 1978, Griesmeier et al. 2010, Somkuti et al. 2012). Parvalbumins can be classified into two evolutionary lineages, the α- and the β-parvalbumins, which share similar architectures. In general, only β-parvalbumins are allergenic. However, an allergenic α-parvalbumin from frog was

Table 1.3 Selected allergenic parvalbumins.

Protein family	Allergen source	Allergen designation
Parvalbumin	Atlantic cod (*Gadus morhua*)	Gad m 1
	Atlantic salmon (*Salmo salar*)	Sal s 1
	Carp (*Cyprinus carpio*)	Cyp c 1
	Rainbow trout (*Oncorhynchus mykiss*)	Onc m 1
	Whiff (*Lepidorhombus whiffagonis*)	Lep w 1

described (Hilger et al. 2002). Gad c 1 was isolated from cod and was the first described allergenic β-parvalbumin (Aas and Jebsen 1967, Elsayed and Bennich 1975). Today, a large number of allergenic β-parvalbumins from a variety of fish species is known (Kuehn et al. 2014, Sharp and Lopata 2014). In addition, two allergenic parvalbumins from red stingray were described (Cai et al. 2010).

1.5 TROPOMYOSIN-LIKE SUPERFAMILY

Tropomyosins are one of three families of the tropomyosin-like superfamily. Tropomyosins are closely related proteins that—together with actin and myosin—are involved in the contraction of muscle fibers. Tropomyosins consist of 40 heptapeptide units and are double stranded, so called coiled-coil, molecules (Li et al. 2002). Tropomyosins are the major allergens of crustaceans and mollusks. Most allergies to shrimps, crabs, lobsters, squids, and shellfish are mediated by tropomyosins. Tropomyosins were originally described as allergenic in shrimps (Daul et al. 1994, Leung et al. 1994, Shanti et al. 1993). Today, tropomyosins are regarded as panallergens of many invertebrate animals (Reese et al. 1999). Tropomyosins of crustaceans and mollusks are highly heat-stable and cross-reactive (Motoyama et al. 2006). Extracts of cooked *Penaeus indicus* shrimps still contained the major allergen Pen i 1 with unchanged IgE-binding capacity (Naqpal et al. 1989). Water-soluble shrimp allergens were also detected in the cooking stock (Lehrer et al. 1990). In seafood processing plants, allergenic tropomyosins are present in aerosols and thus elicit occupational allergies in the work force (Lopata and Jeebhay 2013). Tropomyosins are also inhalant allergens from mites and cockroaches. Although they seem to possess only a

Table 1.4 Selected allergenic tropomyosins.

Protein family	Allergen source	Allergen designation
Tropomyosin: Crustaceans	American lobster (*Homerus americanus*)	Hom a 1
	Crucifix crab (*Charybdis feriatus*)	Cha f 1
	Indian white prawn (*Penaeus indicus*)	Pen i 1
	North Sea shrimp (*Crangon crangon*)	Cra c 1
Tropomyosin: Mollusks	Pacific flying squid (*Todarodes pacificus*)	Tod p 1

limited allergenic potential (Thomas et al. 2010) they are regarded as important for cross-sensitization to tropomyosins of crustaceans and shellfish (Lopata et al. 2010).

1.6 PROFILIN-LIKE SUPERFAMILY

The profilin-like superfamily comprises four member families. One of them, the profilin family, are proteins that are highly conserved in higher plants with sequence identities of at least 75% (Radauer et al. 2006). Profilins are cytoplasmic proteins of 12–15 kDa and are present in all eukaryotic cells. They bind monomeric actin (Schutt et al. 1993) and are involved in the dynamic turnover and restructuring of the actin cytoskeleton (Witke 2004). Profilin from birch pollen was the first profilin that was described as allergenic (Valenta et al. 1991). Subsequently, a large number of cross-reactive profilin allergens were described in pollen of trees, grasses and weeds (Gadermaier et al. 2014, Hauser et al. 2010). As profilin-specific IgE cross-reacts with practically all plant profilins, a profilin sensitization is regarded as a risk factor for allergic reactions to various plant pollen (Mari 2001) and plant foods (Asero et al. 2003, Fernandez-Rivas 2015). However, the clinical relevance of a profilin sensitization is still under discussion (Santos and Van Ree 2011). The clinical relevance of a profilin sensitization was shown for profilins from cantaloupe, watermelon, tomato, banana, pineapple, orange and kaki (Anliker et al. 2001, Asero et al. 2008, Lopez-Torrejon et al. 2005). Recently, profilins were shown to be

Table 1.5 Selected allergenic plant food profilins.

Protein family	Allergen source	Allergen designation
Profilin	Banana (*Musa acuminata*)	Mus a 1
	Cantaloupe (*Cucumis melo*)	Cuc m 2
	Orange (*Citrus sinensis*)	Cit s 2
	Pineapple (*Ananas comosus*)	Ana c 1
	Tomato (*Solanum lycopersicum*)	Sola l 1

complete food allergens capable of eliciting severe reactions in plant food allergic patients that had been exposed to high levels of grass pollen (Alvarado et al. 2014).

1.7 BET V 1-LIKE SUPERFAMILY

Bet v 1, the major birch pollen allergen, is the one member that gave this superfamily its name (Breiteneder et al. 1989). The Bet v 1-like superfamily contains at present 103,375 members from 17,750 species (http://pfam.xfam.org/clan/CL0209, accessed November 2015). Proteins with the typical Bet v 1 architecture can be found in all kingdoms of life and hence belong to the earliest proteins that evolved at the beginning of life (Radauer et al. 2008b). The superfamily consists of 14 families including the Bet v family, which comprises 11 subfamilies. Most of the Bet v 1-homologous allergens known today belong to the PR-10 subfamily (Hoffmann-Sommergruber 2002). The cDNA coding for Bet v 1 was discovered on July 3, 1989 and published as a sequence for the first plant allergen (Breiteneder et al. 1989). Birch belongs to the botanical order Fagales which comprises 8 families, some of which produce allergenic pollen such as hazel (Breiteneder et al. 1993), alder (Breiteneder et al. 1992), oak (Wallner et al. 2009), and beech (Hauser et al. 2011).

The association of a birch pollen allergy with an allergy to diverse plant foods is a frequently observed syndrome, which is due to the presence of homologous allergens in these allergen sources (Katelaris 2010, Vieths et al. 2002). The observed clinical symptoms to the various plant foods are generally elicited by IgE that was induced by exposure to Bet v 1. The known structures of Bet v 1 (Gajhede

Table 1.6 Selected allergens of the Bet v 1 family.

Subfamily of the Bet v 1 family	Allergen source	Allergen designation
PR-10	Apple (*Malus domestica*)	Mal d 1
	Celeriac (*Apium graveolens*)	Api g 1
	Cherry (*Prunus avium*)	Pru av 1
	Mung bean (*Vigna radiata*)	Vig r 1
	Peach (*Prunus persica*)	Pru p 1
	Peanut (*Arachis hypogaea*)	Ara h 8
	Soybean (*Glycine max*)	Gly m 4
RRP	Kiwifruit (*Actinidia deliciosa*)	Act d 11
CSBP	Mung bean (*Vigna radiata*)	Vig r 6

et al. 1996), and its homologs form cherry (Neudecker et al. 2001), celeriac (Markovic-Housley et al. 2009), carrot (Markovic-Housley et al. 2009), soybean (Berkner et al. 2009) and peanut (Hurlburt et al. 2013) clearly illustrate the similarities of these molecules' surfaces that explain the clinically observed cross-reactivities. IgE antibodies bind to Bet v 1-related plant food allergens such as Mal d 1 from apple (Vanek-Krebitz et al. 1995), Api g 1 from celeriac (Breiteneder et al. 1995), Ara h 8 from peanut (Mittag et al. 2004), Vig r 1 from mung bean (Mittag et al. 2005), and Bet v 1 homologs from Sharon fruit (Bolhaar et al. 2005) and jackfruit (Bolhaar et al. 2004). Act d 11 is an allergen of the kiwifruit that belongs to the ripening related protein (RRP) subfamily (D'Avino et al. 2011). Vig r 6 from mung beans is another Bet v 1 homolog that belongs to the cytokinin-specific binding protein (CSBP) family.

1.8 THE CASEIN AND THE CASEIN KAPPA FAMILY

All mammalian milks contain multiple casein proteins characterized as α-, β- and κ-caseins (Oftedal 2012). Caseins are members of the unfolded secretory calcium-binding phosphoproteins called SSCP (Kawasaki and Weiss 2003). The α- and β-caseins evolved from tooth and bone-proteins well before the evolution of lactation

Table 1.7 Allergenic caseins of cow's milk.

Protein family	Allergen source	Allergen designation
Casein	Cow's milk (*Bos domesticus*)	Bos d 9: αS1-casein
		Bos d 10: αS2-casein
		Bos d 11: β-casein
Casein kappa	Cow's milk (*Bos domesticus*)	Bos d 12: κ-casein

(Lenton et al. 2015). In mammalian milks, sequestered nanoclusters of calcium phosphate are substructures in casein micelles which allow the calcium and phosphate concentrations to be far in excess of their solubility. The αS1-, αS2- and β-caseins form a shell around amorphous calcium phosphate to form the nanoclusters. These nanoclustes are then assembled into the casein micelles that are stabilized by κ-casein (ten Grotenhuis et al. 2003). α- and β-caseins are members of the casein family (Kawasaki et al. 2011), while κ-caseins are members of the casein kappa family (Ward et al. 1997). Caseins are major food allergens involved in cow's milk allergy, which affects predominantly young children. In European children, the incidence of challenge-proven cow's milk allergy was 0.54% with national incidences ranging from <0.3% to 1% (Schoemaker et al. 2015). Recently, the official nomenclature of allergenic caseins has been changed (Radauer et al. 2014). The name Bos d 8, as it is widely established, was kept to designate the whole casein fraction. However, based on low sequence similarities, Bos d 8 was demerged into four separate allergens: Bos d 9 (aS1-casein), Bos d 10 (αS2-casein), Bos d 11.0101 (β-casein), and Bos d 12.0101 (κ-casein).

1.9 CALYCIN-LIKE SUPERFAMILY

The calycin structural superfamily includes 20 families. Calycins are an example for a superfamily of proteins, which—although they share structural similarities—have unusually low levels of overall sequence conservation. The calycin architecture is based on an eight-stranded β-barrel which forms an internal ligand binding site for small hydrophobic molecules (Flower et al. 1993).

1.9.1 Lipocalins

Lipocalins form a subset of the calycin superfamily. Lipocalins are small extracellular proteins with a large variety of functions which typically revolves around the binding of small hydrophobic ligands such as retinol (Flower et al. 2000). Most of the allergenic lipocalins are not food allergens but important inhalant allergens from mammals and insects (Hilger et al. 2012, Virtanen et al. 2012). The only lipocalin animal food allergen is β-lactoglobulin (Bos d 5) which is a major allergen in cow's milk (Hochwallner et al. 2014) and is absent from human and camel milk (Restani et al. 2009). Bos d 5 is highly stable to proteolytic degradation and acid hydrolysis (Wal 2004).

1.10 CONCLUSIONS

In 1991, the evolutionary biologist Margie Profet published the toxin hypothesis of allergy (Profet 1991). She proposed that the allergic immune response evolved as a defense mechanism to protect the individual from toxic environmental substances such as venoms and toxic plant compounds. Recently, this hypothesis has found experimental proof for bee and snake venoms (Marichal et al. 2013, Palm et al. 2013, Starkl et al. 2015). It is highly plausible that this hypothesis will be confirmed for allergenic components of other insect venoms. Future experiments will have to be performed for plant food allergens and plant food matrices to explore whether they are as innocuous as they were made out to be. In fact, seed storage proteins which are commonly regarded as inert also have functions in plant defense mechanisms (Candido Ede et al. 2011). 2S albumins from passion fruit seeds have been shown to induce plasma membrane permeabilization (Agizzio et al. 2006) and vicilins from cowpea were discovered to interact with the microvilli of the larval midgut epithelium of the bean-feeding cowpea beetle (Oliveira et al. 2014).

The allergens of the various superfamilies have distinct distributions. Allergenic prolamins and cupins are only present in plants. While the cupin allergens are so far only known as seed storage proteins, allergens of the prolamin superfamily can either be storage proteins or have inhibitory or signal transduction functions. Bet v 1 homologs and profilins are also only known as plant allergens. Allergenic food proteins of the EF-hand superfamily are only known from fish. Likewise, allergenic tropomyosins as food allergens seem to be limited to crustaceans and mollusks. Although lipocalins are also present in plants (Charron et al. 2005), most of them are inhalant animal allergens and only one is an animal food allergen, the β-lactoglobulin from cow's milk. All of these proteins perform a specific biologic function. They become allergenic only when they interact with the immune system of a predisposed individual. It is worth to note, that in general, allergens are restricted to a highly limited number of protein families. That indicates that only a very small number of protein structures are able to induce allergic sensitization or to become involved in such a process. Why this is the case is still unclear. The innate immune system (Herre et al. 2013, Junker et al. 2012, Trompette et al. 2009), binding of ligands to the allergens (Jyonouchi et al. 2011, Mirotti et al. 2013), and adjuvants present in the allergen source seem to play a role (Gilles et al. 2009, Mittag et al. 2013).

When the allergens designated by the WHO/IUIS Allergen Nomenclature Subcommittee (http://www.allergen.org/) are classified by protein families, as was done in this chapter, they become much more manageable. A detailed analysis of the biochemical, structural and immunologic properties of each family of allergens will contribute to the understanding of factors that contribute to the allergenic potential of a protein.

ACKNOWLEDGEMENT

The author acknowledges the support of the Austrian Science Fund (FWF) Grant SFB F4608.

Keywords: Allergen family; molecular phylogeny; superfamily; food allergy

REFERENCES

Aas, K. and J. W. Jebsen (1967). Studies of hypersensitivity to fish. Partial purification and crystallization of a major allergenic component of cod. Int Arch Allergy Appl Immunol. 32: 1–20.

Agizzio, A. P., M. Da Cunha, A. O. Carvalho, M. A. Oliveira, S. F. Ribeiro and V. M. Gomes (2006). The antifungal properties of a 2S albumin-homologous protein from passion fruit seeds involve plasma membrane permeabilization and ultrastructural alterations in yeast cells. Plant Sci. 171: 515–522.

Alcocer, M., L. Rundqvist and G. Larsson (2012). Ber e 1 protein: the versatile major allergen from Brazil nut seeds. Biotechnol Lett. 34: 597–610.

Alvarado, M. I., L. Jimeno, F. De La Torre, P. Boissy, B. Rivas, M. J. Lazaro et al. (2014). Profilin as a severe food allergen in allergic patients overexposed to grass pollen. Allergy. 69: 1610–1616.

Anliker, M. D., J. Reindl, S. Vieths and B. Wuthrich (2001). Allergy caused by ingestion of persimmon (*Diospyros kaki*): detection of specific IgE and cross-reactivity to profilin and carbohydrate determinants. J Allergy Clin Immunol. 107: 718–723.

Asero, R., G. Mistrello, D. Roncarolo, S. Amato, D. Zanoni, F. Barocci et al. (2003). Detection of clinical markers of sensitization to profilin in patients allergic to plant-derived foods. J Allergy Clin Immunol. 112: 427–432.

Asero, R., R. Monsalve and D. Barber (2008). Profilin sensitization detected in the office by skin prick test: a study of prevalence and clinical relevance of profilin as a plant food allergen. Clin Exp Allergy. 38: 1033–1037.

Barciszewski, J., M. Szymanski and T. Haertle (2000). Minireview: analysis of rape seed napin structure and potential roles of the storage protein. J Protein Chem. 19: 249–254.

Beardslee, T. A., M. G. Zeece, G. Sarath and J. P. Markwell (2000). Soybean glycinin G1 acidic chain shares IgE epitopes with peanut allergen Ara h 3. Int Arch Allergy Immunol. 123: 299–307.

Berkner, H., P. Neudecker, D. Mittag, B. K. Ballmer-Weber, K. Schweimer, S. Vieths et al. (2009). Cross-reactivity of pollen and food allergens: soybean Gly m 4 is a member of the Bet v 1 superfamily and closely resembles yellow lupine proteins. Biosci Rep. 29: 183–192.

Bernardi, M. L., I. Giangrieco, L. Camardella, R. Ferrara, P. Palazzo, M. R. Panico et al. (2011). Allergenic lipid transfer proteins from plant-derived foods do not immunologically and clinically behave homogeneously: the kiwifruit LTP as a model. PLoS One. 6: e27856.

Beyer, K., L. Bardina, G. Grishina and H. A. Sampson (2002a). Identification of sesame seed allergens by 2-dimensional proteomics and Edman sequencing: seed storage proteins as common food allergens. J Allergy Clin Immunol. 110: 154–159.

Beyer, K., G. Grishina, L. Bardina, A. Grishin and H. A. Sampson (2002b). Identification of an 11S globulin as a major hazelnut food allergen in hazelnut-induced systemic reactions. J Allergy Clin Immunol. 110: 517–523.

Beyer, K., G. Grishina, L. Bardina and H. A. Sampson (2007). Identification of 2 new sesame seed allergens: Ses i 6 and Ses i 7. J Allergy Clin Immunol. 119: 1554–1556.

Bolhaar, S. T., R. Ree, C. A. Bruijnzeel-Koomen, A. C. Knulst and L. Zuidmeer (2004). Allergy to jackfruit: a novel example of Bet v 1-related food allergy. Allergy. 59: 1187–1192.

Bolhaar, S. T., R. van Ree, Y. Ma, C. A. Bruijnzeel-Koomen, S. Vieths, K. Hoffmann-Sommergruber et al. (2005). Severe allergy to sharon fruit caused by birch pollen. Int Arch Allergy Immunol. 136: 45–52.

Breiteneder, H., K. Pettenburger, A. Bito, R. Valenta, D. Kraft, H. Rumpold et al. (1989). The gene coding for the major birch pollen allergen Betv1, is highly homologous to a pea disease resistance response gene. EMBO J. 8: 1935–1938.

Breiteneder, H., F. Ferreira, A. Reikerstorfer, M. Duchene, R. Valenta, K. Hoffmann-Sommergruber et al. (1992). Complementary DNA cloning and expression in *Escherichia coli* of Aln g I, the major allergen in pollen of alder (*Alnus glutinosa*). J Allergy Clin Immunol. 90: 909–917.

Breiteneder, H., F. Ferreira, K. Hoffmann-Sommergruber, C. Ebner, M. Breitenbach, H. Rumpold et al. (1993). Four recombinant isoforms of Cor a I, the major allergen of hazel pollen, show different IgE-binding properties. Eur J Biochem. 212: 355–362.

Breiteneder, H., K. Hoffmann-Sommergruber, G. O'Riordain, M. Susani, H. Ahorn, C. Ebner et al. (1995). Molecular characterization of Api g 1, the major allergen of celery (Apium graveolens), and its immunological and structural relationships to a group of 17-kDa tree pollen allergens. Eur J Biochem. 233: 484–489.

Bublin, M., T. Eiwegger and H. Breiteneder (2014). Do lipids influence the allergic sensitization process? J Allergy Clin Immunol. 134: 521–529.

Bugajska-Schretter, A., L. Elfman, T. Fuchs, S. Kapiotis, H. Rumpold, R. Valenta et al. (1998). Parvalbumin, a cross-reactive fish allergen, contains IgE-binding epitopes sensitive to periodate treatment and Ca2+ depletion. J Allergy Clin Immunol. 101: 67–74.

Bugajska-Schretter, A., M. Grote, L. Vangelista, P. Valent, W. R. Sperr, H. Rumpold et al. (2000). Purification, biochemical, and immunological characterisation of a major food allergen: different immunoglobulin E recognition of the apo- and calcium-bound forms of carp parvalbumin. Gut. 46: 661–669.

Burks, A. W., L. W. Williams, R. M. Helm, C. Connaughton, G. Cockrell and T. O'Brien (1991). Identification of a major peanut allergen, Ara h I, in patients with atopic dermatitis and positive peanut challenges. J Allergy Clin Immunol. 88: 172–179.

Burks, A. W., L. W. Williams, C. Connaughton, G. Cockrell, T. J. O'Brien and R. M. Helm (1992). Identification and characterization of a second major peanut allergen, Ara h II, with use of the sera of patients with atopic dermatitis and positive peanut challenge. J Allergy Clin Immunol. 90: 962–969.

Cai, Q. F., G. M. Liu, T. Li, K. Hara, X. C. Wang, W. J. Su et al. (2010). Purification and characterization of parvalbumins, the major allergens in red stingray (*Dasyatis akajei*). J Agric Food Chem. 58: 12964–12969.

Candido Ede, S., M. F. Pinto, P. B. Pelegrini, T. B. Lima, O. N. Silva, R. Pogue et al. (2011). Plant storage proteins with antimicrobial activity: novel insights into plant defense mechanisms. FASEB J. 25: 3290–3305.

Charron, J. B., F. Ouellet, M. Pelletier, J. Danyluk, C. Chauve and F. Sarhan (2005). Identification, expression, and evolutionary analyses of plant lipocalins. Plant Physiol. 139: 2017–2028.

D'Avino, R., M. L. Bernardi, M. Wallner, P. Palazzo, L. Camardella, L. Tuppo et al. (2011). Kiwifruit Act d 11 is the first member of the ripening-related protein family identified as an allergen. Allergy. 66: 870–877.

Daul, C. B., M. Slattery, G. Reese and S. B. Lehrer (1994). Identification of the major brown shrimp (*Penaeus aztecus*) allergen as the muscle protein tropomyosin. Int Arch Allergy Immunol. 105: 49–55.

Declercq, J. P., B. Tinant, J. Parello and J. Rambaud (1991). Ionic interactions with parvalbumins. Crystal structure determination of pike 4.10 parvalbumin in four different ionic environments. J Mol Biol. 220: 1017–1039.

Diaz-Perales, A., A. I. Tabar, R. Sanchez-Monge, B. E. Garcia, B. Gomez, D. Barber et al. (2002). Characterization of asparagus allergens: a relevant role of lipid transfer proteins. J Allergy Clin Immunol. 110: 790–796.

Dunwell, J. M. and P. J. Gane (1998). Microbial relatives of seed storage proteins: conservation of motifs in a functionally diverse superfamily of enzymes. J Mol Evol. 46: 147–154.

Dunwell, J. M., S. Khuri and P. J. Gane (2000). Microbial relatives of the seed storage proteins of higher plants: conservation of structure and diversification of function during evolution of the cupin superfamily. Microbiol Mol Biol Rev. 64: 153–179.

Duro, G., P. Colombo, M. A. Costa, V. Izzo, R. Porcasi, R. Di Fiore et al. (1996). cDNA cloning, sequence analysis and allergological characterization of Par j 2.0101, a new major allergen of the Parietaria judaica pollen. FEBS Lett. 399: 295–298.

Edstam, M. M., L. Viitanen, T. A. Salminen and J. Edqvist (2011). Evolutionary history of the non-specific lipid transfer proteins. Mol Plant. 4: 947–964.

Egger, M., M. Hauser, A. Mari, F. Ferreira and G. Gadermaier (2010). The role of lipid transfer proteins in allergic diseases. Curr Allergy Asthma Rep. 10: 326–335.

Elsayed, S. and K. Aas (1971). Characterization of a major allergen (cod). Observations on effect of denaturation on the allergenic activity. J Allergy. 47: 283–291.

Elsayed, S. and H. Bennich (1975). The primary structure of allergen M from cod. Scand J Immunol. 4: 203–208.

Fernandez-Rivas, M (2015). Fruit and vegetable allergy. Chem Immunol Allergy. 101: 162–170.

Filimonov, V. V., W. Pfeil, T. N. Tsalkova and P. L. Privalov (1978). Thermodynamic investigations of proteins. IV. Calcium binding protein parvalbumin. Biophys Chem. 8: 117–122.

Finn, R. D., A. Bateman, J. Clements, P. Coggill, R. Y. Eberhardt, S. R. Eddy et al. (2014). Pfam: the protein families database. Nucleic Acids Res. 42: D222–230.

Flower, D. R., A. C. North and T. K. Attwood (1993). Structure and sequence relationships in the lipocalins and related proteins. Protein Sci. 2: 753–761.

Flower, D. R., A. C. North and C. E. Sansom (2000). The lipocalin protein family: structural and sequence overview. Biochim Biophys Acta. 1482: 9–24.

Franco, O. L., D. J. Rigden, F. R. Melo and M. F. Grossi-De-Sa (2002). Plant alpha-amylase inhibitors and their interaction with insect alpha-amylases. Eur J Biochem. 269: 397–412.

Gadermaier, G., A. Harrer, T. Girbl, P. Palazzo, M. Himly, L. Vogel et al. (2009). Isoform identification and characterization of Art v 3, the lipid-transfer protein of mugwort pollen. Mol Immunol. 46: 1919–1924.

Gadermaier, G., M. Egger, T. Girbl, A. Erler, A. Harrer, E. Vejvar et al. (2011). Molecular characterization of Api g 2, a novel allergenic member of the lipid-transfer protein 1 family from celery stalks. Mol Nutr Food Res. 55: 568–577.

Gadermaier, G., M. Hauser and F. Ferreira (2014). Allergens of weed pollen: an overview on recombinant and natural molecules. Methods. 66: 55–66.

Gajhede, M., P. Osmark, F. M. Poulsen, H. Ipsen, J. N. Larsen, R. J. Joost van Neerven et al. (1996). X-ray and NMR structure of Bet v 1, the origin of birch pollen allergy. Nat Struct Biol. 3: 1040–1045.

Giangrieco, I., C. Alessandri, C. Rafaiani, M. Santoro, S. Zuzzi, L. Tuppo et al. (2015). Structural features, IgE binding and preliminary clinical findings of the 7kDa lipid transfer protein from tomato seeds. Mol Immunol. 66: 154–163.

Gilles, S., V. Mariani, M. Bryce, M. J. Mueller, J. Ring, H. Behrendt et al. (2009). Pollen allergens do not come alone: pollen associated lipid mediators (PALMS) shift the human immune systems towards a T(H)2-dominated response. Allergy Asthma Clin Immunol. 5: 3.

Griesmeier, U., M. Bublin, C. Radauer, S. Vazquez-Cortes, Y. Ma, M. Fernandez-Rivas et al. (2010). Physicochemical properties and thermal stability of Lep w 1, the major allergen of whiff. Mol Nutr Food Res. 54: 861–869.

Hauser, M., A. Roulias, F. Ferreira and M. Egger (2010). Panallergens and their impact on the allergic patient. Allergy Asthma Clin Immunol. 6: 1.

Hauser, M., C. Asam, M. Himly, P. Palazzo, S. Voltolini, C. Montanari et al. (2011). Bet v 1-like pollen allergens of multiple Fagales species can sensitize atopic individuals. Clin Exp Allergy. 41: 1804–1814.

Herre, J., H. Gronlund, H. Brooks, L. Hopkins, L. Waggoner, B. Murton et al. (2013). Allergens as immunomodulatory proteins: the cat dander protein Fel d 1 enhances TLR activation by lipid ligands. J Immunol. 191: 1529–1535.

Hilger, C., F. Grigioni, L. Thill, L. Mertens and F. Hentges (2002). Severe IgE-mediated anaphylaxis following consumption of fried frog legs: definition of alpha-parvalbumin as the allergen in cause. Allergy. 57: 1053–1058.

Hilger, C., A. Kuehn and F. Hentges (2012). Animal lipocalin allergens. Curr Allergy Asthma Rep. 12: 438–447.

Hochwallner, H., U. Schulmeister, I. Swoboda, S. Spitzauer and R. Valenta (2014). Cow's milk allergy: from allergens to new forms of diagnosis, therapy and prevention. Methods. 66: 22–33.

Hoffmann-Sommergruber, K (2002). Pathogenesis-related (PR)-proteins identified as allergens. Biochem Soc Trans. 30: 930–935.

Hurlburt, B. K., L. R. Offermann, J. K. McBride, K. A. Majorek, S. J. Maleki and M. Chruszcz (2013). Structure and function of the peanut panallergen Ara h 8. J Biol Chem. 288: 36890–36901.

Ikura, M. (1996). Calcium binding and conformational response in EF-hand proteins. Trends Biochem Sci. 21: 14–17.

Ivanciuc, O., C. H. Schein and W. Braun (2003). SDAP: database and computational tools for allergenic proteins. Nucleic Acids Res. 31: 359–362.

Junker, Y., S. Zeissig, S. J. Kim, D. Barisani, H. Wieser, D. A. Leffler et al. (2012). Wheat amylase trypsin inhibitors drive intestinal inflammation via activation of toll-like receptor 4. J Exp Med. 209: 2395–2408.

Jyonouchi, S., V. Abraham, J. S. Orange, J. M. Spergel, L. Gober, E. Dudek et al. (2011). Invariant natural killer T cells from children with versus without food allergy exhibit differential responsiveness to milk-derived sphingomyelin. J Allergy Clin Immunol. 128: 102–109 e113.

Kader, J. C. (1996). Lipid-transfer proteins in plants. Annu Rev Plant Physiol Plant Mol Biol. 47: 627–654.

Karp, C. L. (2010). Guilt by intimate association: what makes an allergen an allergen? J Allergy Clin Immunol. 125: 955–960.

Katelaris, C. H. (2010). Food allergy and oral allergy or pollen-food syndrome. Curr Opin Allergy Clin Immunol. 10: 246–251.

Kawasaki, K. and K. M. Weiss (2003). Mineralized tissue and vertebrate evolution: the secretory calcium-binding phosphoprotein gene cluster. Proc Natl Acad Sci USA. 100: 4060–4065.

Kawasaki, K., A. G. Lafont and J. Y. Sire (2011). The evolution of milk casein genes from tooth genes before the origin of mammals. Mol Biol Evol. 28: 2053–2061.

Khuri, S., F. T. Bakker and J. M. Dunwell (2001). Phylogeny, function, and evolution of the cupins, a structurally conserved, functionally diverse superfamily of proteins. Mol Biol Evol. 18: 593–605.

Kleber-Janke, T., R. Crameri, U. Appenzeller, M. Schlaak and W. M. Becker (1999). Selective cloning of peanut allergens, including profilin and 2S albumins, by phage display technology. Int Arch Allergy Immunol. 119: 265–274.

Kreis, M., B. G. Forde, S. Rahman, B. J. Miflin and P. R. Shewry (1985). Molecular evolution of the seed storage proteins of barley, rye and wheat. J Mol Biol. 183: 499–502.

Kuehn, A., I. Swoboda, K. Arumugam, C. Hilger and F. Hentges (2014). Fish allergens at a glance: variable allergenicity of parvalbumins, the major fish allergens. Front Immunol. 5: 179.

Lascombe, M. B., B. Bakan, N. Buhot, D. Marion, J. P. Blein, V. Larue et al. (2008). The structure of "defective in induced resistance" protein of Arabidopsis thaliana, DIR1, reveals a new type of lipid transfer protein. Protein Sci. 17: 1522–1530.

Lauer, I., K. Foetisch, D. Kolarich, B. K. Ballmer-Weber, A. Conti, F. Altmann et al. (2004). Hazelnut (*Corylus avellana*) vicilin Cor a 11: molecular characterization of a glycoprotein and its allergenic activity. Biochem J. 383: 327–334.

Lee, P. W., J. A. Nordlee, S. J. Koppelman, J. L. Baumert and S. L. Taylor (2011). Evaluation and comparison of the species-specificity of 3 antiparvalbumin IgG antibodies. J Agric Food Chem. 59: 12309–12316.

Lehrer, S. B., M. D. Ibanez, M. L. McCants, C. B. Daul and J. E. Morgan (1990). Characterization of water-soluble shrimp allergens released during boiling. J Allergy Clin Immunol. 85: 1005–1013.

Lenton, S., T. Nylander, S. C. Teixeira and C. Holt (2015). A review of the biology of calcium phosphate sequestration with special reference to milk. Dairy Sci Technol. 95: 3–14.

Leung, P. S., K. H. Chu, W. K. Chow, A. Ansari, C. I. Bandea, H. S. Kwan et al. (1994). Cloning, expression, and primary structure of Metapenaeus ensis tropomyosin, the major heat-stable shrimp allergen. J Allergy Clin Immunol. 94: 882–890.

Lewit-Bentley, A. and S. Rety (2000). EF-hand calcium-binding proteins. Curr Opin Struct Biol. 10: 637–643.

Li, Y., S. Mui, J. H. Brown, J. Strand, L. Reshetnikova, L. S. Tobacman et al. (2002). The crystal structure of the C-terminal fragment of striated-muscle alpha-tropomyosin reveals a key troponin T recognition site. Proc Natl Acad Sci USA. 99: 7378–7383.

Liu, F., X. Zhang, C. Lu, X. Zeng, Y. Li, D. Fu et al. (2015). Non-specific lipid transfer proteins in plants: presenting new advances and an integrated functional analysis. J Exp Bot. 66: 5663–5681.

Lopata, A. L., R. E. O'Hehir and S. B. Lehrer (2010). Shellfish allergy. Clin Exp Allergy. 40: 850–858.

Lopata, A. L. and M. F. Jeebhay (2013). Airborne seafood allergens as a cause of occupational allergy and asthma. Curr Allergy Asthma Rep. 13: 288–297.

Lopez-Torrejon, G., G. Salcedo, M. Martin-Esteban, A. Diaz-Perales, C. Y. Pascual and R. Sanchez-Monge (2003). Len c 1, a major allergen and vicilin from lentil seeds: protein isolation and cDNA cloning. J Allergy Clin Immunol. 112: 1208–1215.

Lopez-Torrejon, G., M. D. Ibanez, O. Ahrazem, R. Sanchez-Monge, J. Sastre, M. Lombardero et al. (2005). Isolation, cloning and allergenic reactivity of natural profilin Cit s 2, a major orange allergen. Allergy. 60: 1424–1429.

Marcus, J. P., J. L. Green, K. C. Goulter and J. M. Manners (1999). A family of antimicrobial peptides is produced by processing of a 7S globulin protein in Macadamia integrifolia kernels. Plant J. 19: 699–710.

Mari, A. (2001). Multiple pollen sensitization: a molecular approach to the diagnosis. Int Arch Allergy Immunol. 125: 57–65.

Maria-Neto, S., R. V. Honorato, F. T. Costa, R. G. Almeida, D. S. Amaro, J. T. Oliveira et al. (2011). Bactericidal activity identified in 2S Albumin from sesame seeds and *in silico* studies of structure-function relations. Protein J. 30: 340–350.

Marichal, T., P. Starkl, L. L. Reber, J. Kalesnikoff, H. C. Oettgen, M. Tsai et al. (2013). A beneficial role for immunoglobulin E in host defense against honeybee venom. Immunity. 39: 963–975.

Markovic-Housley, Z., A. Basle, S. Padavattan, B. Maderegger, T. Schirmer and K. Hoffmann-Sommergruber (2009). Structure of the major carrot allergen Dau c 1. Acta Crystallogr D. Biol Crystallogr. 65: 1206–1212.

Mills, E. N., J. Jenkins, N. Marigheto, P. S. Belton, A. P. Gunning and V. J. Morris (2002). Allergens of the cupin superfamily. Biochem Soc Trans. 30: 925–929.

Mirotti, L., E. Florsheim, L. Rundqvist, G. Larsson, F. Spinozzi, M. Leite-de-Moraes et al. (2013). Lipids are required for the development of Brazil nut allergy: the role of mouse and human iNKT cells. Allergy. 68: 74–83.

Mittag, D., J. Akkerdaas, B. K. Ballmer-Weber, L. Vogel, M. Wensing, W. M. Becker et al. (2004). Ara h 8, a Bet v 1-homologous allergen from peanut, is a major allergen in patients with combined birch pollen and peanut allergy. J Allergy Clin Immunol. 114: 1410–1417.

Mittag, D., S. Vieths, L. Vogel, D. Wagner-Loew, A. Starke, P. Hunziker et al. (2005). Birch pollen-related food allergy to legumes: identification and characterization of the Bet v 1 homologue in mungbean (Vigna radiata), Vig r 1. Clin Exp Allergy. 35: 1049–1055.

Mittag, D., N. Varese, A. Scholzen, A. Mansell, G. Barker, G. Rice et al. (2013). TLR ligands of ryegrass pollen microbial contaminants enhance Th1 and Th2 responses and decrease induction of Foxp3(hi) regulatory T cells. Eur J Immunol. 43: 723–733.

Moreno, F. J. and A. Clemente (2008). 2S Albumin Storage Proteins: What Makes them Food Allergens? Open Biochem J. 2: 16–28.

Motoyama, K., S. Ishizaki, Y. Nagashima and K. Shiomi (2006). Cephalopod tropomyosins: identification as major allergens and molecular cloning. Food Chem Toxicol. 44: 1997–2002.

Naqpal, S., L. Rajappa, D. D. Metcalfe and P. V. Rao (1989). Isolation and characterization of heat-stable allergens from shrimp (*Penaeus indicus*). J Allergy Clin Immunol. 83: 26–36.

Neudecker, P., K. Schweimer, J. Nerkamp, S. Scheurer, S. Vieths, H. Sticht et al. (2001). Allergic cross-reactivity made visible: solution structure of the major cherry allergen Pru av 1. J Biol Chem. 276: 22756–22763.

Oda, Y., T. Matsunaga, K. Fukuyama, T. Miyazaki and T. Morimoto (1997). Tertiary and quaternary structures of 0.19 alpha-amylase inhibitor from wheat kernel determined by X-ray analysis at 2.06 Å resolution. Biochemistry. 36: 13503–13511.

Offermann, L. R., M. Bublin, M. L. Perdue, S. Pfeifer, P. Dubiela, T. Borowski et al. (2015). Structural and functional characterization of the Hazelnut allergen cor a 8. J Agric Food Chem. 63: 9150–9158.

Oftedal, O. T. (2012). The evolution of milk secretion and its ancient origins. Animal. 6: 355–368.

Ogawa, T., N. Bando, H. Tsuji, K. Nishikawa and K. Kitamura (1995). Alpha-subunit of beta-conglycinin, an allergenic protein recognized by IgE antibodies of soybean-sensitive patients with atopic dermatitis. Biosci Biotechnol Biochem. 59: 831–833.

Oliveira, G. B., D. Kunz, T. V. Peres, R. B. Leal, A. F. Uchoa, R. I. Samuels et al. (2014). Variant vicilins from a resistant Vigna unguiculata lineage (IT81D-1053) accumulate inside Callosobruchus maculatus larval midgut epithelium. Comp Biochem Physiol B Biochem Mol Biol. 168: 45–52.

Palm, N. W., R. K. Rosenstein and R. Medzhitov (2012). Allergic host defences. Nature. 484: 465–472.

Palm, N. W., R. K. Rosenstein, S. Yu, D. D. Schenten, E. Florsheim and R. Medzhitov (2013). Bee venom phospholipase A2 induces a primary type 2 response that is dependent on the receptor ST2 and confers protective immunity. Immunity. 39: 976–985.

Pastorello, E. A., L. Farioli, V. Pravettoni, M. Ispano, A. Conti, R. Ansaloni et al. (1998). Sensitization to the major allergen of Brazil nut is correlated with the clinical expression of allergy. J Allergy Clin Immunol. 102: 1021–1027.

Pastorello, E. A., L. Farioli, V. Pravettoni, C. Ortolani, M. Ispano, M. Monza et al. (1999). The major allergen of peach (Prunus persica) is a lipid transfer protein. J Allergy Clin Immunol. 103: 520–526.

Pastorello, E. A., L. Farioli, V. Pravettoni, M. Ispano, E. Scibola, C. Trambaioli et al. (2000). The maize major allergen, which is responsible for food-induced allergic reactions, is a lipid transfer protein. J Allergy Clin Immunol. 106: 744–751.

Pastorello, E. A., E. Varin, L. Farioli, V. Pravettoni, C. Ortolani, C. Trambaioli et al. (2001). The major allergen of sesame seeds (Sesamum indicum) is a 2S albumin. J Chromatogr B Biomed Sci Appl. 756: 85–93.

Pastorello, E. A., L. Farioli, V. Pravettoni, C. Ortolani, D. Fortunato, M. G. Giuffrida et al. (2003). Identification of grape and wine allergens as an endochitinase 4, a lipid-transfer protein, and a thaumatin. J Allergy Clin Immunol. 111: 350–359.

Pauls, T. L., J. A. Cox and M. W. Berchtold (1996). The Ca2+(–)binding proteins parvalbumin and oncomodulin and their genes: new structural and functional findings. Biochim Biophys Acta. 1306: 39–54.

Platts-Mills, T. A. and J. A. Woodfolk (2011). Allergens and their role in the allergic immune response. Immunol Rev. 242: 51–68.

Profet, M. (1991). The function of allergy: immunological defense against toxins. Q Rev Biol. 66: 23–62.

Pulendran, B., H. Tang and S. Manicassamy (2010). Programming dendritic cells to induce T(H)2 and tolerogenic responses. Nat Immunol. 11: 647–655.

Quirce, S. and A. Diaz-Perales (2013). Diagnosis and management of grain-induced asthma. Allergy Asthma Immunol Res. 5: 348–356.

Rabjohn, P., E. M. Helm, J. S. Stanley, C. M. West, H. A. Sampson, A. W. Burks et al. (1999). Molecular cloning and epitope analysis of the peanut allergen Ara h 3. J Clin Invest. 103: 535–542.

Radauer, C., M. Willerroider, H. Fuchs, K. Hoffmann-Sommergruber, J. Thalhamer, F. Ferreira et al. (2006). Cross-reactive and species-specific immunoglobulin E epitopes of plant profilins: an experimental and structure-based analysis. Clin Exp Allergy. 36: 920–929.

Radauer, C. and H. Breiteneder (2007). Evolutionary biology of plant food allergens. J Allergy Clin Immunol. 120: 518–525.

Radauer, C., M. Bublin, S. Wagner, A. Mari and H. Breiteneder (2008a). Allergens are distributed into few protein families and possess a restricted number of biochemical functions. J Allergy Clin Immunol. 121: 847–852 e847.

Radauer, C., P. Lackner and H. Breiteneder (2008b). The Bet v 1 fold: an ancient, versatile scaffold for binding of large, hydrophobic ligands. BMC Evol Biol. 8: 286.

Radauer, C., A. Nandy, F. Ferreira, R. E. Goodman, J. N. Larsen, J. Lidholm et al. (2014). Update of the WHO/IUIS Allergen Nomenclature Database based on analysis of allergen sequences. Allergy. 69: 413–419.

Reese, G., R. Ayuso and S. B. Lehrer (1999). Tropomyosin: an invertebrate pan-allergen. Int Arch Allergy Immunol. 119: 247–258.

Restani, P., C. Ballabio, C. Di Lorenzo, S. Tripodi and A. Fiocchi (2009). Molecular aspects of milk allergens and their role in clinical events. Anal Bioanal Chem. 395: 47–56.

Robotham, J. M., F. Wang, V. Seamon, S. S. Teuber, S. K. Sathe, H. A. Sampson et al. (2005). Ana o 3, an important cashew nut (*Anacardium occidentale* L.) allergen of the 2S albumin family. J Allergy Clin Immunol. 115: 1284–1290.

Rodin, J. and L. Rask (1990). Characterization of matteuccin, the 2.2S storage protein of the ostrich fern. Evolutionary relationship to angiosperm seed storage proteins. Eur J Biochem. 192: 101–107.

Ruiter, B. and W. G. Shreffler (2012). Innate immunostimulatory properties of allergens and their relevance to food allergy. Semin Immunopathol. 34: 617–632.

Salcedo, G., R. Sanchez-Monge, D. Barber and A. Diaz-Perales (2007). Plant non specific lipid transfer proteins: an interface between plant defence and human allergy. Biochim Biophys Acta. 1771: 781–791.

Salcedo, G., S. Quirce and A. Diaz-Perales (2011). Wheat allergens associated with Baker's asthma. J Investig Allergol Clin Immunol. 21: 81–92.

Sanchez-Lopez, J., L. Tordesillas, M. Pascal, R. Munoz-Cano, M. Garrido, M. Rueda et al. (2014). Role of Art v 3 in pollinosis of patients allergic to Pru p 3. J Allergy Clin Immunol. 133: 1018–1025.

Sanchez-Monge, R., L. Gomez, D. Barber, C. Lopez-Otin, A. Armentia and G. Salcedo (1992). Wheat and barley allergens associated with baker's asthma. Glycosylated subunits of the alpha-amylase-inhibitor family have enhanced IgE-binding capacity. Biochem J. 281(Pt 2): 401–405.

Santos, A. and R. Van Ree (2011). Profilins: mimickers of allergy or relevant allergens? Int Arch Allergy Immunol. 155: 191–204.

Scanlon, S. T. and A. N. McKenzie (2015). The messenger between worlds: the regulation of innate and adaptive type-2 immunity by innate lymphoid cells. Clin Exp Allergy. 45: 9–20.

Schoemaker, A. A., A. B. Sprikkelman, K. E. Grimshaw, G. Roberts, L. Grabenhenrich, L. Rosenfeld et al. (2015). Incidence and natural history of challenge-proven cow's milk allergy in European children—EuroPrevall birth cohort. Allergy. 70: 963–972.

Schutt, C. E., J. C. Myslik, M. D. Rozycki, N. C. Goonesekere and U. Lindberg (1993). The structure of crystalline profilin-beta-actin. Nature. 365: 810–816.

Shanti, K. N., B. M. Martin, S. Nagpal, D. D. Metcalfe and P. V. Rao (1993). Identification of tropomyosin as the major shrimp allergen and characterization of its IgE-binding epitopes. J Immunol. 151: 5354–5363.

Sharp, M. F. and A. L. Lopata (2014). Fish allergy: in review. Clin Rev Allergy Immunol. 46: 258–271.

Shewry, P. R. and A. S. Tatham (1990). The prolamin storage proteins of cereal seeds: structure and evolution. Biochem J. 267: 1–12.

Shewry, P. R., J. A. Napier and A. S. Tatham (1995). Seed storage proteins: structures and biosynthesis. Plant Cell. 7: 945–956.

Smole, U., C. Radauer, N. Lengger, M. Svoboda, N. Rigby, M. Bublin et al. (2015). The major birch pollen allergen Bet v 1 induces different responses in dendritic cells of birch pollen allergic and healthy individuals. PLoS One. 10: e0117904.

Somkuti, J., M. Bublin, H. Breiteneder and L. Smeller (2012). Pressure-temperature stability, Ca2+ binding, and pressure-temperature phase diagram of cod parvalbumin: Gad m 1. Biochemistry. 51: 5903–5911.

Starkl, P., T. Marichal, N. Gaudenzio, L. L. Reber, R. Sibilano, M. Tsai et al. (2015). IgE antibodies, FcepsilonRIalpha, and IgE-mediated local anaphylaxis can limit snake venom toxicity. J Allergy Clin Immunol.

Svensson, B., K. Fukuda, P. K. Nielsen and B. C. Bonsager (2004). Proteinaceous alpha-amylase inhibitors. Biochim Biophys Acta. 1696: 145–156.

Tatham, A. S. and P. R. Shewry (2008). Allergens to wheat and related cereals. Clin Exp Allergy. 38: 1712–1726.

Tejera, M. L., M. Villalba, E. Batanero and R. Rodriguez (1999). Identification, isolation, and characterization of Ole e 7, a new allergen of olive tree pollen. J Allergy Clin Immunol. 104: 797–802.

ten Grotenhuis, E., R. Tuinier and C. G. de Kruif (2003). Phase stability of concentrated dairy products. J Dairy Sci. 86: 764–769.

Teuber, S. S., A. M. Dandekar, W. R. Peterson and C. L. Sellers (1998). Cloning and sequencing of a gene encoding a 2S albumin seed storage protein precursor from English walnut (Juglans regia), a major food allergen. J Allergy Clin Immunol. 101: 807–814.

Teuber, S. S., K. C. Jarvis, A. M. Dandekar, W. R. Peterson and A. A. Ansari (1999). Identification and cloning of a complementary DNA encoding a vicilin-like proprotein, jug r 2, from english walnut kernel (Juglans regia), a major food allergen. J Allergy Clin Immunol. 104: 1311–1320.

Thomas, W. R., B. J. Hales and W. A. Smith (2010). House dust mite allergens in asthma and allergy. Trends Mol Med. 16: 321–328.

Tordesillas, L., S. Sirvent, A. Diaz-Perales, M. Villalba, J. Cuesta-Herranz, R. Rodriguez et al. (2011). Plant lipid transfer protein allergens: no cross-reactivity between those from foods and olive and Parietaria pollen. Int Arch Allergy Immunol. 156: 291–296.

Trompette, A., S. Divanovic, A. Visintin, C. Blanchard, R. S. Hegde, R. Madan et al. (2009). Allergenicity resulting from functional mimicry of a Toll-like receptor complex protein. Nature. 457: 585–588.

Tsai, M., P. Starkl, T. Marichal and S. J. Galli (2015). Testing the 'toxin hypothesis of allergy': mast cells, IgE, and innate and acquired immune responses to venoms. Curr Opin Immunol. 36: 80–87.

Valenta, R., M. Duchene, K. Pettenburger, C. Sillaber, P. Valent, P. Bettelheim et al. (1991). Identification of profilin as a novel pollen allergen; IgE autoreactivity in sensitized individuals. Science. 253: 557–560.

Vanek-Krebitz, M., K. Hoffmann-Sommergruber, M. Laimer da Camara Machado, M. Susani, C. Ebner, D. Kraft et al. (1995). Cloning and sequencing of Mal d 1, the major allergen from apple (Malus domestica), and its immunological relationship to Bet v 1, the major birch pollen allergen. Biochem Biophys Res Commun. 214: 538–551.

Vejvar, E., M. Himly, P. Briza, S. Eichhorn, C. Ebner, W. Hemmer et al. (2013). Allergenic relevance of nonspecific lipid transfer proteins 2: Identification and characterization of Api g 6 from celery tuber as representative of a novel IgE-binding protein family. Mol Nutr Food Res. 57: 2061–2070.

Vieths, S., S. Scheurer and B. Ballmer-Weber (2002). Current understanding of cross-reactivity of food allergens and pollen. Ann N Y Acad Sci. 964: 47–68.

Virtanen, T., T. Kinnunen and M. Rytkonen-Nissinen (2012). Mammalian lipocalin allergens—insights into their enigmatic allergenicity. Clin Exp Allergy. 42: 494–504.

Wal, J. M. (2004). Bovine milk allergenicity. Ann Allergy Asthma Immunol. 93: S2–11.

Wallner, M., A. Erler, M. Hauser, E. Klinglmayr, G. Gadermaier, L. Vogel et al. (2009). Immunologic characterization of isoforms of Car b 1 and Que a 1, the major hornbeam and oak pollen allergens. Allergy. 64: 452–460.

Wang, F., J. M. Robotham, S. S. Teuber, P. Tawde, S. K. Sathe and K. H. Roux (2002). Ana o 1, a cashew (*Anacardium occidental*) allergen of the vicilin seed storage protein family. J Allergy Clin Immunol. 110: 160–166.

Wang, F., J. M. Robotham, S. S. Teuber, S. K. Sathe and K. H. Roux (2003). Ana o 2, a major cashew (*Anacardium occidentale* L.) nut allergen of the legumin family. Int Arch Allergy Immunol. 132: 27–39.

Ward, T. J., R. L. Honeycutt and J. N. Derr (1997). Nucleotide sequence evolution at the kappa-casein locus: evidence for positive selection within the family Bovidae. Genetics. 147: 1863–1872.

Willart, M. A. and H. Hammad (2010). Alarming dendritic cells for allergic sensitization. Allergol Int. 59: 95–103.

Willison, L. N., S. K. Sathe and K. H. Roux (2014). Production and analysis of recombinant tree nut allergens. Methods. 66: 34–43.

Wills-Karp, M. (2010). Allergen-specific pattern recognition receptor pathways. Curr Opin Immunol. 22: 777–782.

Witke, W. (2004). The role of profilin complexes in cell motility and other cellular processes. Trends Cell Biol. 14: 461–469.

Zuidmeer, L., W. A. van Leeuwen, I. K. Budde, J. Cornelissen, I. Bulder, I. Rafalska et al. (2005). Lipid transfer proteins from fruit: cloning, expression and quantification. Int Arch Allergy Immunol. 137: 273–281.

2

Nomenclature of Food Allergens

Christian Radauer

CONTENTS

Department of Pathophysiology and Allergy Research, Medical University of Vienna, Austria.
E-mail: christian.radauer@muv.ac.at

2.1 INTRODUCTION

In the past few decades, great progress in identification and characterization of food allergens has been achieved. Major food allergens from a small number of sources were identified and purified as early as the 1970s. A boost in the number of newly identified allergens was elicited by the general availability of recombinant DNA technology in the late 1980s. Currently (September 2015), the IUIS Allergen Nomenclature Database contains 835 allergens from 252 sources and their isoforms and variants. Of these, 299 allergens from 101 sources are food allergens.

Originally, researchers were free in naming the allergens they had identified and each researcher used a different naming scheme. For instance, the major allergen from cod, beta-parvalbumin, was named allergen M (now termed Gad m 1 and Gad c 1 for the allergens from Atlantic and Baltic cod, respectively). In some instances, different groups even used different names for the same allergens. In order to establish a uniform system for the nomenclature of allergens, the IUIS Allergen Nomenclature Sub-Committee was founded in 1984 under the auspices of the World Health Organization (WHO) and the International Union of Immunological Societies (IUIS). It comprises leading experts in allergen characterization, structure, function, molecular biology, and bioinformatics and is responsible for maintaining and developing a unique, unambiguous and systematic nomenclature for all proteins inducing IgE-mediated allergies in humans, including food allergens. No other form or system of allergen nomenclature is recognized by the WHO or the IUIS. The committee also maintains the database of approved allergen names (www.allergen.org), which has developed from a plain text list to a fully functional, searchable database.

In order to maintain a consistent allergen nomenclature that complies with the guidelines established by the sub-committee, researchers are required to submit newly described allergens to the Allergen Nomenclature Sub-Committee before submitting their manuscript to a journal for consideration for publication. Submissions are kept confidential, and no specific information other than the name of the new allergen will be disclosed on the web site before publication. The submission form is available at www. allergen.org.

This chapter summarizes the current official nomenclature of allergens as established in previous publications (King, Hoffman et al. 1994a, King, Hoffman et al. 1994b, King, Hoffman et al. 1995a, King, Hoffman et al. 1995b, King, Hoffman et al. 1995c) and informs the reader about the guidelines to be followed if submitting new allergens to the Allergen Nomenclature Sub-Committee.

2.2 ALLERGEN NOMENCLATURE

The official nomenclature of allergenic proteins is based on the Linnaean binominal nomenclature identifying genus and species of all organisms. It was first published in 1986 (Marsh et al. 1986) and revised in 1994 (King, Hoffman et al. 1994a, King, Hoffman et al. 1994b, King, Hoffman et al. 1995a, King, Hoffman et al. 1995b, King, Hoffman et al. 1995c). Allergen names are usually kept consistent over time, but might be occasionally revised following new developments in allergy research (Chapman et al. 2007, Radauer et al. 2014).

The general format of allergen names is presented in Table 2.1. Allergen names are composed of a small letter indicating the origin of the allergen, the abbreviated genus name, a space, the abbreviated species name, a space, the allergen number, a period, and a four-digit number comprising isoallergen and variant numbers. In contrast to

Table 2.1 Allergen nomenclature using rAra h 3.0201 from peanut (*Arachis hypogaea*) as an example.

Element	Example	Explanation
Origin	r	n: natural; r: recombinant; s: synthetic
Genus	Ara	The first 3 or 4 letters of the genus name
Species	h	The first or the first two letters of the species name
Allergen number	3	Mostly assigned in chronological order of identification
Isoallergen number	02	Groups of allergen sequences from the same species with > 67% but < 90% sequence identity are defined as isoallergens
Variant number	01	Variants of an isoallergen have > 90% sequence identity

species names in biology, which are written in italics, allergen names are written in plain letters. Italicized letters are used to designate allergen-encoding genes.

The following guidelines apply when naming new allergens:

2.2.1 Origin

A small letter may be added to specify the origin of the allergens with n, r and s indicating natural, recombinant and synthetic allergens or allergen-derived peptides.

2.2.2 Genus and Species Names

Usually the first three letters of the genus and the first letter of the species are used. In cases of ambiguity, a fourth and second letter, respectively, may be added. For example, in the genus *Prunus* (stone fruits), allergens from apricot (*Prunus armeniaca*) and cherry (*P. avium*) are designated Pru ar and Pru av, respectively. The allergen nomenclature database follows the taxonomic system used in the NCBI (http://www.ncbi.nlm.nih.gov/taxonomy) and UniProt (http://www.uniprot.org/taxonomy/) sequence databases.

2.2.3 Allergen Numbers

Originally, allergen numbers were assigned in the order of identification of the allergens. Hence, Ara h 1 was the first identified peanut (*Arachis hypogaea*) allergen. With the increasing number of available allergen sequences and the advances in bioinformatics, the classification of allergens into families of evolutionary related proteins showed that most allergens can be grouped into a strikingly small number of protein families (Radauer et al. 2008). Homologous allergens from the same protein family have similar structures and sequences and are often, but not always, cross-reactive. In order to enhance the clarity of allergen names, the Allergen Nomenclature Sub-committee aims at assigning corresponding numbers to homologous allergens from related species (usually from the same family or order). For instance, in the rose family (Rosaceae), numbers 1 through 4 were consistently assigned to Bet v 1-related allergens (e.g., Mal d 1, Pru p 1, Fra a 1), thaumatin-like proteins (Mal d 2, Pru av 2), non-specific lipid transfer proteins (Mal d 3, Pru p 3) and profilins (Mal d 4, Pru av 4, Pyr c 4). Established names will not be changed in order to keep the literature consistent.

2.2.4 Isoallergens and Variants

Homologous allergens from one species with similar molecular masses, similar biochemical functions (if known) and sequence identities of more than 67% are called isoallergens (Figure 2.1). They receive the same allergen number and are distinguished by the first two digits of the four-digit number following the period. Closely-related sequences with more than 90% identity are called variants and distinguished using the third and fourth digit. The 67% and 90%

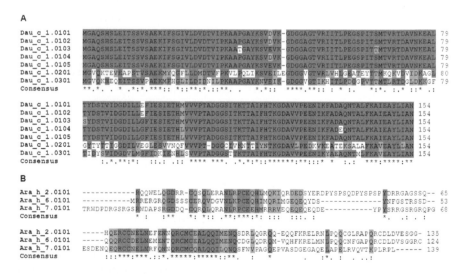

Figure 2.1 Examples for naming allergens, isoallergens and variants. **A.** Isoallergens and variants of the major carrot allergen, Dau c 1. Isoallergen sequences show identities between 49% and 71%, while the variants of Dau c 1.01 are 96–99% identical. **B.** Peanut allergens from the 2S albumin seed storage protein family. Ara h 2, Ara h 6 and Ara h 7 show low sequence identities between 31% and 58% and contain large sequence insertions and deletions when compared with each other. Hence they received different allergen numbers.

sequence identity thresholds serve only as guidelines and may be adjusted in certain cases to reflect the degree of sequence conservation in specific groups of sequences. For instance, Dau c 1, the major allergen from carrot (*Daucus carota*) is composed of three isoallergens. Dau c 1.02 has only 49–51% sequence identity to variants of Dau c 1.01 and Dau c 1.03, but due to their similar sequence lengths all those sequences were classified as isoallergens of a single allergen (Figure 2.1A). In contrast, Ara h 2 and Ara h 6 from peanut both belong to the 2S albumin family and show 58% sequence identity but aligning their sequences revealed insertions and deletion of considerable extent. Therefore they received different allergen numbers (Figure 2.1B).

Not all elements of an allergen name have to be specified. For example, nMal d 1 denotes the mixture of isoallergens as isolated from its natural source, Mal d 1.01 may indicate the first isoallergen without differentiating between variants, while rMal d 1.0101 specifies a particular variant produced as a recombinant protein. The nomenclature also contains a system for naming fragments, peptides, mutants and other allergen derivatives. For instance sMal d 1.0101 (120–132) names a synthetic peptide derived from the sequence of Mal d 1.0101. For further details, the reader is referred to the official allergen nomenclature publications (King, Hoffman et al. 1994a, King, Hoffman et al. 1994b, King, Hoffman et al. 1995a, King, Hoffman et al. 1995b, King, Hoffman et al. 1995c).

2.3 SUBMITTING NEW ALLERGENS TO THE WHO/IUIS ALLERGEN DATABASE

Researchers are requested to fill in an allergen submission form to be downloaded from the allergen nomenclature website (www.allergen. org). This form should contain as much relevant information about the proposed new allergens as possible. It should be submitted as soon as collecting data on the allergen has been completed and prior to submitting these data for publication. The completed form should be submitted by e-mail to the chair of the Allergen Nomenclature Sub-Committee whose contact details are indicated on the website.

Two (or more) members of the sub-committee will review the submission and assess whether the allergen fulfills the requirements for inclusion into the allergen nomenclature. The reviewers will also check for homologies with other known allergens and for any conflicts with already assigned allergen names. The review process will take approximately one month after the date of submission.

The chair of the sub-committee will notify investigators about the status of the proposed new allergen. If the allergen is deemed to fulfill the requirements, the investigators can use the new allergen nomenclature in publications, presentations, etc. The full committee will confirm the nomenclature at the next Sub-Committee meeting. Should the allergen not fulfill the requirements, the reviewers may ask for more data or for further clarification about the allergen from the investigator. Once the allergen name has been approved, the data will be included in the Allergen Nomenclature database at www. allergen.org.

Data to be filled into the submission form are summarized in Table 2.2. The following guidelines will be applied by the Allergen Nomenclature Sub-Committee in its evaluation procedure.

2.3.1 Allergen Source

Only allergens from sources causing IgE-mediated allergies in humans are included in the allergen nomenclature database. Hence, neither proteins allergenic for animals, such as dogs and horses, nor IgE binding homologs of allergens from organisms to which no allergic reactions has been described will be considered. For example, sequences of tropomyosins from invertebrates are highly conserved, which leads to a high extent of IgE cross-reactivity between invertebrate tropomyosins. They were identified as food allergens in crustaceans, mollusks and the fish parasite *Anisakis simplex* as well as inhalant allergens in mites and cockroaches (Jeong et al. 2006). A tropomyosin from fruit fly (*Drosophila melanogaster*) was also shown to bind IgE but received no official allergen names, as fruit flies have not been recognized as allergen sources (Leung et al. 1996).

Table 2.2 Data requested from submitters of new allergens to the IUIS allergen database.

Data	Notes
Source organism	
Name	Scientific name, common name and taxonomic classification
Tissue in which the allergen is expressed	Allergens are accepted only if the protein is expressed in a clinically relevant tissue or organ of the source organism. Specific sequences are accepted provided that their expression is shown at least at the mRNA level
Biochemical data of the allergen	
Proposed allergen name	The final name will be assigned by the allergen nomenclature sub-committee
Biochemical name	
Molecular mass	Including methods of determination
Protein sequence and accession number	At least a partial protein sequence is required. Submitters are required to submit their sequence data to the NCBI or UniProt sequence database
Nucleotide sequence and accession number	For cloned allergens. Submitters are required to submit their sequences to the GenBank/EMBL/DDBJ nucleotide sequence databases
Protein Data Bank accession number	If the three-dimensional structure is known
Post-translational modifications	Glycosylation, cleavage of signal peptides, etc., if known
Sequence-related publication	
Allergenicity data	
Route of exposure	Inhalation, ingestion, contact, etc.
Number of tested patients	
Methods of diagnosing the patients' allergies	Patients have to be allergic to the source of the submitted allergen
Methods of determining sensitization to the submitted allergen	Measurement of allergen-specific serum IgE, skin tests, basophil activation, etc.
Number of patients sensitized to the submitted allergen	Allergens are accepted only if at least 5 patients or 5% of the patients allergic to the respective source have IgE binding to the submitted allergen
Allergenicity related publication	

2.3.2 Sequence Data

At least partial sequences are required for submitted allergens. For allergens purified from their sources, these data may be obtained, for instance, by Edman degradation or tandem mass spectrometry. The allergen should be purified to an extent that allows the unambiguous connection of IgE binding with the sequenced protein. It is not sufficient to analyze a crude extract by SDS-PAGE and excise and sequence those bands which correspond to IgE binding bands in a Western blot. For allergen sequences amplified from genomic DNA, the expression of that particular isoform in the tissue relevant for allergic reaction has to be proved at least on the mRNA level. Researchers should submit their allergen sequence data to the respective nucleotide or protein sequence databases. Allergens submitted without associated accession numbers will not be accepted by the sub-committee.

2.3.3 Tested Patient Population

IgE binding to the submitted allergen had to be tested using sera from patients allergic to the source from which the submitted allergen has been obtained. For instance, when submitting a new apple allergen, IgE binding has to be tested with sera from apple allergic patients. It is not sufficient to analyze sera from birch pollen allergic patients instead, some of whom might have apple allergy while others might not.

2.3.4 Sensitization to the Submitted Allergen

Submitters are required to evaluate to IgE binding capability of the new allergen in a sufficiently large population of patients fulfilling the criteria described above. Allergen will receive an official allergen name only if either at least five patients or at least 5% of the tested patients have IgE binding to that allergen.

2.4 CONCLUSIONS

The official WHO/IUIS allergen nomenclature is the only recognized system of naming allergenic proteins. Researchers are requested to use exclusively official allergen names in their publications. In addition, creating IUIS-like names for new allergens without prior consultation of the Allergen Nomenclature Sub-Committee is strongly discouraged, because such names might create inconsistencies with future official names. Furthermore, it is important to always specify the exact sequence used when publishing allergenicity data. As an example, it has been shown that isoallergens of the Bet v 1-related allergens from celeriac (Api g 1) and carrot (Dau c 1) showed considerably different IgE binding capacities (Wangorsch et al. 2007, Wangorsch et al. 2012). Hence specifying only the allergen name without adding the isoallergen number would provide insufficient information.

The WHO/IUIS allergen nomenclature database is continuously updated and supplemented with newly submitted allergens as well as new data of already published allergens. The IUIS Allergen Nomenclature Sub-Committee encourages users to notify the committee of missing or inconsistent records in the database and thereby aid in providing a reliable and up-to-date resource of unambiguous allergen names as well as isoallergen and variant sequence information for the scientific community. Updates and error reports may be sent to the sub-committee by using the submission form or by contacting one of the committee members, whose contact details are published at www.allergen.org.

Keywords: Allergen family; WHO/IUIS; allergen nomenclature; allergen amino acid sequence; food allergy

REFERENCES

Chapman, M. D., A. Pomes, H. Breiteneder and F. Ferreira (2007). Nomenclature and structural biology of allergens. J Allergy Clin Immunol. 119(2): 414–420.

Jeong, K. Y., C. S. Hong and T. S. Yong (2006). Allergenic tropomyosins and their cross-reactivities. Protein Pept Lett. 13(8): 835–845.

King, T. P., D. Hoffman, H. Lowenstein, D. G. Marsh, T. A. Platts-Mills and W. Thomas (1994a). Allergen nomenclature. Int Arch Allergy Immunol. 105(3): 224–233.

King, T. P., D. Hoffman, H. Lowenstein, D. G. Marsh, T. A. Platts-Mills and W. Thomas (1994b). Allergen nomenclature. Bull World Health Organ. 72(5): 797–806.

King, T. P., D. Hoffman, H. Lowenstein, D. G. Marsh, T. A. Platts-Mills and W. Thomas (1995a). Allergen nomenclature. Clin Exp Allergy. 25(1): 27–37.

King, T. P., D. Hoffman, H. Lowenstein, D. G. Marsh, T. A. Platts-Mills and W. Thomas (1995b). Allergen nomenclature. Allergy. 50(9): 765–774.

King, T. P., D. Hoffman, H. Lowenstein, D. G. Marsh, T. A. Platts-Mills and W. Thomas (1995c). Allergen Nomenclature. J Allergy Clin Immunol. 96(1): 5–14.

Leung, P. S., W. K. Chow, S. Duffey, H. S. Kwan, M. E. Gershwin and K. H. Chu (1996). IgE reactivity against a cross-reactive allergen in crustacea and mollusca: evidence for tropomyosin as the common allergen. J Allergy Clin Immunol. 98(5 Pt 1): 954–961.

Marsh, D. G., L. Goodfriend, T. P. King, H. Lowenstein and T. A. Platts-Mills (1986). Allergen nomenclature. Bull World Health Organ. 64(5): 767–774.

Radauer, C., M. Bublin, S. Wagner, A. Mari and H. Breiteneder (2008). Allergens are distributed into few protein families and possess a restricted number of biochemical functions. J Allergy Clin Immunol. 121(4): 847–852 e847.

Radauer, C., A. Nandy, F. Ferreira, R. E. Goodman, J. N. Larsen, J. Lidholm, A. Pomes, M. Raulf-Heimsoth, P. Rozynek, W. R. Thomas and H. Breiteneder (2014). Update of the WHO/IUIS Allergen Nomenclature Database based on analysis of allergen sequences. Allergy. 69(4): 413–419.

Wangorsch, A., B. K. Ballmer-Weber, P. Rosch, T. Holzhauser and S. Vieths (2007). Mutational epitope analysis and cross-reactivity of two isoforms of Api g 1, the major celery allergen. Mol Immunol. 44(10): 2518–2527.

Wangorsch, A., D. Weigand, S. Peters, V. Mahler, K. Fotisch, A. Reuter, J. Imani, A. M. Dewitt, K. H. Kogel, J. Lidholm, S. Vieths and S. Scheurer (2012). Identification of a Dau c PRPlike protein (Dau c 1.03) as a new allergenic isoform in carrots (cultivar Rodelika). Clin Exp Allergy. 42(1): 156–166.

3

Nut Allergy

Dwan Price,[1] *Wesley Burks*[2] and *Cenk Suphioglu*[1],*

CONTENTS

[1] Neuro Allergy Research Laboratory (NARL), School of Life and Environmental Sciences, Deakin University, Victoria, Australia.
[2] Department of Pediatrics, The University of North Carolina, NC, USA.
* Corresponding author: cenk.suphioglu@deakin.edu.au

3.1 INTRODUCTION

The prevalence of nut allergy in western countries is a growing trend. Australian food allergy rates are the highest in the world with up to 10% of 12 month old children having a food allergy (Osborne et al. 2011). A severe peanut allergy is typically life threatening and persists through to adulthood (Fleischer et al. 2003). The safest means of managing the condition is strict avoidance from the diet, and access to emergency medications, such as a epinephrine autoinjector (Boyce et al. 2011).

In fact, peanut allergy rates are reported to have doubled since 1995, according to a small population study involving children living in the Australian Capital Territory, who were also referred to a specialist allergy clinic (Mullins et al. 2009). Also, allergy rates differ demografically within Australia. Challenge proven peanut allergy in a regional Australian city was lower at 0.8% (Molloy et al. 2015), compared to the closest major city, 100 Km away, where peanut allergy occurs in 3% of 12 month old infants (Osborne et al. 2011).

While Australia has some of the highest rates of peanut allergy, it is increasing globally. A study by Venter et al. (2010) conducted in the United Kingdom in 2009, demonstrated that peanut allergy had increased from 1.3% in 1989 to 2% in 2001-2002. Meanwhile, in a Canadian cohort, peanut allergy was estimated to affect 2.24%, while tree nut allergy affected 1.73% of children, these estimates were based on a random telephone survey of self reported and clinical diagnoses in 2008-2009 (Ben-Shoshan et al. 2012).

Also, a 2014 report by Bunyavanich and collegues (2014), confirms this increasing trend. Where, peanut specific IgE confirmed allergy, with prescription of epinephrine autoinjector, occurs at a rate of 2 to 4.9%, with peanut specific IgE at ≥ 0.35 kU/L and ≥ 14 kU/L respectively for children aged 7–10 years old in the northeast US.

While one study has shown that the rate of peanut allergy has doubled according to diagnoses by general practitioners (Kotz et al. 2011), another study has indicated that diagnosis rates of peanut allergy have tripled since 1999 (Rinaldi et al. 2012). Tree nut allergy in particular seems to be rising more in children (Sicherer et al. 2010). These findings are not uncommon throughout western countries (Grundy et al. 2002, Sicherer et al. 2003), where allergy rates to peanuts are lower in developing countries (Ahn et al. 2012, Ho et al. 2012).

Lastly, children of peanut farmers are just as likely to acquire a peanut allergy as the general population (Jagdis et al. 2015). In contrast, higher environmental exposure appears to enhance the likelihood of peanut allergy (Fox et al. 2009). Allergy to specific nuts also varies demographically, since hazelnut allergy is more common among Europeans (Burney et al. 2014). Suggesting, conflicting evidence as to whether possible increase in peanut exposure in westernized countries is a key factor in peanut allergy acquisition.

Collectively, this ongoing rising international trend emphasises the significance of therapeutic intervention and research directed at stopping the rising frequency of nut allergy.

3.2 WHY ARE NUT ALLERGENS SO ALLERGENIC?

There are several explanations as to why nuts are particularly more allergenic compared to other foods, and only a trace amount is capable of eliciting an immune response (Hourihane et al. 1997). Peanuts in particular are the major cause of an anaphylactic death after consuming a food allergen (Bock et al. 2007). These include:

3.2.1 Allergen Abundance

Most nuts contain more than one major allergen. According to World Health Organisation and International Union of Immunological Societies (WHO/IUIS) Allergen Nomenclature Sub-committee, the following allergens have been detected within the following nuts: Peanuts (*Arachis hypogaea*: Ara h 1 through to Ara h 17), Cashew nut (*Anacardium occidentale*: Ana o 1 through to Ana o 3), Hazelnut (*Corylus avellana*: Cor a 1 through to Cor a 14), walnut (*Juglans regia*: Jug r 1 through to Jug r 4), Pistachio (*Pistacia vera*: Pis v 1 through to Pis v 5, Brasil nut (*Bertholletia excelsa*: Ber e 1 through to Ber e 3), Almond (*Prunus dulcis*: Pru du 3 through to Pru du 6).

3.2.2 Complex Structural Integrity

Most nut allergens are structurally resilient and resistant to thermal treatment and degradation during gastrointestinal digestion (Astwood et al. 1996, Kopper et al. 2005). Many allergenic proteins which are functionally similar also share similar allergenic structural integrity. For example, Hazelnut allergen Cor a 14, a 2S albumin, is thermostable and digestion resistant (Pfeifer et al. 2015). Much like the 2S albumins from peanut, the major allergen Ara h 2 (Koppelman et al. 2010) and Ara h 6 (Hazebrouck et al. 2012). Nut allergens also, commonly have many binding sites for IgE (Burks et al. 1997, Stanley et al. 1997, Rabjohn et al. 1999). These IgE binding sites are often internalized within the allergens core (Maleki et al. 2000) and hence, unreachable by various digestive enzymes.

3.2.3 Special Allergen Attributes

Many nut allergens have been observed to also act in other ways to have increased allergenic potential. For example, Ara h 1 acts as an immune adjuvant through inducing monocyte-derived dendritic cells through attachment to Dendritic cell (DC)-specific ICAM-grabbing nonintegrin (Shreffler et al. 2006). Possible adjuvant activity has also been observed with peanut extract, which has been observed to disrupt tight junctions in human intestinal epithelial cells (Price et al. 2014), which seal the intestinal barrier from the external lumen, therefore allowing increased allergen contact with the intestinal immune system. Additionally, Ara h 2 is an effective inhibitor of trypsin (Maleki et al. 2003). This suggests that there is an increased chance that peanut allergens will remain intact and thus be exposed to the intestinal epithelium, from where they unleash their potential immune response.

3.3 WHAT THERAPIES ARE CURRENTLY ADDRESSING NUT ALLERGY?

Current therapies used to mitigate peanut allergy are mucosally targeted. Since, subcutaneous immunotherapies are not successful for treating peanut allergies, as they are associated with higher numbers of adverse reactions (Oppenheimer et al. 1992). Mucosally targeted therapies include oral therapies such as Oral Immunotherapy (OIT) and sublingual Immunotherapy (SLIT).

OIT shows promising success rates. A Randomised, double-blind placebo-controlled Food challenge (DBPCFC) trial, testing for the efficacy of orally administered peanut, observed that 62% of peanut recipients achieved increased tolerance to peanut compared to 0% of those assigned to placebo challenge during phase 1 of the study (Anagnostou et al. 2014). At the end of the study, 91% of individuals tolerated 800 mg peanut, and 54% of subjects tolerated a 1400 mg dose. Other OIT trials have demonstrated similar successful findings (Kim et al. 2011). Also, simultaneous dosing of probiotics in conjunction with OIT is currently showing promise as a successful treatment for peanut allergy (Tang et al. 2015). Preliminary data indicates that early intervention OIT in children aged less than 3 years old, demonstrates better success and sustained unresponsiveness to peanut (Vickery et al. 2015). Suggesting that treating peanut allergy early is more successful.

One study, comparing OIT and SLIT, demonstrated that OIT was far more effective at inducing unresponsiveness to peanut compared with SLIT (Chin et al. 2013). However, adverse reactions were also more likely to occur using OIT in addition to participant withdrawal prior to study end (Narisety et al. 2015). Indicating that participants are more likely to endure SLIT for safety reasons, despite the clinical success of OIT.

Another DBPCFC trial, examining the efficiency and safety of SLIT, observed that 70% of those receiving peanut sublingually demonstrated increased tolerance to peanut compared to 15% of placebo recipients. Among those responders, average peanut tolerance increased by more than 140% to 496 mg peanut (Fleischer et al. 2013). Long term follow up of this study, observed that responders generally maintained desensitisation to peanut, and its use as a potential therapy for peanut allergy is promising given its high safety rate (Burks et al. 2015). However, since this therapy is conducted on a daily basis over a number of years, participant adherence rates are considerably low, making it an unsuitable therapy for some.

Collectively, current available therapies all have their limitations. As yet, there is no safe, effective, practical therapy to treat peanut allergy other than strict dietary avoidance and emergency medication. This remains a major area of ongoing clinical and biological investigation.

3.4 EXPLORING CAUSES OF NUT ALLERGY

3.4.1 Breaking Down Barriers

3.4.1.1 Increased intestinal permeability

The specific role that a more permeable intestinal barrier plays in food allergy is yet undefined. A 2005 review by Liu et al. (Liu et al. 2005), summarises the paediatric diseases, which may be augmented via an ongoing leaky intestinal epithelium. These include diabetes, asthma, inflammatory bowel, allergy, celiac disease and even autism. Although there is a correlation between these diseases and intestinal permeability, additional data is required for a concise conclusion to be made about the subject.

Early studies assessing the macromolecular absorption of food particles in term and pre term infants, observed that premature babies had significantly higher absorption rates of β-lactoglobulin compared to term infants (Axelsson et al. 1989). On another note, increased intestinal permeability has been associated with babies fed with formula milk compared to babies fed with breast milk before 1 month of age (Taylor et al. 2009).

Collectively, if we consider the mode of action of the common immune adjuvant: cholera toxin, and the intestinal barrier response to peanut extract (Price et al. 2014). That is, they both act to disrupt tight junctions to increase intestinal permeability and consequent immune system exposure. Then we can apply this mechanism to infantile intestinal permeability with regards to nut allergen immune responses. For example, if increased infant intestinal permeability is associated with increased risk of allergy, then we could assume that premature infants, that have increased intestinal permeability compared to term infants, should be more likely to suffer from allergic disease.

Currently, evidence shows mixed views on whether prematurity is a risk factor for allergy. Earlier studies observed a connection between prematurity and atopy (Lucas et al. 1990, Kuehr et al. 1992). However, more recent studies have shown no association between the incidence of atopic disease and premature birth (Liem et al. 2007, Kvenshagen et al. 2009). Other studies indicate that premature birth may actually decrease the incidence of allergic disease (Pekkanen et al. 2001, Crump et al. 2011, Siltanen et al. 2011). In fact, prematurity may actually decrease the occurrence of atopy in the long term (Siltanen et al. 2001). The differences within these studies may be accredited partially to variation in study design. However, it must be acknowledged that the recent evidence indicates zero association and even a decreased risk of atopy for preterm individuals. Thus, an extra 'leaky' gut may not only be beneficial from a nutritional point of view for premature infants but may help to desensitise the intestinal immune system to potential allergens.

Collectively, it is likely that infant intestinal permeability and the infant immune system are developing concurrently, and that the developments of the two are intrinsically linked. Perhaps the increased permeability observed in preterm, and to a lesser degree term, neonates may allow for an increase in immune exposure to many dietary antigens. Additionally, if it is less likely for a preterm neonate to develop allergy despite increased intestinal permeability, then perhaps this extra 'leaky' gut leads to increased antigen exposure, and therefore may be protective against food allergy, through the promotion of antigen tolerance. This would therefore be dissimilar to the immune response we see in animal atopic studies to immune adjuvants like cholera toxin, which are usually adult animals with mature immune systems, and where intestinal barrier permeability is artificially amplified to increase allergen entry. Therefore, perhaps increased intestinal permeability is beneficial in promoting immune tolerance only when the immune system is developing during infancy.

Lastly, it is interesting to note that an infant's gastrointestinal tract is immature at birth, and continues to develop postnatal. For example, the acidity of the stomach does not appear to reach adult levels until the age of 2 (Deren 1971). Also, various proteolytic enzymes are below adult levels, as reviewed by Lebenthal and colleagues (1983). In addition, food allergens, especially nuts, are extremely robust regarding food processing, heat treatment, and digestion leaving relatively intact allergens or peptides with IgE-binding epitopes able to sensitise and elicit an immune reaction (described above). These characteristics, combined with a more permeable intestinal barrier in infants, favour the exposure of semi-digested/intact allergens to the mucosal immune system.

Infant gut digestive maturity and intestinal permeability may be critical in acclimatizing the developing infant's immune system to common environmental and dietary proteins, including nut allergens. The timing of allergen introduction, discussed below, during infant digestive and immune maturity may be a key player in allergic sensitization process, as infant gut maturity could ultimately dictate the allergic potential and intestinal absorption pathway of particular allergens, especially peanuts (Price et al. 2013).

3.4.1.2 Dermal barrier failure

Unlike the possible protective benefits of increased intestinal permeability, described above. Breakdown of the dermal barrier appears to increase the risk of food allergy. Where, loss of function mutations in the filaggrin gene is associated with peanut allergy (Brown et al. 2011) as well as other food allergies (Linneberg et al. 2013, Venkataraman et al. 2014, Ginkel et al. 2015). Where filaggrin plays a crucial role in maintaining dermal barrier function (Irvine et al. 2011). This indicates that dermal barrier function can play a critical role in the sensitization process, and that the sensitization process is not route specific. This supports the notion that food intolerances are initiated through the skin. Where, it has been shown that the dermal barrier is capable of sensitizing the immune system when an adjuvant is present (Dunkin et al. 2011), where the adjuvant promotes barrier breakdown. Exposure through skin has also shown to illicit a Th2 immune response to peanut allergens on areas of undamaged skin in mice, where adjuvant activity by peanut is suspected (Tordesillas et al. 2014). Also, unintentional exposure is likely to occur through the skin, since exposure is uncontrolled, unlike that of the oral route.

3.4.2 Initial Allergen Encounters—Is the Timing of Allergen Introduction Important?

3.4.2.1 In utero

It has been long recommended to delay the introduction of certain foods to infants at risk of developing atopic disease (Fiocchi et al. 2006), and antigen avoidance during pregnancy, to prevent atopic disease. Other studies with contradicting or inconclusive outcomes regarding maternal and infant avoidance diets prompted the re-evaluation of the dietary management strategies to prevent the incidence of allergic disease (Poole et al. 2006, Snijders et al. 2008). A report by Greer et al. (Greer et al. 2008) on early nutritional intervention, summarised that there is inconclusive evidence to suggest that maternal antigen avoidance poses an increased risk for the development of atopic disease.

In fact, A Danish population study indicated that peanut and tree nut consumption during pregnancy may actually reduce the occurrence of allergic disease in children (Maslova et al. 2012). Moreover, for non-nut allergic mothers that consumed peanuts during and just prior to pregnancy, there was decreased risk of peanut allergy in their children. These mothers were however, more likely to give their children nuts at an early age, so tolerance may have been achieved through early childhood introduction, discussed later, and not maternal diet (Frazier et al. 2014). Also, children in the United States were 47% less likely to have a peanut allergy if their mothers consumed peanuts in the first trimester of their pregnancy (Bunyavanich et al. 2014). However, infants with a pre-existing milk or egg intolerance were more likely to be sensitized to peanuts, as per elevated peanut specific IgE, if their mothers consumed peanuts during pregnancy (Sicherer et al. 2010). Also, peanut and hazelnut specific IgE have been detected in cord blood indicating that sensitization may in fact occur in utero (Pfefferle et al. 2008). Higher levels of total IgE have been observed in the cord blood of babies from non atopic mothers who were exposed to pesticides (Hernández et al. 2013), air pollution and cigarette smoke (Herr et al. 2011). However, IgE of sole feotal origin remains questionable (Bergmann et al. 1995). Thus, currently it appears debatable whether nut allergen exposure during pregnancy is a key risk factor for acquiring a nut allergy in childhood. However, current evidence may suggest that maternal consumption is protective against peanut allergy.

3.4.2.2 Breast milk

Recently, breastfed infants in the Australian Capital Territory where more likely to develop a parental reported nut allergy by the age of 5 if they were exclusively breastfed for the first 6 months of life, compared to combined feeding with breast milk and other foods, or other foods alone (Paton et al. 2012). This is not surprising since peanut allergens are present in breast milk upon maternal consumption (Vadas et al. 2001, Schocker et al. 2015), and therefore available for immune sensitisation. However, another study comparing breastfed children from Jewish mothers in the United Kingdom and Israel, observed a protective effect from peanut consumption. Where, peanut consumption is higher, and peanut allergy lower, in Israeli mothers compared to UK mothers (Du Toit et al. 2008). Also, peanut sensitised mice have demonstrated tolerance rather that sensitisation to peanut allergens contained within human milk, further suggesting a protective benefit (Bernard et al. 2014).

3.4.2.3 Early foods

There is lack of evidence to support delaying the introduction of certain foods, such as, those deemed highly allergenic (peanuts, fish and eggs) after the recommended time to introduce complimentary solid foods into the infant's diet, ~ 4 to 6 months of age. Other studies show similar findings (Lilja et al. 1989, Falth-Magnusson and Max Kjeltman 1992, Kramer 2000, Kramer and Kakuma 2006). In fact, in the same study mentioned above comparing peanut allergy among Israeli and UK Jewish children, there is a tenfold increase in the prevalence of peanut allergy among children who do not consume peanuts before the age of 1 (Du Toit et al. 2008).

Also, a recent randomized trial study, also demonstrated overall protective benefits against peanut allergy in infants with a pre-existing egg allergy, severe eczema or both combined. When peanuts were consumed compared to a peanut-free diet, the occurrence of peanut allergy was 3.2% and 17.2% respectively by the time the children where 5 years old. This same study aimed at assessing high-risk individuals who had a mild skin prick test to peanut in addition to other atopic conditions, compared to individuals with a negative skin prick test. Findings revealed that individuals at high risk for peanut allergy who avoided peanuts were more than 3 times as likely to have a peanut allergy by the time they were 5 years old. Also, individuals with negative skin prick test to peanut at the commencement of the study where 7 times more likely to have a peanut allergy at age 5 if they had avoided peanut protein (Du Toit et al. 2015). Overall, Peanut consumption rather than avoidance during infancy appears to be protective against peanut allergy compared to prior knowledge.

3.4.3 Immune System Development—Preparing the Gut for Nut Allergen Contact

3.4.3.1 The mucosal response to microbe colonization and gut development

The specific roles gut flora play in gut development is less understood. Hooper and colleagues (Hooper et al. 2001) investigated the effect of inoculating germ free mice with a single microflora *Bacterioides theaiotaomicron*. This resulted in the increased expression of genes related to nutrient absorption, gut motility, barrier integrity and toxin metabolism. Interestingly, germ free mice are also more easily sensitised to the peanut allergen Ara h 1 compared to mice with normal intestinal microflora (Stefka et al. 2014). Indicating that microflora colonisation is necessary to protect against allergy, at least in mice. Also, treatment of two intestinal cell lines, LS-123 and IEC-6 respectively, with varying bacterial components (lipopolysaccharide from *E. coli*, *K. pneuminia* and *P. aeruginosa*. Lipid A monophosphoryl from *E. coli*, lipothecoid acid from *S. faecalis* and peptidoglycan from *S. aures*) increased DNA expression in both cell types to variable degrees (Olaya et al. 2001). However, colonisation of the gut by certain bacteria have demonstrated direct interaction with tight junctions and thus modified barrier function by increasing intestinal permeability. *Clostridium perfringens* enterotoxin causes the delocalisation of the tight junction protein occludin in caco-2 tight junctions into larger more complex species within the cell (Singh et al. 2000). The same group also showed that the interaction caused cellular toxicity and that this toxicity permitted the access to additional tight junction proteins (Singh et al. 2001).

Also, specific populations of bacteria are more common than others in infants with food allergy, specifically Clostridium from cluster 1 (*Clostridium sensu stricto*) (Ling et al. 2014). This indicates that not all microbes promote healthy gut and immune development. Thus various mechanisms are likely in place to prevent the growth of pathogenic microbes while supporting the growth of others. It may also suggest that specific bacterial populations dominate in disease states, such as food allergy, as either cause or effect of gut/immune disequilibrium.

In response to luminal antigens, intestinal epithelial cells secrete a range of products. These control microbial overgrowth, preventing infection and promote microbial equilibrium within the intestinal environment. Meyer-Hoffert and collegues (2008), suggest that the ability to actively inhibit intestinal bacterial growth occurs predominately after weaning, and therefore the establishment of the neonatal microbiota may not be influenced by the presence of enteric derived antimicrobials. Antimicrobial peptides present within extracts of mucus obtained from mouse intestine actively killed both healthy commensal bacteria as well as pathogenic bacteria, significantly more than extracts obtained from the luminal contents. Here, the particular known microbials identified included defensins 1–4 and 6, thymosin-b, ubiquicidin, ribosomal proteins L29 and L35, lysozyme P (Meyer-Hoffert et al. 2008). Thus, proper colonisation of the gut by commensal flora is necessary for its proper physiological development and function as well as immune tolerance induction.

Given that our gut flora are a major stimulus for our developing gut and immune system, it is likely that an intricate combination of specific flora are required for the timely development and proper functioning of our gut and associated immune system. Also, it is highly possible that pre stimulation of the intestinal epithelium by commensal flora may play a role in epithelial conditioning to antigen sensitisation. For example, by improving barrier integrity, which controls allergen transport or acclimatising the immune response to foreign antigens. Many variables involved with the establishment of a healthy gut microbiome are discussed in further detail below.

3.4.3.2 *Normal establishment of the microbiome*

Changes to child rearing practices and lifestyle in the last century including birthing techniques, living conditions, diet and antibiotics all affect the intestinal microflora, but whether these alterations have been detrimental is unknown. This concept of decreased exposure to the microbial environment forms the basis of the hygiene hypothesis, where less atopic disease was observed among larger family sizes (Strachan 1989). Below we discuss the major influencing factors that may affect the intestinal flora. In doing so we can observe a correlation between deviations to intestinal health and development of nut allergy.

3.4.3.3 Living conditions

The immune system is constantly receiving information about external environmental antigens, through contact between the external environment and the various epithelia throughout the body. How our environment affects our sensitivity to specific antigens is not well understood. Environmental stimulation of the immune system by bacteria can be reflected by comparative epidemiological studies on individuals living anthroposophic or farming lifestyles, where these lifestyles are typically associated with lower prevalence of atopic disease. Those individuals who live an anthroposophic lifestyle, where there is a lack of immunisation and occurrence of childhood diseases are higher (e.g., measles), have a significantly reduced prevalence of allergy (Alm et al. 1999). This study also observed that those children who lived a anthroposophic lifestyle were typically also breast fed longer and consumed more fermented vegetables, which may directly affect the intestinal microflora (Alm et al. 1999). Similarly, a cross-sectional survey conducted by Douwes et al. (Douwes et al. 2008), on farmers in New Zealand, has demonstrated that exposure to a farming environment during pregnancy correlates positively with reduced childhood eczema, hay fever and asthma. An earlier study conducted in Germany demonstrating similar findings, observed a positive correlation between exposure to a farming environment and reduced asthma and hay fever (Ehrenstein et al. 2000). This may very well be attributed to endotoxin exposure as a result of a farming environment compared to a non-farming environment (Braun-Fahrländer 2003).

3.4.3.4 Birth type

Colonisation of the sterile intestinal tract by intestinal microflora begins immediately after birth. The type of birth influences the initial bacterial species colonising the newborns intestinal tract. Vaginal delivery more commonly promotes the initial colonisation of Bifidobacterium,

B. catenulatum and *B. longum* (Biasucci et al. 2008), as well as the presence of *Bacterioides fragilis* (Grölund et al. 1999). Interestingly, vaginal delivery also causes significant fluctuations to circulating cytokine levels compared to caesarean delivery (Malamitsi-Puchner et al. 2005). Caesarean delivery on the other hand, delays the colonisation of Bifidobacteria in the newborn gut for up to 6 months (Biasucci et al. 2008). These infants also typically have significantly higher numbers of *Clostridium perfringes* compared to vaginally delivered infants (Grölund et al. 1999). Interestingly, the number of caesarean births has risen from 19.1 to 29.5 per 100 births, as reported by a New South Wales study from 1998 to 2008 (Stavrou et al. 2011), with similar trends observed in the United States (MacDorman et al. 2008). Also, general sterility of the birthing environment has increased with the utilisation of hospital grade cleaners.

3.4.3.5 Infant feeding practices

Feeding practices also influence the composition of the intestinal microflora. Breastfeeding promotes the growth of Bifidobacteria (Harmsen et al. 2000). Also recently, the discovery of Ruminococci at equally high levels as Bifidobacteria, has been observed using molecular techniques (Favier et al. 2002, Favier et al. 2003). This observation of Ruminococci has lead to the finding of Ruminococci A, which has shown to prevent Clostridium (Dabard et al. 2001). Interestingly, breast fed infants are less frequently colonised by Clostridium before weaning (Stark and Lee 1982). Thus, the role that Ruminococci plays in interbacterial inhibition may be of interest when analysing the development of the intestinal mucosal immune system, especially regarding Clostridium colonisation.

3.4.3.6 Antibiotics

Antibiotics has been shown to induce sensitization to the major peanut allergen Ara h 1 in wild type mice much like the response observed in mice lacking a functional Toll-Like receptor 4, which is a receptor involved in bacterial lipopolysaccharide signaling (Bashir et al. 2004). This suggests that signaling between the gut immune system and the commensal flora is key in regulating immune responses to luminal antigens. Similar allergic responses were observed in antibiotic treated mice as well as germ free mice exposed to peanut and cholera toxin compared to normal colonized mice (Stefka et al. 2014). Here, it was shown that antibiotics effectively removed bacterial populations belonging to the phyla Firmicutes and Bacterioidetes. Where, these populations were replaced by those belonging to Lactobacilliacae. Interestingly, recolonizing the mice with Clostridia clusters XIVa, XIVb, and IV but not Bacterioides appeared to reduce the allergic response to peanut (Stefka et al. 2014). This observation could be attributed to the loss of induction and decreased expansion of T regulatory cells (Treg), where clostridium from clusters IV, XIVa and XVIII are assumed to provide an abundance of bacterial antigens to increase the bacterial antigen repertoire for the immune system (Atarashi et al. 2011), specifically CD4$^+$ CD25$^+$ Foxp3$^+$ Tregs (Russell et al. 2012). However, Clostridium clusters XIVa and XVIII are dominate in food allergic infant fecal samples (Ling et al. 2014), suggesting that their role in immune suppression is not as strong as previously thought.

As mentioned previously, Clostridium colonisation has been shown to be protective against peanut allergy (Stefka et al. 2014). However, Clostridia are also actively inhibited in breast fed infants, indicating that clostridia are opportunistic pathogens and their growth is suppressed by that of healthy commensals. However, despite Clostridium's pathogenic tendencies, perhaps their presence in the gut is also required to guard against immune overreaction. Since, breast fed infants are more likely to be sensitized to peanuts (Paton et al. 2012). It is more than likely however, that immune modulation is species specific regarding Clostridium.

An earlier study observed that prior to the development of atopy, distinct differences in the gut microflora are evident. Thus, a specific microflora composition may be crucial for the proper development of non-atopic immunity. This particular study used fluorescence *in situ* hybridisation, which is more sensitive than classical culturing techniques, to identify that atopic individuals had more Clostridia and less Bifidobacteria than their non-atopic counterparts (Kalliomaki et al. 2001). Similarly, in a more recent study, atopic individuals were also observed to have higher numbers of Clostridium difficile compared to non-atopic individuals using RT-PCR (Penders et al. 2007).

3.4.3.7 Probiotics

In recent years, the use of probiotics has been utilised for the potential treatment and prevention of atopic disease, but are subject to debate. Beneficial effects have been observed with the supplementation of *Lactobacillus casei* in the form of fermented milk, with a 33% reduction in the occurrence of allergic rhinitis, but little alteration to asthmatic episodes in children aged 2 to 5 in northern Italy (Giovannini et al. 2007). Also, maternal supplementation with *Lactobacillus rhamnosus* and *Bifidobacterium lactis* helped to prevent atopic disease in infants who were exclusively breast fed for more than 2.5 months by atopic mothers (Huurre et al. 2008). In addition, supplementation with a combined mixture of Lactobacillus, Bifidobacterium and Propionibacterium only helped to prevent IgE-associated allergy in infants delivered by caesarean up to the age of 5 years (Kuitunen et al. 2009). Conversely, another study observed no protective effect of probiotic supplementation with *Lactobacillus acidophilus* for the first 6 months of life, as observed at 1 year of age (Taylor et al. 2007). Similar results were obtained, where no collective preventative effect was observed for all allergies. However, borderline significance was observed for IgE-associated atopic disease, as well as a significant reduction in eczema (Kukkonen et al. 2007). Recently, a study comparing the bacterial species diversity of food allergic and non food allergic Chinese infants, demonstrated that food allergic infants had increased levels of the probiotic bacteria Lactobacillus and Bifidobacterium (Ling et al. 2014).

Given that antibiotic use can modify intestinal microflora populations and consequently increase the risk of nut sensitivity. Then we can apply this concept to the use of probiotics to act as immune adjuvants to stimulate desensitization to existing nut allergies. Since, in peanut allergic mice, Foxp3+ Tregs were induced and allergic response reduced after probiotic treatment (Barletta et al. 2013). Interestingly, another Treg and T helper cell inducing probiotic, *Lactobacillus rhamnosus* has shown promise at modulating the immune response to cow's milk (Pohjavuori et al. 2004) and more recently peanut allergy (Tang et al. 2015). Specifically, *L. Rhamnosus* was administered combined with peanut challenge in a double-blind placebo-controlled trial, to children aged between 1 and 10 for 1.5 years. Here, 80% of allergic individuals were unresponsive to peanut challenge at the end of the trial (Tang et al. 2015). It remains unclear whether prolonged unresponsiveness is achieved after 5.3 weeks, the study conclusion. Also recently, *Bacillus subtilis* expressing a Ara h 2 and Cholera toxin B fusion protein, was administered as a probiotic to peanut allergic mice. Here, peanut tolerance was induced through production of protective IgA instead of IgE (Zhou et al. 2015). In fact, probiotics have been shown to encourage the degradation of the peanut allergen Ara h 2 in human intestinal epithelial cells, through mechanisms increasing their barrier function (Song et al. 2012).

3.5 CONCLUSIONS

The incidence of nut allergy is increasing. High allergen content, robustness to degradation by heat and digestive processes, as well as immunological adjuvant attributes, makes nut allergens, particularly peanut, potent allergenic foods. Current therapies aimed at mitigating nut allergies have their limitations, and is an area of ongoing research and development. Current explanations for the increased rates of nut allergies, particularly peanut, are centralised around timing of exposure, barrier dysfunction, and microbial desensitisation of Th2 immune responses by environmental pathogen exposure. The escalation of nut allergy prevalence remains an enigma, and will continue to drive sufferers, scientists and clinicians alike, nutty for many years to come, until safe and effective therapeutics and prevention strategies are developed in the future.

Keywords: Nut allergy; allergen sensitisation; allergy therapy; food allergy

REFERENCES

Ahn, K., J. Kim, M. -I. Hahm, S. -Y. Lee, W. K. Kim, Y. Chae, Y. M. Park, M. Y. Han, K. -J. Lee, J. K. Kim, E. S. Yang and H. -J. Kwon (2012). Prevalence of immediate-type food allergy in Korean schoolchildren: a population-based study. Allergy and Asthma Proceedings: The Official Journal of Regional and State Allergy Societies. 33(6): 481–487.

Alm, J. S., J. Swartz, G. Lilja, A. Scheynius and G. r. Pershagen (1999). Atopy in children of families with an anthroposophic lifestyle. The Lancet. 353(9163): 1485–1488.

Anagnostou, K., S. Islam, Y. King, L. Foley, L. Pasea, C. Palmer, S. Bond, P. Ewan and A. Clark (2014). Study of induction of Tolerance to Oral Peanut: a randomised controlled trial of desensitisation using peanut oral immunotherapy in children (STOP II).

Astwood, J. D., J. N. Leach and R. L. Fuchs (1996). Stability of food allergens to digestion *in vitro*. Nature Biotechnology. 14(10): 1269–1273.

Atarashi, K., T. Tanoue, T. Shima, A. Imaoka, T. Kuwahara, Y. Momose, G. Cheng, S. Yamasaki, T. Saito and Y. Ohba (2011). Induction of colonic regulatory T cells by indigenous Clostridium species. Science. 331(6015): 337–341.

Axelsson, I., I. Jakobsson, T. Lindberg, S. Polberger, B. Benediktsson and N. Räihä (1989). Macromolecular absorption in Preterm and term infants. Acta Pœdiatrica. 78(4): 532–537.

Barletta, B., G. Rossi, E. Schiavi, C. Butteroni, S. Corinti, M. Boirivant and G. Di Felice (2013). Probiotic VSL# 3-induced TGF-β ameliorates food allergy inflammation in a mouse model of peanut sensitization through the induction of regulatory T cells in the gut mucosa. Molecular Nutrition & Food Research. 57(12): 2233–2244.

Bashir, M. E. H., S. Louie, H. N. Shi and C. Nagler-Anderson (2004). Toll-like receptor 4 signaling by intestinal microbes influences susceptibility to food allergy. The Journal of Immunology. 172(11): 6978–6987.

Ben-Shoshan, M., D. W. Harrington, L. Soller, J. Fragapane, L. Joseph, Y. S. Pierre, S. B. Godefroy, S. J. Elliott and A. E. Clarke (2012). Demographic predictors of Peanut, Tree Nut, Fish, Shellfish, and Sesame Allergy in Canada. Journal of Allergy. 2012: 6.

Bergmann, R., J. Schulz, S. Günther, J. Dudenhausen, K. Bergmann, C. Bauer, W. Dorsch, E. Schmidt, W. Luck and S. Lau (1995). Determinants of cord-blood IgE concentrations in 6401 German neonates. Allergy. 50(1): 65–71.

Bernard, H., S. Ah-Leung, M. F. Drumare, C. Feraudet-Tarisse, V. Verhasselt, J. M. Wal, C. Créminon and K. Adel-Patient (2014). Peanut allergens are rapidly transferred in human breast milk and can prevent sensitization in mice. Allergy. 69(7): 888–897.

Biasucci, G., B. Benenati, L. Morelli, E. Bessi and G. n. Boehm (2008). Cesarean delivery may affect the early biodiversity of intestinal bacteria. The Journal of Nutrition. 138(9): 1796S–1800S.

Bock, S. A., A. Muñoz-Furlong and H. A. Sampson (2007). Further fatalities caused by anaphylactic reactions to food, 2001–2006. Journal of Allergy and Clinical Immunology. 119(4): 1016–1018.

Boyce, J. A., A. Assa'ad, A. Burks, S. M. Jones, H. A. Sampson, R. A. Wood, M. Plaut, S. F. Cooper, M. J. Fenton and S. H. Arshad (2011). Guidelines for the diagnosis and management of food allergy in the United States: summary of the NIAID-sponsored expert panel report. Nutrition Research. 31(1): 61–75.

Braun-Fahrländer, C. (2003). Environmental exposure to endotoxin and other microbial products and the decreased risk of childhood atopy: evaluating developments since April 2002. Current Opinion in Allergy and Clinical Immunology. 3(5): 325–329.

Brown, S. J., Y. Asai, H. J. Cordell, L. E. Campbell, Y. Zhao, H. Liao, K. Northstone, J. Henderson, R. Alizadehfar and M. Ben-Shoshan (2011). Loss-of-function variants in the filaggrin gene are a significant risk factor for peanut allergy. Journal of Allergy and Clinical Immunology. 127(3): 661–667.

Bunyavanich, S., S. L. Rifas-Shiman, T. A. Platts-Mills, L. Workman, J. E. Sordillo, C. A. Camargo, M. W. Gillman, D. R. Gold and A. A. Litonjua (2014). Peanut, milk, and wheat intake during pregnancy is associated with reduced allergy and asthma in children. Journal of Allergy and Clinical Immunology. 133(5): 1373–1382.

Bunyavanich, S., S. L. Rifas-Shiman, T. A. E. Platts-Mills, L. Workman, J. E. Sordillo, M. W. Gillman, D. R. Gold and A. A. Litonjua (2014). Peanut allergy prevalence among school-age children in a US cohort not selected for any disease. Journal of Allergy and Clinical Immunology. 134(3): 753–755.

Burks, A. W., D. Shin, G. Cockrell, J. S. Stanley, R. M. Helm and G. A. Bannon (1997). Mapping and mutational analysis of the IgE-binding epitopes on Ara h 1, a legume vicilin protein and a major allergen in peanut hypersensitivity. European Journal of Biochemistry. 245(2): 334–339.

Burks, A. W., R. A. Wood, S. M. Jones, S. H. Sicherer, D. M. Fleischer, A. M. Scurlock, B. P. Vickery, A. H. Liu, A. K. Henning and R. Lindblad (2015). Sublingual immunotherapy for peanut allergy: Long-term follow-up of a randomized multicenter trial. Journal of Allergy and Clinical Immunology. 135(5): 1240–1248. e1243.

Burney, P. G. J., J. Potts, I. Kummeling, E. N. C. Mills, M. Clausen, R. Dubakiene, L. Barreales, C. Fernandez-Perez, M. Fernandez-Rivas, T. M. Le, A. C. Knulst, M. L. Kowalski, J. Lidholm, B. K. Ballmer-Weber, C. Braun-Fahlander, T. Mustakov, T. Kralimarkova, T. Popov, A. Sakellariou, N. G. Papadopoulos, S. A. Versteeg, L. Zuidmeer, J. H. Akkerdaas, K. Hoffmann-Sommergruber and R. van Ree

(2014). The prevalence and distribution of food sensitization in European adults. Allergy. 69(3): 365–371.

Chin, S. J., B. P. Vickery, M. D. Kulis, E. H. Kim, P. Varshney, P. Steele, J. Kamilaris, A. M. Hiegel, S. K. Carlisle and P. B. Smith (2013). Sublingual versus oral immunotherapy for peanut-allergic children: a retrospective comparison. Journal of Allergy and Clinical Immunology. 132(2): 476–478. e472.

Crump, C., K. Sundquist, J. Sundquist and M. A. Winkleby (2011). Gestational age at birth and risk of allergic rhinitis in young adulthood. Journal of Allergy and Clinical Immunology. 127(5): 1173–1179.

Dabard, J., C. Bridonneau, C. Phillipe, P. Anglade, D. Molle, M. Nardi, M. Ladire, H. Girardin, F. Marcille, A. Gomez and M. Fons (2001). Ruminococcin A, a new Lantibiotic produced by a Ruminococcus gnavus strain isolated from human feces. Appl Environ Microbiol. 67(9): 4111–4118.

Deren, J. S. (1971). Development of structure and function in the fetal and newborn stomach. Am J Clin Nutr. 24(1): 144–159.

Douwes, J., S. Cheng, N. Travier, C. Cohet, A. Niesink, J. McKenzie, C. Cunningham, G. Le Gros, E. von Mutius and N. Pearce (2008). Farm exposure in utero may protect against asthma, hay fever and eczema. Eur Respir J. 32(3): 603–611.

Du Toit, G., Y. Katz, P. Sasieni, D. Mesher, S. J. Maleki, H. R. Fisher, A. T. Fox, V. Turcanu, T. Amir, G. Zadik-Mnuhin, A. Cohen, I. Livne and G. Lack (2008). Early consumption of peanuts in infancy is associated with a low prevalence of peanut allergy. Journal of Allergy and Clinical Immunology. 122(5): 984–991.

Du Toit, G., G. Roberts, P. H. Sayre, H. T. Bahnson, S. Radulovic, A. F. Santos, H. A. Brough, D. Phippard, M. Basting and M. Feeney (2015). Randomized trial of peanut consumption in infants at risk for peanut allergy. New England Journal of Medicine. 372(9): 803–813.

Dunkin, D., M. C. Berin and L. Mayer (2011). Allergic sensitization can be induced via multiple physiologic routes in an adjuvant-dependent manner. Journal of Allergy and Clinical Immunology. 128(6): 1251–1258. e1252.

Ehrenstein, V., V. Mutius, Illi Baumann, Böhm and V. Kries (2000). Reduced risk of hay fever and asthma among children of farmers. Clinical & Experimental Allergy. 30(2): 187–193.

Falth-Magnusson, K. and N. I. Max Kjeltman (1992). Allergy prevention by maternal elimination diet during late pregnancy—a 5-year follow-up of a randomized study. Journal of Allergy and Clinical Immunology. 89(3): 709–713.

Favier, C. F., W. M. de Vos and A. D. L. Akkermans (2003). Development of bacterial and bifidobacterial communities in feces of newborn babies. Anaerobe. 9(5): 219–229.

Favier, C. F., E. E. Vaughan, W. M. De Vos and A. D. L. Akkermans (2002). Molecular monitoring of succession of bacterial communities in human neonates. Appl Environ Microbiol. 68(1): 219–226.

Fiocchi, A., A. Assa'ad and S. Bahna (2006). Food allergy and the introduction of solid foods to infants: a consensus document. Annals of Allergy, Asthma & Immunology. 97(1): 10–21.

Fleischer, D. M., A. W. Burks, B. P. Vickery, A. M. Scurlock, R. A. Wood, S. M. Jones, S. H. Sicherer, A. H. Liu, D. Stablein and A. K. Henning (2013). Sublingual immunotherapy for peanut allergy: a randomized, double-blind, placebo-controlled multicenter trial. Journal of Allergy and Clinical Immunology. 131(1): 119–127. e117.

Fleischer, D. M., M. K. Conover-Walker, L. Christie, A. W. Burks and R. A. Wood (2003). The natural progression of peanut allergy: resolution and the possibility of recurrence. Journal of Allergy and Clinical Immunology. 112(1): 183–189.

Fox, A. T., P. Sasieni, G. du Toit, H. Syed and G. Lack (2009). Household peanut consumption as a risk factor for the development of peanut allergy. Journal of Allergy and Clinical Immunology. 123(2): 417–423.

Frazier, A. L., C. A. Camargo, S. Malspeis, W. C. Willett and M. C. Young (2014). Prospective study of peripregnancy consumption of peanuts or tree nuts by mothers and the risk of peanut or tree nut allergy in their offspring. JAMA Pediatrics. 168(2): 156–162.

Ginkel, C., B. Flokstra-de Blok, B. Kollen, J. Kukler, G. Koppelman and A. Dubois (2015). Loss-of-function variants of the filaggrin gene associated with clinical reactivity to foods. Allergy. 70(4): 461–464.

Giovannini, M., C. Agostoni, E. Riva, F. Salvini, A. Ruscitto, G. V. Zuccotti, G. Radaelli and Felicita Study Group (2007). A randomized prospective double blind controlled trial on effects of long-term consumption of fermented milk containing Lactobacillus casei in pre-school children with allergic asthma and/or rhinitis. Pediatric Research. 62(2): 215–220 210.1203/PDR.1200b1013e3180a1276d1294.

Greer, F. R., S. H. Sicherer, A. W. Burks, a. t. C. o. Nutrition, S. o. Allergy and Immunology (2008). Effects of early nutritional interventions on the development of atopic disease in infants and children: The role of maternal dietary restriction, breastfeeding, timing of introduction of complementary foods, and hydrolyzed formulas. Pediatrics. 121(1): 183–191.

Grölund, M. -M., O. -P. Lehtonen, E. Eerola and P. Kero (1999). Fecal microflora in healthy infants born by different methods of delivery: Permanent changes in intestinal flora after cesarean delivery. Journal of Pediatric Gastroenterology and Nutrition. 28(1): 19–25.

Grundy, J., S. Matthews, B. Bateman, T. Dean and S. H. Arshad (2002). Rising prevalence of allergy to peanut in children: Data from 2 sequential cohorts. Journal of Allergy and Clinical Immunology. 110(5): 784–789.

Harmsen, H. J. M., A. C. M. Wildeboer-Veloo, G. C. Raangs, A. A. Wagendorp, N. Klijn, J. G. Bindels and G. W. Welling (2000). Analysis of intestinal flora development in breast-fed and formula-fed infants by using molecular identification and detection methods. Journal of Pediatric Gastroenterology and Nutrition. 30(1): 61–67.

Hazebrouck, S., B. Guillon, M. F. Drumare, E. Paty, J. M. Wal and H. Bernard (2012). Trypsin resistance of the major peanut allergen Ara h 6 and allergenicity of the digestion products are abolished after selective disruption of disulfide bonds. Molecular Nutrition & Food Research. 56(4): 548–557.

Hernández, E., A. Barraza-Villarreal, M. C. Escamilla-Nunez, L. Hernández-Cadena, P. D. Sly, L. M. Neufeld, U. Ramakishnan and I. Romieu (2013). Prenatal determinants of cord blood total immunoglobulin E levels in Mexican newborns. Allergy and Asthma Proceedings, Ocean Side Publications.

Herr, C. E., R. Ghosh, M. Dostal, V. Skokanova, P. Ashwood, M. Lipsett, J. P. Joad, K. E. Pinkerton, P. S. Yap and J. D. Frost (2011). Exposure to air pollution in critical prenatal time windows and IgE levels in newborns. Pediatric Allergy and Immunology. 22(1 Part I): 75–84.

Ho, M. H. K., S. L. Lee, W. H. S. Wong, P. Ip and Y. L. Lau (2012). Prevalence of self-reported food allergy in Hong Kong children and teens—a population survey. Asian Pacific Journal of Allergy and Immunology/Launched by the Allergy and Immunology Society of Thailand. 30(4): 275–284.

Hooper, L. V., M. H. Wong, A. Thelin, L. Hansson, P. G. Falk and J. I. Gordon (2001). Molecular analysis of commensal host-microbial relationships in the intestine. Science. 291(5505): 881–884.

Hourihane, J. O. B., S. A. Kilburn, J. A. Nordlee, S. L. Hefle, S. L. Taylor and J. O. Warner (1997). An evaluation of the sensitivity of subjects with peanut allergy to very low doses of peanut protein: A randomized, double-blind, placebo-controlled food challenge study. Journal of Allergy and Clinical Immunology. 100(5): 596–600.

Huurre, A., K. Laitinen, S. Rautava, M. Korkeamäki and E. Isolauri (2008). Impact of maternal atopy and probiotic supplementation during pregnancy on infant sensitization: a double-blind placebo-controlled study. Clinical & Experimental Allergy. 38(8): 1342–1348.

Irvine, A. D., W. I. McLean and D. Y. Leung (2011). Filaggrin mutations associated with skin and allergic diseases. New England Journal of Medicine. 365(14): 1315–1327.

Jagdis, A., G. Liss, S. Maleki and P. Vadas (2015). Prevalence of peanut allergy in offspring of peanut farmers. Food and Nutrition Sciences. 6(01): 29.

Kalliomaki, M., P. Kirjavainen, E. Eerola, P. Kero, S. Salminen and E. Isolauri (2001). Distinct patterns of neonatal gut microflora in infants in whom atopy was and was not developing. Journal of Allergy and Clinical Immunology. 107(1): 129–134.

Kim, E. H., J. A. Bird, M. Kulis, S. Laubach, L. Pons, W. Shreffler, P. Steele, J. Kamilaris, B. Vickery and A. W. Burks (2011). Sublingual immunotherapy for peanut allergy: clinical and immunologic evidence of desensitization. Journal of Allergy and Clinical Immunology. 127(3): 640–646. e641.

Koppelman, S. J., S. L. Hefle, S. L. Taylor and G. A. de Jong (2010). Digestion of peanut allergens Ara h 1, Ara h 2, Ara h 3, and Ara h 6: A comparative *in vitro* study and partial characterization of digestion-resistant peptides. Molecular Nutrition & Food Research. 54(12): 1711–1721.

Kopper, R. A., N. J. Odum, M. Sen, R. M. Helm, J. S. Stanley and A. W. Burks (2005). Peanut protein allergens: the effect of roasting on solubility and allergenicity. International Archives of Allergy and Immunology. 136(1): 16–22.

Kotz, D., C. R. Simpson and A. Sheikh (2011). Incidence, prevalence, and trends of general practitioner-recorded diagnosis of peanut allergy in England, 2001 to 2005. Journal of Allergy and Clinical Immunology. 127(3): 623–630. e621.

Kramer, M. S. (2000). Maternal antigen avoidance during pregnancy for preventing atopic disease in infants of women at high risk. Cochrane Database Syst Rev. (2): CD000133.

Kramer, M. S. and R. Kakuma (2006). Maternal dietary antigen avoidance during pregnancy or lactation, or both, for preventing or treating atopic disease in the child. Cochrane Database Syst Rev. 3: CD000133.

Kuehr, J., T. Frischer, W. Karmaus, R. Meinert, R. Barth, E. Herrmann-Kunz, J. Forster and R. Urbanek (1992). Early childhood risk factors for sensitization at school age. J Allergy Clin Immunol. 90(3 Pt 1): 358–363.

Kuitunen, M., K. Kukkonen, K. Juntunen-Backman, R. Korpela, T. Poussa, T. Tuure, T. Haahtela and E. Savilahti (2009). Probiotics prevent IgE-associated allergy until age 5 years in cesarean-delivered children but not in the total cohort. Journal of Allergy and Clinical Immunology. 123(2): 335–341.

Kukkonen, K., E. Savilahti, T. Haahtela, K. Juntunen-Backman, R. Korpela, T. Poussa, T. Tuure and M. Kuitunen (2007). Probiotics and prebiotic galacto-oligosaccharides in the prevention of allergic diseases: A randomized, double-blind, placebo-controlled trial. Journal of Allergy and Clinical Immunology. 119(1): 192–198.

Kvenshagen, B., M. Jacobsen and R. Halvorsen (2009). Atopic dermatitis in premature and term children. Archives of Disease in Childhood. 94(3): 202-205.

Lebenthal, E., P. C. Lee and L. A. Heitlinger (1983). "Impact of development of the gastrointestinal tract on infant feeding." The Journal of Pediatrics. 102(1): 1–9.

Liem, J. J., A. L. Kozyrskyj, S. I. Huq and A. B. Becker (2007). The risk of developing food allergy in premature or low-birth-weight children. New York, NY, ETATS-UNIS, Elsevier.

Lilja, G., A. Dannaeus, T. Foucard, V. Graff-Lonnevig, S. G. O. Johansson and H. ÖMan (1989). Effects of maternal diet during late pregnancy and lactation on the development of atopic diseases in infants up to 18 months of age—*in-vivo* results. Clinical & Experimental Allergy. 19(4): 473–479.

Ling, Z., Z. Li, X. Liu, Y. Cheng, Y. Luo, X. Tong, L. Yuan, Y. Wang, J. Sun, L. Li and C. Xiang (2014). Altered fecal microbiota composition associated with food allergy in infants. Applied and Environmental Microbiology. 80(8): 2546–2554.

Linneberg, A., R. V. Fenger, L. -L. Husemoen, B. H. Thuesen, T. Skaaby, A. Gonzalez-Quintela, C. Vidal, B. C. Carlsen, J. D. Johansen and T. Menné (2013). Association between loss-of-function mutations in the filaggrin gene and self-reported food allergy and alcohol sensitivity. International Archives of Allergy and Immunology. 161(3): 234–242.

Liu, Z., N. Li and J. Neu (2005). Tight junctions, leaky intestines, and pediatric diseases. Acta Paediatr. 94(4): 386–393.

Lucas, A., O. G. Brooke, T. J. Cole, R. Morley and M. F. Bamford (1990). Food and drug reactions, wheezing, and eczema in preterm infants. Arch Dis Child. 65(4): 411–415.

MacDorman, M. F., F. Menacker and E. Declercq (2008). Cesarean birth in the United States: Epidemiology, trends, and outcomes. Clinics in Perinatology. 35(2): 293–307.

Malamitsi-Puchner, A., E. Protonotariou, T. Boutsikou, E. Makrakis, A. Sarandakou and G. Creatsas (2005). The influence of the mode of delivery on circulating cytokine concentrations in the perinatal period. Early Human Development. 81(4): 387–392.

Maleki, S. J., R. A. Kopper, D. S. Shin, P. ChunWook, C. M. Compadre, H. Sampson, A. W. Burks and G. A. Bannon (2000). Structure of the major peanut allergen Ara h 1 may protect IgE-binding epitopes from degradation. Journal of Immunology (Baltimore). 164(11): 5844–5849.

Maleki, S. J., O. Viquez, T. Jacks, H. Dodo, E. T. Champagne, S. Y. Chung and S. J. Landry (2003). The major peanut allergen, Ara h 2, functions as a trypsin inhibitor, and roasting enhances this function. The Journal of Allergy and Clinical Immunology. 112(1): 190–195.

Maslova, E., C. Granström, S. Hansen, S. B. Petersen, M. Strøm, W. C. Willett and S. F. Olsen (2012). Peanut and tree nut consumption during pregnancy and allergic disease in children—should mothers decrease their intake? Longitudinal evidence from the Danish National Birth Cohort. Journal of Allergy and Clinical Immunology. 130(3): 724–732.

Meyer-Hoffert, U., M. W. Hornef, B. Henriques-Normark, L. -G. Axelsson, T. Midtvedt, K. Pütsep and M. Andersson (2008). Secreted enteric antimicrobial activity localises to the mucus surface layer. Gut. 57(6): 764–771.

Molloy, J., J. Koplin, A. -L. Ponsonby, M. L. K. Tang, F. Collier, K. Allen and P. Vuillermin (2015). Prevalence of challenge-proven IgE-mediated food allergy in infants in the Barwon Region, Victoria, Australia. Clinical and Translational Allergy. 5(Suppl. 3): P89–P89.

Mullins, R. J., K. B. Dear and M. L. Tang (2009). Characteristics of childhood peanut allergy in the Australian Capital Territory, 1995 to 2007. Journal of Allergy and Clinical Immunology. 123(3): 689–693.

Narisety, S. D., P. A. Frischmeyer-Guerrerio, C. A. Keet, M. Gorelik, J. Schroeder, R. G. Hamilton and R. A. Wood (2015). A randomized, double-blind, placebo-controlled pilot study of sublingual versus oral immunotherapy for the treatment of peanut allergy. Journal of Allergy and Clinical Immunology. 135(5): 1275–1282. e1276.

Olaya, J. H., V. Neopikhanov, C. Söderman and A. Uribe (2001). Bacterial wall components such as Lipothecoid acid, Peptidoglycan, Liposaccharide and lipid A stimulate cell proliferation in intestinal epithelial cells. Microbial Ecology in Health and Disease. 13(2): 124–128.

Oppenheimer, J. J., H. S. Nelson, S. A. Bock, F. Christensen and D. Y. Leung (1992). Treatment of peanut allergy with rush immunotherapy. Journal of Allergy and Clinical Immunology. 90(2): 256–262.

Osborne, N. J., J. J. Koplin, P. E. Martin, L. C. Gurrin, A. J. Lowe, M. C. Matheson, A. -L. Ponsonby, M. Wake, M. L. K. Tang, S. C. Dharmage and K. J. Allen (2011). Prevalence of challenge-proven IgE-mediated food allergy using population-based sampling and predetermined challenge criteria in infants. Journal of Allergy and Clinical Immunology. 127(3): 668–676.e662.

Paton, J., M. Kljakovic, K. Ciszek and P. Ding (2012). Infant feeding practices and nut allergy over time in Australian school entrant children. International Journal of Pediatrics 2012.

Pekkanen, J., B. Xu and M. R. Järvelin (2001). Gestational age and occurrence of atopy at age 31—a prospective birth cohort study in Finland. Clinical & Experimental Allergy. 31(1): 95–102.

Penders, J., C. Thijs, P. A. van den Brandt, I. Kummeling, B. Snijders, F. Stelma, H. Adams, R. van Ree and E. E. Stobberingh (2007). Gut microbiota composition and development of atopic manifestations in infancy: the KOALA Birth Cohort Study. Gut. 56(5): 661–667.

Pfefferle, P. I., S. Sel, M. J. Ege, G. Büchele, N. Blümer, S. Krauss-Etschmann, I. Herzum, C. E. Albers, R. P. Lauener and M. Roponen (2008). Cord blood allergen-specific IgE is associated with reduced IFN-γ production by cord blood cells: The protection against allergy—study in rural environments (PASTURE) study. Journal of Allergy and Clinical Immunology. 122(4): 711–716.

Pfeifer, S., M. Bublin, P. Dubiela, K. Hummel, J. Wortmann, G. Hofer, W. Keller, C. Radauer and K. Hoffmann-Sommergruber (2015). Cor a 14, the allergenic 2S albumin from hazelnut, is highly thermostable and resistant to gastrointestinal digestion. Molecular Nutrition & Food Research.

Pohjavuori, E., M. Viljanen, R. Korpela, M. Kuitunen, M. Tiittanen, O. Vaarala and E. Savilahti (2004). Lactobacillus GG effect in increasing IFN-γ production in infants with cow's milk allergy. Journal of Allergy and Clinical Immunology. 114(1): 131–136.

Poole, J. A., K. Barriga, D. Y. M. Leung, M. Hoffman, G. S. Eisenbarth, M. Rewers and J. M. Norris (2006). Timing of initial Eexposure to cereal grains and the risk of wheat allergy. Pediatrics. 117(6): 2175–2182.

Price, D., L. Ackland and C. Suphioglu (2013). Nuts'n'guts: transport of food allergens across the intestinal epithelium. Asia Pacific Allergy. 3(4): 257.

Price, D., M. Ackland, W. Burks, M. Knight and C. Suphioglu (2014). Peanut allergens alter intestinal barrier permeability and tight junction localisation in caco-2 cell cultures 1. Cellular Physiology and Biochemistry. 33(6): 1758–1777.

Rabjohn, P., E. M. Helm, J. S. Stanley, C. M. West, H. A. Sampson, A. W. Burks and G. A. Bannon (1999). Molecular cloning and epitope analysis of the peanut allergen Ara h 3. The Journal of Clinical Investigation. 103(4): 535–542.

Rinaldi, M., L. Harnack, C. Oberg, P. Schreiner, J. St. Sauver and L. L. Travis (2012). Peanut allergy diagnoses among children residing in Olmsted County, Minnesota. Journal of Allergy and Clinical Immunology. 130(4): 945–950.

Russell, S. L., M. J. Gold, M. Hartmann, B. P. Willing, L. Thorson, M. Wlodarska, N. Gill, M. R. Blanchet, W. W. Mohn and K. M. McNagny (2012). Early life antibiotic-driven changes in microbiota enhance susceptibility to allergic asthma. EMBO Reports. 13(5): 440–447.

Schocker, F., J. Baumert, S. Kull, A. Petersen and U. Jappe (2015). Maternal transfer of the most potent peanut allergen Ara h 2 into human breast milk in a German cohort. Clinical and Translational Allergy. 5(Suppl. 3): P24.

Shreffler, W. G., R. R. Castro, Z. Y. Kucuk, Z. Charlop-Powers, G. Grishina, S. Yoo, A. Burks and H. A. Sampson (2006). The major glycoprotein allergen from Arachis hypogaea, Ara h 1, is a ligand of dendritic cell-specific ICAM-grabbing nonintegrin and acts as a Th2 adjuvant *in vitro*. The Journal of Immunology. 177(6): 3677.

Sicherer, S. H., A. Muñoz-Furlong, J. H. Godbold and H. A. Sampson (2010). US prevalence of self-reported peanut, tree nut, and sesame allergy: 11-year follow-up. Journal of Allergy and Clinical Immunology. 125(6): 1322–1326.

Sicherer, S. H., A. Muñoz-Furlong and H. A. Sampson (2003). Prevalence of peanut and tree nut allergy in the United States determined by means of a random digit dial telephone survey: A 5-year follow-up study. Journal of Allergy and Clinical Immunology. 112(6): 1203–1207.

Sicherer, S. H., R. A. Wood, D. Stablein, R. Lindblad, A. W. Burks, A. H. Liu, S. M. Jones, D. M. Fleischer, D. Y. Leung and H. A. Sampson (2010). Maternal consumption of peanut during pregnancy is associated with peanut sensitization in atopic infants. Journal of Allergy and Clinical Immunology. 126(6): 1191–1197.

Siltanen, M., M. Kajosaari, M. Pohjavuori and E. Savilahti (2001). Prematurity at birth reduces the long-term risk of atopy. Journal of Allergy and Clinical Immunology. 107(2): 229–234.

Siltanen, M., K. Wehkalampi, P. Hovi, J. G. Eriksson, S. Strang-Karlsson, A. -L. Järvenpää, S. Andersson and E. Kajantie (2011). Preterm birth reduces the incidence of atopy in adulthood. Journal of Allergy and Clinical Immunology. 127(4): 935–942.

Singh, U., L. L. Mitic, E. U. Wieckowski, J. M. Anderson and B. A. McClane (2001). Comparative biochemical and immunocytochemical studies reveal differences in the effects of Clostridium perfringens enterotoxin on polarized caco-2 cellsversus vero cells. Journal of Biological Chemistry. 276(36): 33402–33412.

Singh, U., C. M. Van Itallie, L. L. Mitic, J. M. Anderson and B. A. McClane (2000). CaCo-2 cells treated with Clostridium perfringens enterotoxin form multiple large complex species, one of which contains the tight junction protein occludin. Journal of Biological Chemistry. 275(24): 18407–18417.

Snijders, B. E. P., C. Thijs, R. van Ree and P. A. van den Brandt (2008). Age at first introduction of cow milk products and other food products in relation to infant atopic manifestations in the first 2 years of life: The KOALA birth cohort study. Pediatrics. 122(1): e115–e122.

Song, C. -H., Z. -Q. Liu, S. Huang, P. -Y. Zheng and P. -C. Yang (2012). Probiotics promote endocytic allergen degradation in gut epithelial cells. Biochemical and Biophysical Research Communications.

Stanley, J. S., N. King, A. W. Burks, S. K. Huang, H. Sampson, G. Cockrell, R. M. Helm, C. M. West and B. GA (1997). Identification and mutational analysis of the immunodominant IgE binding epitopes of the major peanut allergen ara h 2. Archives of Biochemistry and Biophysics. 342: 244–253.

Stark, P. L. and A. Lee (1982). The microbial ecology of the large bowel of breastfed and formula-fed infants during the first year of life. J Med Microbiol. 15(2): 189–203.

Stavrou, E., J. Ford, A. Shand, J. Morris and C. Roberts (2011). Epidemiology and trends for Caesarean section births in New South Wales, Australia: A population-based study. BMC Pregnancy and Childbirth. 11(1): 8.

Stefka, A. T., T. Feehley, P. Tripathi, J. Qiu, K. McCoy, S. K. Mazmanian, M. Y. Tjota, G. -Y. Seo, S. Cao and B. R. Theriault (2014). Commensal bacteria protect against food allergen sensitization. Proceedings of the National Academy of Sciences. 111(36): 13145–13150.

Strachan, D. P. (1989). Hay fever, hygiene, and household size. Bmj. 299(6710): 1259–1260.

Tang, M. L., A. -L. Ponsonby, F. Orsini, D. Tey, M. Robinson, E. L. Su, P. Licciardi, W. Burks and S. Donath (2015). Administration of a probiotic with peanut oral immunotherapy: A randomized trial. Journal of Allergy and Clinical Immunology. 135(3): 737–744. e738.

Taylor, A. L., J. A. Dunstan and S. L. Prescott (2007). Probiotic supplementation for the first 6 months of life fails to reduce the risk of atopic dermatitis and increases the risk of allergen sensitization in high-risk children: A randomized controlled trial. Journal of Allergy and Clinical Immunology. 119(1): 184–191.

Taylor, S. N., L. A. Basile, M. Ebeling and C. L. Wagner (2009). Intestinal permeability in preterm infants by feeding type: Mother's milk versus formula. Breastfeeding Medicine. 4(1): 11–15.

Tordesillas, L., R. Goswami, S. Benedé, G. Grishina, D. Dunkin, K. M. Järvinen, S. J. Maleki, H. A. Sampson and M. C. Berin (2014). Skin exposure promotes a Th2-dependent sensitization to peanut allergens. The Journal of Clinical Investigation. 124(11): 4965.

Vadas, P., Y. Wai, W. Burks and B. Perelman (2001). Detection of peanut allergens in breast milk of lactating women. JAMA: The Journal of the American Medical Association. 285(13): 1746–1748.

Venkataraman, D., N. Soto-Ramírez, R. J. Kurukulaaratchy, J. W. Holloway, W. Karmaus, S. L. Ewart, S. H. Arshad and M. Erlewyn-Lajeunesse (2014). Filaggrin loss-of-function mutations are associated with food allergy in childhood and adolescence. Journal of Allergy and Clinical Immunology. 134(4): 876–882. e874.

Venter, C., S. Hasan Arshad, J. Grundy, B. Pereira, C. Bernie Clayton, K. Voigt, B. Higgins and T. Dean (2010). Time trends in the prevalence of peanut allergy: three cohorts of children from the same geographical location in the UK. Allergy. 65(1): 103–108.

Vickery, B. P., A. Beavers, J. Berglund, J. P. French, D. K. Hamilton, L. Herlihy, E. H. Kim, M. D. Kulis, N. Szczepanski and P. H. Steele (2015). High rate of sustained unresponsiveness with early-intervention peanut oral immunotherapy. Journal of Allergy and Clinical Immunology. 135(2): AB155.

Zhou, Z., Y. Song, C. Mao, K. D. Srivastava, C. Liu, N. Yang, Z. Liu and X. -M. Li (2015). Recombinant probiotic *Bacillus subtilis* spores with surface expression of Ara h2 reduce peanut-induced anaphylaxis in mice. Journal of Allergy and Clinical Immunology. 135(2): AB29.

4

Egg Allergy

Paul J. Turner[1] and *Dianne E. Campbell*[2,*]

CONTENTS

[1] Section of Paediatrics (Allergy and Infectious Diseases), Imperial College London, London, United Kingdom.
[2] Department of Allergy and Immunology, Children's Hospital Westmead, Sydney, Australia.
* Corresponding author: dianne.campbell1@health.nsw.gov.au

4.1 INTRODUCTION

Allergy to hen's egg is perhaps the most common paediatric food allergy worldwide. This is certainly true for Australia and North America. It is unclear what makes this otherwise nutritious, innocuous and widely consumed food quite so allergenic. In this chapter we will examine the molecular characteristics and biochemistry of hen's egg and the epidemiology, diagnosis, management and prevention of both IgE- and non-IgE mediated egg allergies.

4.2 EGG PROTEIN ALLERGENS: COMPOSITION AND CHEMISTRY

The proteins responsible for the majority of allergic reactions to egg are thought to be present within egg white, as opposed to the yolk (Caubet et al. 2011, Everberg et al. 2011). The predominant allergens have been characterised and are described in Table 4.1.

Table 4.1 Major allergens present in egg.

	Allergen	% w/w total protein	Mol wt (kDa)	IgE-binding activity	Heat stability	Digestion-resistance
Egg White						
Ovalbumin	Gal d 2	54%	45	++	−	+
Ovomucoid	Gal d 1	11%	28	+++	+++	+++
Ovotransferrin (conalbumin)	Gal d 3	12%	76.6	+	−	−
Lysozyme	Gal d 4	3.4%	14.3	++	−	+
Riboflavin binding protein (RfBP)		0.8%	30–35	?	++	+
Egg Yolk						
Alpha livetin (chicken serum albumin)	Gal d 5	~9%	65–70	++	+/−	?
Vitellogenin-1 precursor	Gal d 6 (YGP42)		35	?	++	−
Riboflavin binding protein (RfBP)		0.3%	30–35	?	++	+
Apovitellenin I, VI		~5%	~10+	?	?	?

4.2.1 Egg White

Egg white is made up of approximately 90% water and 10% protein; this protein fraction constitutes just over half of the total protein content of an egg. Approximately 40 different proteins have been identified within egg white, but the literature in terms of allergenicity is generally limited to the four main constituents: ovalbumin, ovomucoid, ovotransferrin and lysozyme. The relative importance of these proteins in IgE-mediated allergy has at times proven controversial. Bernhisel-Broadbent et al. proposed that in most egg-allergic individuals, IgE-binding to ovalbumin may in fact be due to contamination of commercial ovalbumin with ovomucoid. They suggested that "ovomucoid is the immunodominant protein fraction in egg white, and that the use of commercially purified ovalbumin has led to an overestimation of the dominance of ovalbumin as a major egg allergen" (Bernhisel-Broadbent et al. 1994). However, a subsequent study using purified commercial preparations of these four protein constituents (by chromatography based on affinity to monoclonal antibodies) found contamination was as much of an issue with commercial ovomucoid as for ovalbumin, with significant lysozyme contamination in the former, an observation reported elsewhere (Jacobsen et al. 2008). Both studies confirmed IgE to ovalbumin as a major determinant of egg allergy, in addition to ovomucoid (Table 4.2).

Ovalbumin (Gal d 2) is the most abundant glycoprotein in egg, constituting approximately 54% of the total egg protein by weight.

Table 4.2 Comparison of IgE binding to commercially available (crude) and purified preparations of egg-white allergen components in various studies.

Study	Allergen type	Ovomucoid (Gal d 1)	Ovalbumin (Gal d 2)	Ovotransferrin (Gal d 3)	Lysozyme (Gal d 4)
D'Urbano et al. (2010)	Commercial	24/46 (52%)	20/46 (44%)	6/46 (13%)	17/46 (37%)
Jacobsen et al. (2008)	Purified	10/10 (100%)	7/10 (70%)	10/10 (100%)	3/10 (30%)
Everberg et al. (2011)	Commercial	75/83 (90%)	75/83 (90%)	73/83 (88%)	48/83 (58%)
	Purified	60/83 (72%)	72/83 (87%)	57/83 (69%)	Purification not needed

A number of IgE-binding epitopes have been identified (Mine and Rupa 2003), many of which retain their ability to bind IgE from sera of egg-allergic individuals following simulated digestion (Benede et al. 2014). Ovalbumin is heat-labile and undergoes conformational changes which might reduce IgE-binding (Joo and Kato 2006). This has been proposed as one mechanism through which egg-allergic individuals are able to tolerate extensively-heated egg, for example in baked foods (Martos et al. 2011). However, IgE-binding to linear epitopes in ovalbumin, which are not modified through heat-treatment, have also been reported (Mine and Zhang 2002).

Ovomucoid (Gal d 1) is the second most abundant glycoprotein by weight, and may be the dominant allergen in IgE-mediated egg allergy, primarily due to the resistance of its IgE-binding epitopes to heat and digestion (Urisu et al. 1997). However, ovomucoid does undergo a degree of gastric-modification, which affects allergenicity; it has been suggested that this observation might explain the phenomenon of contact reactions to raw egg in children who otherwise tolerate egg ingestion (Urisu et al. 1997). Ovomucoid comprises 186 amino acids arranged in three tandem domains (Gal d 1.1, 1.2, and 1.3), the second of which may be most associated with IgE binding (Cooke and Sampson 1997). A number of studies have reported IgE to ovomucoid is predictive for lack of tolerance to heat-modified egg (Ando et al. 2008, Lemon-Mule et al. 2008, Tan et al. 2013) and persistence of egg allergy (Bernhisel-Broadbent et al. 1994, Jarvinen et al. 2007). Epitope mapping, using a peptide microarray immunoassay, has been used to study IgE-binding to ovomucoid; interestingly, sera from 17 of 50 patients studied did not demonstrate IgE binding to linear epitopes (Martinez-Botas et al. 2013).

Ovotransferrin (Gal d 3) is an iron-binding protein, which exhibits significant lability to heat and digestion. Although the 3-D structure has been elucidated, the IgE-binding epitopes remain unidentified (reviewed in Matsuo et al. 2015).

Lysozyme (Gal d 4) is not considered to be a major allergen, however at least one third of egg-allergic patients appear to be sensitised to lysozyme (Fremont et al. 1997) (Table 4.1). It has been well-characterised, primarily due to its use as a food preservative (D'Urbano et al. 2010) and a number of IgE-binding epitopes have been identified, which are digestion resistant (Jimenez-Saiz et al. 2014).

Egg white also contains a number of other proteins, of which 2 (egg white cystatin and lipocalin-type prostaglandin D synthase (L-PGDS)) have recently been identified as possible allergens (Suzuki et al. 2010). Riboflavin binding protein (RfBP), present in small quantities in both egg white and yolk, has also been reported to bind IgE from sera of egg-allergic children (Martos et al. 2012). However, the study used pooled sera, so the clinical relevance of their findings is difficult to ascertain.

4.2.2 Egg Yolk

Chicken serum albumin (Gal d 5), also known as alpha-livetin, is not considered to be a major determinant of conventional IgE-mediated egg allergy (D'Urbano et al. 2010). However, it has been identified as the causative allergen of bird-egg syndrome: where affected individuals are thought to be primarily sensitised to airborne bird allergens, with cross-reactivity to albumin in egg yolk causing predominantly respiratory symptoms (both upper and lower airways) with egg ingestion (Mandallaz et al. 1988, Quirce et al. 2001).

A 35-kDa fragment of the vitellogenin-1 precursor, known as the **YGP42 protein** (Gal d 6), has also been identified as an egg yolk allergen (Amo et al. 2010). This protein is also heat-stable, but susceptible to gastric modification. The protein does not appear to have a major role in conventional egg allergy, with a minority of patient (5/27) showing specific binding to the protein. It may, however, explain the presence of bird-egg syndrome in patients with negative testing to Gal d 5.

Other potential allergens within egg yolk include apovitellenins I and VI and phosvitin, with 25% of 40 egg-allergic children demonstrating IgE to these proteins in one report (Walsh et al. 2005). However, these 'minor' yolk allergens remain poorly investigated, and it is difficult to draw conclusions as to the relative importance of these allergens.

4.3 PRIMARY PREVENTION OF EGG ALLERGY

Given the rise in prevalence of food allergy in both the developed and developing world (Prescott et al. 2013), attention is now keenly focused on strategies that will prevent the development or establishment of food allergy. One key question in this area relates to the best timing for introduction of egg (and other allergic foods) into the diet of both infants considered at high risk of allergy and low risk (based on heredity). Two decades of recommendations to avoid early dietary exposure to egg have been accompanied by increasing rather than decreasing incidence of egg allergy in children. Findings from several large observational and cross-sectional cohorts now suggest that introduction of egg between 4–8 months may be associated with a reduced incidence of IgE-mediated egg allergy (Koplin et al. 2010, Nwaru et al. 2013). Moreover, other observational cohorts have been unable to demonstrate a protective effect from delayed introduction of egg. Level 1 evidence from randomised controlled trials is currently lacking, however several early egg introduction primary prevention trials are currently underway. Based on the epidemiological evidence to date, and level 1 evidence from the LEAP peanut allergy trial (Du Toit et al. 2015), and encouraging immunological data from smaller studies (Palmer et al. 2013) most feeding guidelines from specialist national allergy societies recommend introduction of egg into the diet of infants from 4 months of age, irrespective of allergic heredity (Fleischer et al. 2013, Muraro et al. 2014).

There is no evidence that maternal avoidance of egg in the diet during pregnancy or lactation reduces the risk of egg allergy and for this reason, maternal dietary restrictions on the basis of prevention of food allergy are not recommended (Kramer and Kakuma 2012).

There is much interest in the possible allergy prevention effects of a range of macro- and micro-nutrient supplements including vitamin D, vitamin A, pre- and pro-biotics and fish oils, both as maternal and infant supplements (reviewed in (Rueter et al. 2015)). To date, egg allergy as an outcome has only been examined in terms of allergen sensitization, not clinical or challenge proven allergy. Current meta-analyses of these nutritional strategies have not supported their use to date, however higher quality randomised controlled trials with clinically relevant food allergy outcomes are required to fully answer this question.

4.4 IgE-MEDIATED EGG ALLERGY

4.4.1 Prevalence and Natural History

Egg allergy most commonly presents in the first year of life. In many regions is the most common cause of food allergy in infants and preschool-aged children. Estimating the true prevalence of egg allergy (and other food allergies) is generally hindered by the lack of studies that have utilized food challenge methodology in large unselected populations. A further confounder is the form of egg used in food challenges (OFC). IgE-mediated egg allergy has been robustly demonstrated to be most common food allergy in the infant population in Australia, with prevalence of 8.9% (Osborne et al. 2011). This study used raw egg (potentially the most allergenic form of egg) for diagnostic food challenges, which might explain, at least in part, the lower rates reported for Europe, with a meta-analysis estimating a lifetime prevalence of egg allergy at 2.5% and challenge proven prevalence of 0.2% (Nwaru et al. 2014). The mean incidence of egg allergy within the first two years of life from EuroPrevall cohorts across Europe is estimated at 1.2% (Xepapadaki et al. 2015). Data from other regions, including North America, is less robust, with 3.4% over 6 and 13% under six years self-reporting egg allergy (Salo et al. 2011). In Asia, where food allergy appears to be increasing in-line with westernised lifestyles and eating habits (Hu et al. 2010, Prescott et al. 2013), egg allergy appears to be amongst the most common food allergens in the paediatric population but has lower rates of estimated prevalence. Egg allergy was the most common reported food allergy in South Korean children 0–6 year (Park et al. 2014) and 6–7 years (Ahn et al. 2012) and in Chinese infants at 2.5% (Chen et al. 2011), whereas prevalence in Singapore, Taiwan and Hong Kong is reported at <1% (reviewed in (Lee and Shek 2014)). Selected allergic cohorts have also reported high rates of sensitisation to egg in children in Malaysia (Yadav and Naidu 2015) and India (Dey et al. 2014). Risk factors for the development of egg allergy include a family history of allergy and, in the case of infants in Australia, having parents born in East Asia (Koplin et al. 2012).

As a trigger for anaphylaxis presentation to the emergency department in children or adults, an egg trigger is reported in several case series including US (1% of all anaphylaxis admissions (Ma et al. 2014)), Australia (9% of all food-related anaphylaxis admissions (Liew et al. 2009)), Singapore (12% of paediatric food-related anaphylaxis admissions (Liew et al. 2013)). Egg is an uncommon cause of fatal anaphylaxis, with rare reports of egg anaphylaxis fatalities in the UK (Colver et al. 2005, Turner et al. 2015).

It is traditionally believed that IgE-mediated egg is nearly always outgrown and that this occurs early in life. Recent evidence suggests that this may not actually be the case. Median age of outgrowing egg allergy has been reported to vary between 6–9 years from more recent observational cohorts (Savage et al. 2007, Sicherer et al. 2014). Moreover in the later cohort, only 68% of egg allergic children had attained resolution by 16 years. The risk factors in this cohort for the persistence of egg allergy were high initial egg-specific IgE, other food allergies and co-existing allergic disease. Other factors which have been reported to be associated with earlier acquisition of natural tolerance include tolerance to extensively heated egg (Peters et al. 2014) severity of initial allergic reaction, allergy skin test wheal size, rate of decline in sensitisation and age at onset (reviewed in (Tan and Joshi 2014)).

Up to 70% of infants with IgE-mediated egg allergy can tolerate extensively heated/baked egg in their diet (Lemon-Mule et al. 2008, Turner et al. 2013). Although production of IgE directed against ovomucoid is associated with an increased chance of reacting to baked egg (as discussed above), there are no current tests or clinical predictors which can safely determine whether a given egg allergic individual will be baked egg tolerant. Because of this, many experts recommend that baked egg challenges in children with egg allergy be performed under medical supervision (Turner et al. 2013, Leonard et al. 2015), because of the risk of anaphylaxis (reported to occur in up to 14% of children reacting to baked egg on OFC (Turner et al. 2013). Liberalisation of the diet for these children is likely to improve quality of life, however, a follow-up study of children who underwent "successful" baked egg and milk OFC suggest that ongoing symptoms occur commonly and only two-thirds of the cohort maintained longer-term adherence incorporation of dietary baked egg (Lee et al. 2015). Furthermore evidence from a quasi-controlled study that examined introduction of baked egg in the diet of children with established IgE-mediated egg allergy and tolerance suggests that regular ingestion of baked egg may actually accelerate time to egg tolerance (Leonard et al. 2012).

4.4.2 Diagnosis

The gold-standard for diagnosis remains the double-blind, placebo-controlled food challenges (DBPCFC). However, in the vast majority of cases, diagnosis can be achieved through a typical clinical history of IgE-mediated symptoms and evidence of sensitisation or egg-specific IgE, either through skin testing (SPT) or blood test.

Egg is generally thought to cause more mild symptoms than other allergens (Clark et al. 2010), but the evidence for this is limited. A series review of 2304 food challenges at a single paediatric centre over a 10 year period, of which 30% were positive, reported egg was more likely to cause gastrointestinal symptoms than other allergens (Gupta et al. 2015). Most series report the potential for anaphylaxis following food challenge to egg, with a frequency of 15–58% of positive reactions associated with more severe symptoms such as lower respiratory involvement (Benhamou et al. 2008, Clark et al. 2011, Rolinck-Werninghaus et al. 2012).

Cut-offs for 95% PPV for clinical egg allergy have been published, although these have mostly been derived from selected populations, and may vary depend on the type of egg allergy being assessed, e.g., raw egg versus lightly cooked or boiled egg. Hill et al. reported a diagnostic cut-off level for SPT in egg-allergic children of 7 mm or greater (5 mm in children under 2 years) to lightly boiled egg (Hill et al. 2004), a cut-off confirmed elsewhere for raw egg (Monti et al. 2002). Similar cut-offs have been reported for serum-specific IgE to egg, typically in the region >6kU/L (Sampson 2001, Celik-Bilgili et al. 2005,

Ando et al. 2008). While these cut-offs are only validated for the populations in which they are derived, a lack of data has resulted in their widespread application. More recently, a large, unselected population cohort study provided a 95% PPV cut-off of 4 mm to raw egg for infants aged 12 months (Peters et al. 2013). The same study reported a 95% cut-off for ssIgE of 1.7 kUA/L or greater for allergy to raw egg. Whether this lower cut-off reflects the younger age of the cohort or that challenge testing was performed to raw egg rather than lightly-cooked egg is unclear. A recent meta-analysis of 5 cohorts reported pooled sensitivities of 92% and 93% and specificities of 58% and 49% for SPTs and sIgE, respectively for the diagnosis of egg allergy (using the mixed cut-offs). However, it is not possible to predict those at risk of more severe reactions to egg. Some authors have reported that those with more severe reactions tend to have higher degrees of sensitisation (Benhamou et al. 2008), but not others. Either way, the degree of sensitisation does not allow discrimination of those likely to have severe reactions with a level of certainty that is of clinical utility.

The utility of component resolved diagnostics (CRD) in diagnosis of egg allergy, where IgE to single allergen components are measured, has been examined. For standard whole-egg and raw-egg OFC, there appears little evidence that CRD provides significant improvement over standard testing for predicting challenge outcome. Ovomucoid is relatively heat-stable, and it has been suggested that individuals with low level sensitisation to ovomucoid may be more tolerant to heated forms of egg. Indeed, ssIgE to ovomucoid has been reported to be predictive of tolerance to hard-boiled egg (Haneda et al. 2012). An association between reactivity to extensively-heated egg white and ssIgE to ovomucoid is reported, but results are not sufficiently discriminatory to be of clinical use (Ando et al. 2008). It has also been proposed that the CRD approach, particularly the level of ovomucoid sensitisation, might be promising in predicting those tolerant to extensively-heated egg, such as in cakes and other baked foods. As discussed above, in the context of ovomucoid, several studies have reported very high cut-off of IgE to OVM of >50 kUA/l (Lemon-Mule et al. 2008) and ovomucoid-SPT >11 mm (Tan et al. 2013) predicted baked egg reactivity. However, the majority of egg-allergic individuals have levels of sensitisation below these high cut-offs, which limits the utility of the test to a relatively small proportion of the egg allergic population.

The basophil activation test (BAT), in which the expression of activation markers on basophils present in whole blood is assessed following incubation with a food allergen, has been found to predict clinical reactivity in peanut-allergic subjects (Santos et al. 2015). There is only one report of the BAT being used in the diagnosis of egg allergy, in which the test did not confer a significant advantage over conventional diagnostic techniques (Ocmant et al. 2009).

4.4.3 Treatment

Standard avoidance of allergen, dietary advice and education, provision of emergency actions plans for allergic reactions/ anaphylaxis and provision of adrenaline autoinjectors (AAI) (where deemed appropriate) all form the basis for management of individuals with IgE-mediated egg allergy. As discussed above, it may be possible to provide some dietary liberalisation with incorporation of baked egg into the diet of egg allergic individuals. It those individuals, special attention should be paid to the potential development of symptoms suggestive of EoE, as this has been rarely reported as a complication of baked egg incorporation (Maggadottir et al. 2014).

Many individuals with egg allergy may have other food allergies, and for all individuals who are required to avoid multiple foods, assessment of dietary sufficiency, including total caloric requirements for growth, protein and calcium requirements, micro and macronutrient intake should be undertaken and monitored by a dietician familiar with food exclusion diets.

4.4.3.1 Immunotherapy

Oral immunotherapy (OIT) for egg allergy has been the subject of many observational studies and RCTs (Staden et al. 2007, Burks et al. 2012, Dello Iacono et al. 2013, Fuentes-Aparicio et al. 2013,

Meglio et al. 2013) and a Cochrane review (Romantsik et al. 2014). It is increasingly used in some countries (predominantly Spain and Italy) for the treatment of egg allergy, despite lack of standardised protocols or agreed international guidelines. Although most reports suggest that the majority of children can be successfully desensitised to egg, significant side effects during therapy, both in the updosing and maintenance phases, and low rates of long term sustained unresponsiveness, and intensity of medical/hospital visits for updosing limit its current clinical utility.

An alternative approach has been the introduction of foods containing extensively-heated egg into the diet where tolerated. A multicentre randomised controlled trial is currently underway comparing the relative efficacy of "baked egg" consumption versus conventional egg-OIT in egg-allergic children (clinicaltrials.gov registration NCT01846208).

4.4.3.2 *Vaccinations and medications containing egg*

Egg allergy is often incorrectly cited as a contraindication for immunisation with the measles/measles-mumps-rubella vaccines, despite that fact that these vaccinations are produced in chick-fibroblast cell lines that contain negligible amounts of egg protein. Egg-allergic individuals can receive these vaccines without any additional precautions, irrespective of the severity of their egg allergy (Clark et al. 2010). Three vaccines continue to be produced in chick embryos (and thus contain small amounts of ovalbumin and other egg proteins); influenza, yellow fever and rabies. An egg-free rabies vaccine is available and recommended for use in egg-allergic individuals. Although a recombinant egg-free influenza vaccine is available, it is not licensed for use in children. However, the available data indicates that current influenza vaccination is probably no more likely to cause anaphylaxis in an egg-allergic individual than someone without egg allergy (Des Roches et al. 2012). Parenteral influenza vaccines with less than 1 ug of ovalbumin per dose are recommended for egg allergic individuals (Mullins and Gold 2012). The intranasal, live attenuated influenza vaccine also appears safe in egg allergic children (Turner et al. 2015). Allergic reactions in egg-allergic individuals may occur with yellow fever vaccine and for egg allergic individuals who must receive yellow fever vaccination a desensitization regime has been published (Rutkowski et al. 2013).

The anaesthetic agent propofol is variably cited as being contraindicated in individuals with egg allergy. There are no prospective studies of this agent in egg allergic patients, however a retrospective case series in egg allergic children suggested that the vast majority of children with egg allergy received propofol without incident (Murphy et al. 2011). Case reports suggest that those individuals with a history of egg anaphylaxis may be at risk, and in the absence of high quality data, avoidance in the setting of known previous egg anaphylaxis may be prudent (Baombe and Parvez 2013). Similar concerns exist for the use of intralipid in individuals with egg allergy, with case reports of anaphylaxis in egg allergic individuals and no large case series or prospective studies.

4.5 NON IGE-MEDIATED FOOD ALLERGY

4.5.1 Food Protein Induced Enteropathy Syndrome (FPIES)

The clinical presentation, natural history and immunobiology of FPIES is reviewed in detail elsewhere. Egg is a known trigger food in FPIES. In the largest population prevalence survey to date, reactions to egg accounted for 8% of all FPIES episodes in Australian infants less than two years old (Frith et al. 2013). Similarly, FPIES to egg was reported in 6% of infants in a retrospective Italian case series (Sopo et al. 2012). Age of onset of in egg FPIES is later than that of rice or cow's milk FPIES, however this is likely related to age of initial exposure. Egg FPIES may occur as an isolated allergy, or may be associated with allergy to other foods, however no specific pattern of pairing food allergens has been described consistently for egg induced FPIES. The natural history of egg FPIES is not well characterised, however it appears that egg FPIES may persist for longer than FPIES to grains or cow's milk (Nowak-Wegrzyn et al. 2015). Rare FPIES reactions to allergens in breast milk following maternal ingestion of cows milk and soy but not egg have been reported. There are no *in vitro* tests which can currently identify which allergens may cause on FPIES reaction, or which predict the acquisition of tolerance. Similarly both allergy skin tests and allergy patch testing are not informative. For this reason, observed food challenges, usually conducted with intravenous access and observation for at least 4 hours following a single portion size allergen ingestion remain the gold standard for confirming a diagnosis and for determining if remission has occurred (Nowak-Wegrzyn et al. 2015).

4.5.2 Eosinophilic Oesophagitis (EoE)

EoE is a relatively uncommon disorder characterised by eosinophil infiltration of the oesophageal mucosa and diagnosed on the basis of histological criteria. It is likely to under-diagnosed, and the true incidence is unknown. It affects both adults and children, with typical clinical manifestations that vary according to age. Young infants typically present with vomiting, food refusal and failure to thrive, school age children with dysphagia, slow eating, and abdominal pain, and older teenagers and adults with dysphagia and food bolus obstruction. Although its pathogenesis is unclear, evidence for food allergy as a primary disease driver comes from its clinical response and remission to various food elimination diets. Whilst there is a high rate of co-existent IgE mediated food allergy of around 20% in populations with EoE, non-IgE mediated mechanism are thought to underlie the food reactivity. Egg, cow's milk and nut allergies are the most commonly associated food allergies, and overall rates of allergic disease such as asthma and allergic rhinitis in EoE population are very high, reported at over 70%. mRNA analysis of tissue biopsies in active disease suggests activation of a disease specific *transciptome*, with increased expression of eoxatin S (Blanchard et al. 2006) and periostin (Blanchard et al. 2008).

Elimination diets in the management of EoE are varied, but most commonly consist of an empiric 8, 6 or 4 food free diet, all of which include dietary elimination of egg. As an EoE trigger, egg has been reported to account for 21–36% of diet-responsive EoE in adult EoE sufferers populations (Molina-Infante et al. 2014, Rodriguez-Sanchez et al. 2014) and 13%–17% in paediatric cohorts (Spergel et al. 2012) (Kagalwalla et al. 2011). Overall egg appears to be the third most commonly implicated food allergen in food-induced EoE, after cow's milk and wheat.

Diets based upon positive allergy patch tests, allergy skin tests and ss-IgE have examined for efficacy. In general diets based upon testing have not shown markedly improved performed over empiric exclusion of egg, cow's milk, wheat, soy, fish and nuts. Where egg allergy testing has been specifically examined for test performance characteristics in children with EoE, egg SPT were was found to have a negative predictive value (NPV) of 90% and a positive predictive value (PPV) of 62%. Similarly egg APTs were found to have 91% NPV, but only 51% PPV (Spergel et al. 2012).

The development of EoE following either traditional egg oral immunotherapy (Lucendo et al. 2014), or liberalisation of the diet of egg allergic individuals with introduction of baked/extensively heated egg (Maggadottir et al. 2014) have been reported.

4.5.3 Eczema

Eczema (atopic dermatitis) is common in children with egg allergy, and is an established risk factor for food allergies. Infants with eczema are six times more likely to have egg allergy by age 12 months than those without (Martin et al. 2015) and approximately two thirds of children with egg allergy have eczema (Turner et al. 2013). Many of these individuals will develop both IgE- and non-IgE-mediated (including eczema) following exposure to egg (and other foods to which they are allergic), however it is difficult to determine the frequency egg can cause a flare in eczema flares as part of non-IgE-mediated egg symptoms in these individuals, as they avoid egg in their diet.

Where an individual has sensitisation to egg, but no history of typical IgE mediated reaction on known exposure, the likelihood that egg is playing a role in triggering or flaring eczema is quite low (reviewed in (Campbell 2012)). Overall, it is estimated that only 30–40% of children with at least moderate AD attending a specialist service with a positive IgE to a particular food will have a positive challenge to that food under double-blind, placebo-controlled conditions (Sampson 2003). A Cochrane review which specifically examined milk and/or egg exclusion diets for eczema severity outcomes found only limited evidence to support dietary exclusion of egg (Bath-Hextall et al. 2008). Particular care on reintroduction must be taken if egg is removed from the diet in individuals with evidence of sensitisation on SPT or egg specific-IgE, as anaphylaxis on re-exposure after prolonged exclusion has been reported in individuals who previously tolerated food to which they were sensitised (Flinterman et al. 2006).

Keywords: Egg allergy; egg allergens; food allergy diagnosis; FPIES; EoE; food allergy

REFERENCES

Ahn, K., J. Kim, M. I. Hahm, S. Y. Lee, W. K. Kim, Y. Chae, Y. M. Park, M. Y. Han, K. J. Lee, J. K. Kim, E. S. Yang and H. J. Kwon (2012). Prevalence of immediate-type food allergy in Korean schoolchildren: a population-based study. Allergy Asthma Proc. 33(6): 481–487. doi: 410.2500/aap.2012.2533.3598.

Amo, A., R. Rodriguez-Perez, J. Blanco, J. Villota, S. Juste, I. Moneo and M. L. Caballero (2010). Gal d 6 is the second allergen characterized from egg yolk. J Agric Food Chem. 58(12): 7453–7457. doi: 7410.1021/jf101403h.

Ando, H., R. Moverare, Y. Kondo, I. Tsuge, A. Tanaka, M. P. Borres and A. Urisu (2008). Utility of ovomucoid-specific IgE concentrations in predicting symptomatic egg allergy. J Allergy Clin Immunol. 122(3): 583–588. doi: 510.1016/j.jaci.2008.1006.1016. Epub 2008 Aug. 1019.

Baombe, J. P. and K. Parvez (2013). Towards evidence-based emergency medicine: best BETs from the Manchester Royal Infirmary. BET 1: is propofol safe in patients with egg anaphylaxis? Emerg Med J. 30(1): 79–80. doi: 10.1136/emermed-2012-202183.202182.

Bath-Hextall, F., F. M. Delamere and H. C. Williams (2008). Dietary exclusions for established atopic eczema. Cochrane Database Syst Rev. (1): CD005203. doi: 005210.001002/14651858.CD14005203.pub14651852.

Benede, S., I. Lopez-Exposito, R. Lopez-Fandino and E. Molina (2014). Identification of IgE-binding peptides in hen egg ovalbumin digested *in vitro* with human and simulated gastroduodenal fluids. J Agric Food Chem. 62(1): 152–158. doi: 110.1021/jf404226w. Epub 402013 Dec. 404227.

Benhamou, A. H., S. A. Zamora and P. A. Eigenmann (2008). Correlation between specific immunoglobulin E levels and the severity of reactions in egg allergic patients. Pediatr Allergy Immunol. 19(2): 173–179. doi: 110.1111/j.1399-3038.2007.00602.x.

Bernhisel-Broadbent, J., H. M. Dintzis, R. Z. Dintzis and H. A. Sampson (1994). Allergenicity and antigenicity of chicken egg ovomucoid (Gal d III) compared with ovalbumin (Gal d I) in children with egg allergy and in mice. J Allergy Clin Immunol. 93(6): 1047–1059.

Blanchard, C., M. K. Mingler, M. McBride, P. E. Putnam, M. H. Collins, G. Chang, K. Stringer, J. P. Abonia, J. D. Molkentin and M. E. Rothenberg (2008). Periostin facilitates eosinophil tissue infiltration in allergic lung and esophageal responses. Mucosal Immunol. 1(4): 289–296. doi: 210.1038/mi.2008.1015. Epub 2008 May 1037.

Blanchard, C., N. Wang, K. F. Stringer, A. Mishra, P. C. Fulkerson, J. P. Abonia, S. C. Jameson, C. Kirby, M. R. Konikoff, M. H. Collins, M. B. Cohen, R. Akers, S. P. Hogan, A. H. Assa'ad, P. E. Putnam, B. J. Aronow and M. E. Rothenberg (2006). Eotaxin-3 and a uniquely conserved gene-expression profile in eosinophilic esophagitis. J Clin Invest. 116(2): 536–547.

Burks, A. W., S. M. Jones, R. A. Wood, D. M. Fleischer, S. H. Sicherer, R. W. Lindblad, D. Stablein, A. K. Henning, B. P. Vickery, A. H. Liu, A. M. Scurlock, W. G. Shreffler, M. Plaut and H. A. Sampson (2012). Oral immunotherapy for treatment of egg allergy in children. N Engl J Med. 367(3): 233–243. doi: 210.1056/NEJMoa1200435.

Campbell, D. E. (2012). Role of food allergy in childhood atopic dermatitis. J Paediatr Child Health. 48(12): 1058–1064. doi: 1010.1111/j.1440-1754.2011.02125.x. Epub 02011 Jun. 02117.

Caubet, J. C., Y. Kondo, A. Urisu and A. Nowak-Wegrzyn (2011). Molecular diagnosis of egg allergy. Curr Opin Allergy Clin Immunol. 11(3): 210–215. doi: 210.1097/ACI.1090b1013e3283464d3283461b.

Celik-Bilgili, S., A. Mehl, A. Verstege, U. Staden, M. Nocon, K. Beyer and B. Niggemann (2005). The predictive value of specific immunoglobulin E levels in serum for the outcome of oral food challenges. Clin Exp Allergy. 35(3): 268–273.

Chen, J., Y. Hu, K. J. Allen, M. H. Ho and H. Li (2011). The prevalence of food allergy in infants in Chongqing, China. Pediatr Allergy Immunol. 22(4): 356–360. doi: 310.1111/j.1399-3038.2011.01139.x. Epub 02011 Jan. 01125.

Clark, A., S. Islam, Y. King, J. Deighton, S. Szun, K. Anagnostou and P. Ewan (2011). A longitudinal study of resolution of allergy to well-cooked and uncooked egg. Clin Exp Allergy. 41(5): 706–712. doi: 710.1111/j.1365-2222.2011.03697.x.

Clark, A. T., I. Skypala, S. C. Leech, P. W. Ewan, P. Dugue, N. Brathwaite, P. A. Huber and S. M. Nasser (2010). British society for allergy and clinical immunology guidelines for the management of egg allergy. Clin Exp Allergy. 40(8): 1116–1129. doi: 1110.1111/j.1365-2222.2010.03557.x.

Colver, A. F., H. Nevantaus, C. F. Macdougall and A. J. Cant (2005). Severe food-allergic reactions in children across the UK and Ireland, 1998–2000. Acta Paediatr. 94(6): 689–695.

Cooke, S. K. and H. A. Sampson (1997). Allergenic properties of ovomucoid in man. J Immunol. 159(4): 2026–2032.

D'Urbano, L. E., K. Pellegrino, M. C. Artesani, S. Donnanno, R. Luciano, C. Riccardi, A. E. Tozzi, L. Rava, F. De Benedetti and G. Cavagni (2010). Performance of a component-based allergen-microarray in the diagnosis of cow's milk and hen's egg allergy. Clin Exp Allergy. 40(10): 1561–1570. doi: 1510.1111/j.1365-2222.2010.03568.x.

Dello Iacono, I., S. Tripodi, M. Calvani, V. Panetta, M. C. Verga and S. Miceli Sopo (2013). Specific oral tolerance induction with raw hen's egg in children with very severe egg allergy: a randomized controlled trial. Pediatr Allergy Immunol. 24(1): 66–74. doi: 10.1111/j.1399-3038.2012.01349.x. Epub 02012 Sep. 01349.

Des Roches, A., L. Paradis, R. Gagnon, C. Lemire, P. Begin, S. Carr, E. S. Chan, J. Paradis, L. Frenette, M. Ouakki, M. Benoit and G. De Serres (2012). Egg-allergic patients can be safely vaccinated against influenza. J Allergy Clin Immunol. 130(5): 1213–1216.e1211. doi: 1210.1016/j.jaci.2012.1207.1046. Epub 2012 Sep. 1227.

Dey, D., N. Ghosh, N. Pandey and S. Gupta Bhattacharya (2014). A hospital-based survey on food allergy in the population of Kolkata, India. Int Arch Allergy Immunol. 164(3): 218–221. doi: 210.1159/000365629. Epub 000362014 Aug. 000365616.

Du Toit, G., G. Roberts, P. H. Sayre, H. T. Bahnson, S. Radulovic, A. F. Santos, H. A. Brough, D. Phippard, M. Basting, M. Feeney, V. Turcanu, M. L. Sever, M. Gomez Lorenzo, M. Plaut and G. Lack (2015). Randomized trial of peanut consumption in infants at risk for peanut allergy. N Engl J Med. 372(9): 803–813. doi: 810.1056/NEJMoa1414850. Epub 1412015 Feb. 1414823.

Everberg, H., P. Brostedt, H. Oman, S. Bohman and R. Moverare (2011). Affinity purification of egg-white allergens for improved component-resolved diagnostics. Int Arch Allergy Immunol. 154(1): 33–41. doi: 10.1159/000319206. Epub 000312010 Jul. 000319224.

Fleischer, D. M., J. M. Spergel, A. H. Assa'ad and J. A. Pongracic (2013). Primary prevention of allergic disease through nutritional interventions. J Allergy Clin Immunol Pract. 1(1): 29–36. doi: 10.1016/j.jaip.2012.1009.1003. Epub 2012 Nov. 1022.

Flinterman, A. E., A. C. Knulst, Y. Meijer, C. A. Bruijnzeel-Koomen and S. G. Pasmans (2006). Acute allergic reactions in children with AEDS after prolonged cow's milk elimination diets. Allergy. 61(3): 370–374.

Fremont, S., G. Kanny, J. P. Nicolas and D. A. Moneret-Vautrin (1997). Prevalence of lysozyme sensitization in an egg-allergic population. Allergy. 52(2): 224–228.

Frith, C. J. P., D. E. Campbell and S. S. Mehr (2013). The first 12 months of FPIES surveillance in Australia. Intern Med J. 43(s4): 1–21.

Fuentes-Aparicio, V., A. Alvarez-Perea, S. Infante, L. Zapatero, A. D'Oleo and E. Alonso-Lebrero (2013). Specific oral tolerance induction in paediatric patients with persistent egg allergy. Allergol Immunopathol (Madr). 41(3): 143–150. doi: 110.1016/j.aller.2012.1002.1007. Epub 2012 Jul. 1024.

Gupta, M., L. D. Grossmann, J. M. Spergel and A. Cianferoni (2015). Egg food challenges are associated with more gastrointestinal reactions. Children. 2(3): 371–381.

Haneda, Y., N. Kando, M. Yasui, T. Kobayashi, T. Maeda, A. Hino, S. Hasegawa, T. Ichiyama and K. Ito (2012). Ovomucoids IgE is a better marker than egg white-specific IgE to diagnose boiled egg allergy. J Allergy Clin Immunol. 129(6): 1681–1682. doi: 1610.1016/j.jaci.2012.1603.1041. Epub 2012 Apr. 1630.

Hill, D. J., R. G. Heine and C. S. Hosking (2004). The diagnostic value of skin prick testing in children with food allergy. Pediatr Allergy Immunol. 15(5): 435–441.

Hu, Y., J. Chen and H. Li (2010). Comparison of food allergy prevalence among Chinese infants in Chongqing, 2009 versus 1999. Pediatr Int. 52(5): 820–824. doi: 810.1111/j.1442-1200X.2010.03166.x.

Jacobsen, B., K. Hoffmann-Sommergruber, T. T. Have, N. Foss, P. Briza, C. Oberhuber, C. Radauer, S. Alessandri, A. C. Knulst, M. Fernandez-Rivas and V. Barkholt (2008). The panel of egg allergens, Gal d 1-Gal d 5: Their improved purification and characterization. Mol Nutr Food Res. 52(Suppl. 2): S176–185. doi: 110.1002/mnfr.200700414.

Jarvinen, K. M., K. Beyer, L. Vila, L. Bardina, M. Mishoe and H. A. Sampson (2007). Specificity of IgE antibodies to sequential epitopes of hen's egg ovomucoid as a marker for persistence of egg allergy. Allergy. 62(7): 758–765.

Jimenez-Saiz, R., S. Benede, B. Miralles, I. Lopez-Exposito, E. Molina and R. Lopez-Fandino (2014). Immunological behavior of *in vitro* digested egg-white lysozyme. Mol Nutr Food Res. 58(3): 614–624. doi: 610.1002/mnfr.201300442. Epub 201302013 Oct. 201300441.

Joo, K. and Y. Kato (2006). Assessment of allergenic activity of a heat-coagulated ovalbumin after *in vivo* digestion. Biosci Biotechnol Biochem. 70(3): 591–597.

Kagalwalla, A. F., A. Shah, B. U. Li, T. A. Sentongo, S. Ritz, M. Manuel-Rubio, K. Jacques, D. Wang, H. Melin-Aldana and S. P. Nelson (2011). Identification of specific foods responsible for inflammation in children with eosinophilic esophagitis successfully treated with empiric elimination diet. J Pediatr Gastroenterol Nutr. 53(2): 145–149. doi: 110.1097/MPG.1090b1013e31821cf31503.

Koplin, J. J., S. C. Dharmage, A. L. Ponsonby, M. L. Tang, A. J. Lowe, L. C. Gurrin, N. J. Osborne, P. E. Martin, M. N. Robinson, M. Wake, D. J. Hill and K. J. Allen (2012). Environmental and demographic risk factors for egg allergy in a population-based study of infants. Allergy. 67(11): 1415–1422. doi: 1410.1111/all.12015. Epub 12012 Sep. 12017.

Koplin, J. J., N. J. Osborne, M. Wake, P. E. Martin, L. C. Gurrin, M. N. Robinson, D. Tey, M. Slaa, L. Thiele, L. Miles, D. Anderson, T. Tan, T. D. Dang, D. J. Hill, A. J. Lowe, M. C. Matheson, A. L. Ponsonby, M. L. Tang, S. C. Dharmage and K. J. Allen (2010). Can early introduction of egg prevent egg allergy in infants? A population-based study. J Allergy Clin Immunol. 126(4): 807–813. doi: 810.1016/j.jaci.2010.1007.1028.

Kramer, M. S. and R. Kakuma (2012). Maternal dietary antigen avoidance during pregnancy or lactation, or both, for preventing or treating atopic disease in the child. Cochrane Database Syst Rev. 9: CD000133. doi: 10.1002/14651858. CD14000133.pub14651853.

Lee, A. J. and L. P. Shek (2014). Food allergy in Singapore: Opening a new chapter. Singapore Med J. 55(5): 244–247.

Lee, E., S. Mehr, P. J. Turner, P. Joshi and D. E. Campbell (2015). Adherence to extensively heated egg and cow's milk after successful oral food challenge. J Allergy Clin Immunol Pract. 3(1): 125–127. e124. doi: 110.1016/j. jaip.2014.1008.1013. Epub 2014 Oct. 1029.

Lemon-Mule, H., H. A. Sampson, S. H. Sicherer, W. G. Shreffler, S. Noone and A. Nowak-Wegrzyn (2008). Immunologic changes in children with egg allergy ingesting extensively heated egg. J Allergy Clin Immunol. 122(5): 977–983. e971. doi: 910.1016/j.jaci.2008.1009.1007. Epub 2008 Oct. 1011.

Leonard, S. A., J. C. Caubet, J. S. Kim, M. Groetch and A. Nowak-Wegrzyn (2015). Baked milk- and egg-containing diet in the management of milk and egg allergy. J Allergy Clin Immunol Pract. 3(1): 13–23; quiz 24. doi: 10.1016/j. jaip.2014.1010.1001.

Leonard, S. A., H. A. Sampson, S. H. Sicherer, S. Noone, E. L. Moshier, J. Godbold and A. Nowak-Wegrzyn (2012). Dietary baked egg accelerates resolution of egg allergy in children. J Allergy Clin Immunol. 130(2): 473–480. e471. doi: 410.1016/j.jaci.2012.1006.1006.

Liew, W. K., W. C. Chiang, A. E. Goh, H. H. Lim, O. M. Chay, S. Chang, J. H. Tan, E. Shih and M. Kidon (2013). Paediatric anaphylaxis in a Singaporean children cohort: changing food allergy triggers over time. Asia Pac Allergy. 3(1): 29–34. doi: 10.5415/apallergy.2013.5413.5411.5429. Epub 2013 Jan. 5430.

Liew, W. K., E. Williamson and M. L. Tang (2009). Anaphylaxis fatalities and admissions in Australia." J Allergy Clin Immunol. 123(2): 434–442. doi: 410.1016/j.jaci.2008.1010.1049. Epub 2008 Dec. 1030.

Lucendo, A. J., A. Arias and J. M. Tenias (2014). Relation between eosinophilic esophagitis and oral immunotherapy for food allergy: a systematic review with meta-analysis. Ann Allergy Asthma Immunol. 113(6): 624–629. doi: 610.1016/j. anai.2014.1008.1004. Epub 2014 Sep. 1010.

Ma, L., T. M. Danoff and L. Borish (2014). Case fatality and population mortality associated with anaphylaxis in the United States. J Allergy Clin Immunol. 133(4): 1075–1083. doi: 1010.1016/j.jaci.2013.1010.1029. Epub 2013 Dec. 1014.

Maggadottir, S. M., D. A. Hill, K. Ruymann, T. F. Brown-Whitehorn, A. Cianferoni, M. Shuker, M. L. Wang, K. Chikwava, R. Verma, C. A. Liacouras and J. M. Spergel (2014). Resolution of acute IgE-mediated allergy with development of eosinophilic esophagitis triggered by the same food. J Allergy Clin Immunol. 2014 May; 133(5): 1487–1489.

Mandallaz, M. M., A. L. de Weck and C. A. Dahinden (1988). Bird-egg syndrome. Cross-reactivity between bird antigens and egg-yolk livetins in IgE-mediated hypersensitivity. Int Arch Allergy Appl Immunol. 87(2): 143–150.

Martin, P. E., J. K. Eckert, J. J. Koplin, A. J. Lowe, L. C. Gurrin, S. C. Dharmage, P. Vuillermin, M. L. Tang, A. L. Ponsonby, M. Matheson, D. J. Hill and K. J. Allen (2015). Which infants with eczema are at risk of food allergy? Results from a population-based cohort. Clin Exp Allergy. 45(1): 255–264. doi: 210.1111/ cea.12406.

Martinez-Botas, J., I. Cerecedo, J. Zamora, C. Vlaicu, M. C. Dieguez, D. Gomez-Coronado, V. de Dios, S. Terrados and B. de la Hoz (2013). Mapping of the IgE and IgG4 sequential epitopes of ovomucoid with a peptide microarray immunoassay. Int Arch Allergy Immunol. 161(1): 11–20. doi: 10.1159/000343040. Epub 000342012 Dec. 000343013.

Martos, G., I. Lopez-Exposito, R. Bencharitiwong, M. C. Berin and A. Nowak-Wegrzyn (2011). Mechanisms underlying differential food allergy response to heated egg. J Allergy Clin Immunol. 127(4): 990–997. e991–992. doi: 910.1016/j. jaci.2011.1001.1057. Epub 2011 Mar. 1015.

Martos, G., C. Pineda-Vadillo, B. Miralles, E. Alonso-Lebrero, R. Lopez-Fandino, E. Molina and J. Belloque (2012). Identification of an IgE reactive peptide in hen egg riboflavin binding protein subjected to simulated gastrointestinal digestion. J Agric Food Chem. 60(20): 5215–5220. doi: 5210.1021/jf3001586. Epub 3002012 May 3001514.

Matsuo, H., T. Yokooji and T. Taogoshi (2015). Common food allergens and their IgE-binding epitopes. Allergol Int. 64(4): 332–343. doi: 310.1016/j.alit.2015.1006.1009. Epub 2015 Jul. 1029.

Meglio, P., P. G. Giampietro, R. Carello, I. Gabriele, S. Avitabile and E. Galli (2013). Oral food desensitization in children with IgE-mediated hen's egg allergy: a new protocol with raw hen's egg. Pediatr Allergy Immunol. 24(1): 75–83. doi: 10.1111/j.1399-3038.2012.01341.x. Epub 02012 Aug. 01313.

Mine, Y. and P. Rupa (2003). Fine mapping and structural analysis of immunodominant IgE allergenic epitopes in chicken egg ovalbumin. Protein Eng. 16(10): 747–752.

Mine, Y. and J. W. Zhang (2002). Comparative studies on antigenicity and allergenicity of native and denatured egg white proteins. J Agric Food Chem. 50(9): 2679–2683.

Molina-Infante, J., A. Arias, J. Barrio, J. Rodriguez-Sanchez, M. Sanchez-Cazalilla and A. J. Lucendo (2014). Four-food group elimination diet for adult eosinophilic esophagitis: A prospective multicenter study. J Allergy Clin Immunol. 134(5): 1093–1099. e1091. doi: 1010.1016/j.jaci.2014.1007.1023. Epub 2014 Aug. 1028.

Monti, G., M. C. Muratore, A. Peltran, G. Bonfante, L. Silvestro, R. Oggero and G. C. Mussa (2002). High incidence of adverse reactions to egg challenge on first known exposure in young atopic dermatitis children: predictive value of skin prick test and radioallergosorbent test to egg proteins. Clin Exp Allergy. 32(10): 1515–1519.

Mullins, R. J. and M. S. Gold (2012). Influenza vaccination of the egg-allergic individual: 2012 update. Med J Aust. 196(11): 682.

Muraro, A., S. Halken, S. H. Arshad, K. Beyer, A. E. Dubois, G. Du Toit, P. A. Eigenmann, K. E. Grimshaw, A. Hoest, G. Lack, L. O'Mahony, N. G. Papadopoulos, S. Panesar, S. Prescott, G. Roberts, D. de Silva, C. Venter, V. Verhasselt, A. C. Akdis and A. Sheikh (2014). EAACI food allergy and anaphylaxis guidelines. Primary prevention of food allergy. Allergy. 69(5): 590–601. doi: 510.1111/all.12398. Epub 12014 Apr. 12393.

Murphy, A., D. E. Campbell, D. Baines and S. Mehr (2011). Allergic reactions to propofol in egg-allergic children. Anesth Analg. 113(1): 140–144. doi: 110.1213/ANE.1210b1013e31821b31450f. Epub 32011 Apr. 31825.

Nowak-Wegrzyn, A., Y. Katz, S. S. Mehr and S. Koletzko (2015). Non-IgE-mediated gastrointestinal food allergy. J Allergy Clin Immunol. 135(5): 1114–1124. doi: 1110.1016/j.jaci.2015.1103.1025.

Nwaru, B. I., L. Hickstein, S. S. Panesar, G. Roberts, A. Muraro and A. Sheikh (2014). Prevalence of common food allergies in Europe: a systematic review and meta-analysis. Allergy. 69(8): 992–1007. doi: 1010.1111/all.12423. Epub 12014 May 12410.

Nwaru, B. I., H. M. Takkinen, O. Niemela, M. Kaila, M. Erkkola, S. Ahonen, H. Tuomi, A. M. Haapala, M. G. Kenward, J. Pekkanen, R. Lahesmaa, J. Kere, O. Simell, R. Veijola, J. Ilonen, H. Hyoty, M. Knip and S. M. Virtanen (2013). Introduction of complementary foods in infancy and atopic sensitization at the age of 5 years: timing and food diversity in a Finnish birth cohort. Allergy. 68(4): 507–516. doi: 510.1111/all.12118. Epub 12013 Feb. 12115.

Ocmant, A., S. Mulier, L. Hanssens, M. Goldman, G. Casimir, F. Mascart and L. Schandene (2009). Basophil activation tests for the diagnosis of food allergy in children. Clin Exp Allergy. 39(8): 1234–1245. doi: 1210.1111/j.1365-2222.2009.03292.x. Epub 02009 Jun. 03222.

Osborne, N. J., J. J. Koplin, P. E. Martin, L. C. Gurrin, A. J. Lowe, M. C. Matheson, A. L. Ponsonby, M. Wake, M. L. Tang, S. C. Dharmage and K. J. Allen (2011). Prevalence of challenge-proven IgE-mediated food allergy using population-based sampling and predetermined challenge criteria in infants. J Allergy Clin Immunol. 127(3): 668–676. e661–662. doi: 610.1016/j.jaci.2011.1001.1039.

Palmer, D. J., J. Metcalfe, M. Makrides, M. S. Gold, P. Quinn, C. E. West, R. Loh and S. L. Prescott (2013). Early regular egg exposure in infants with eczema: A randomized controlled trial. J Allergy Clin Immunol. 132(2): 387–392. e381. doi: 310.1016/j.jaci.2013.1005.1002. Epub 2013 Jun. 1026.

Park, M., D. Kim, K. Ahn, J. Kim and Y. Han (2014). Prevalence of immediate-type food allergy in early childhood in seoul. Allergy Asthma Immunol Res. 6(2): 131–136. doi: 110.4168/aair.2014.4166.4162.4131. Epub 2013 Nov. 4128.

Peters, R. L., K. J. Allen, S. C. Dharmage, M. L. Tang, J. J. Koplin, A. L. Ponsonby, A. J. Lowe, D. Hill and L. C. Gurrin (2013). Skin prick test responses and allergen-specific IgE levels as predictors of peanut, egg, and sesame allergy in infants. J Allergy Clin Immunol. 132(4): 874–880. doi: 810.1016/j.jaci.2013.1005.1038. Epub 2013 Jul. 1024.

Peters, R. L., S. C. Dharmage, L. C. Gurrin, J. J. Koplin, A. L. Ponsonby, A. J. Lowe, M. L. Tang, D. Tey, M. Robinson, D. Hill, H. Czech, L. Thiele, N. J. Osborne and K. J. Allen (2014). The natural history and clinical predictors of egg allergy in the first 2 years of life: a prospective, population-based cohort study. J Allergy Clin Immunol. 133(2): 485–491. doi: 410.1016/j.jaci.2013.1011.1032. Epub 2013 Dec. 1025.

Prescott, S. L., R. Pawankar, K. J. Allen, D. E. Campbell, J. Sinn, A. Fiocchi, M. Ebisawa, H. A. Sampson, K. Beyer and B. W. Lee (2013). A global survey of changing patterns of food allergy burden in children. World Allergy Organ J. 6(1): 21. doi: 10.1186/1939-4551-1186-1121.

Quirce, S., F. Maranon, A. Umpierrez, M. de las Heras, E. Fernandez-Caldas and J. Sastre (2001). Chicken serum albumin (Gal d 5) is a partially heat-labile inhalant and food allergen implicated in the bird-egg syndrome. Allergy. 56(8): 754–762.

Rodriguez-Sanchez, J., E. Gomez Torrijos, B. Lopez Viedma, E. de la Santa Belda, F. Martin Davila, C. Garcia Rodriguez, F. Feo Brito, J. Olmedo Camacho, P. Reales Figueroa and J. Molina-Infante (2014). Efficacy of IgE-targeted vs empiric six-food elimination diets for adult eosinophilic oesophagitis. Allergy. 69(7): 936–942. doi: 910.1111/all.12420. Epub 12014 May 12429.

Rolinck-Werninghaus, C., B. Niggemann, L. Grabenhenrich, U. Wahn and K. Beyer (2012). Outcome of oral food challenges in children in relation to symptom-eliciting allergen dose and allergen-specific IgE. Allergy. 67(7): 951–957. doi: 910.1111/j.1398-9995.2012.02838.x. Epub 02012 May 02814.

Romantsik, O., M. Bruschettini, M. A. Tosca, S. Zappettini, O. Della Casa Alberighi and M. G. Calevo (2014). Oral and sublingual immunotherapy for egg allergy. Cochrane Database Syst Rev. 11: CD010638. doi: 10.1002/14651858.CD14010638. pub14651852.

Rueter, K., A. Haynes and S. L. Prescott (2015). Developing primary intervention strategies to prevent allergic disease. Curr Allergy Asthma Rep. 15(7): 40. doi: 10.1007/s11882-11015-10537-x.

Rutkowski, K., P. W. Ewan and S. M. Nasser (2013). Administration of yellow fever vaccine in patients with egg allergy. Int Arch Allergy Immunol. 161(3): 274–278. doi: 210.1159/000346350. Epub 000342013 Mar. 000346315.

Salo, P. M., A. Calatroni, P. J. Gergen, J. A. Hoppin, M. L. Sever, R. Jaramillo, S. J. Arbes, Jr. and D. C. Zeldin (2011). Allergy-related outcomes in relation to serum IgE: results from the National Health and Nutrition Examination Survey 2005–2006. J Allergy Clin Immunol. 127(5): 1226–1235. e1227. doi: 1210.1016/j.jaci.2010.1212.1106. Epub 2011 Feb. 1212.

Sampson, H. A. (2001). Utility of food-specific IgE concentrations in predicting symptomatic food allergy. J Allergy Clin Immunol. 107(5): 891–896.

Sampson, H. A. (2003). The evaluation and management of food allergy in atopic dermatitis. Clin Dermatol. 21(3): 183–192.

Santos, A. F., G. Du Toit, A. Douiri, S. Radulovic, A. Stephens, V. Turcanu and G. Lack (2015). Distinct parameters of the basophil activation test reflect the severity and threshold of allergic reactions to peanut. J Allergy Clin Immunol. 135(1): 179–186. doi: 110.1016/j.jaci.2014.1009.1001.

Savage, J. H., E. C. Matsui, J. M. Skripak and R. A. Wood (2007). The natural history of egg allergy. J Allergy Clin Immunol. 120(6): 1413–1417.

Sicherer, S. H., R. A. Wood, B. P. Vickery, S. M. Jones, A. H. Liu, D. M. Fleischer, P. Dawson, L. Mayer, A. W. Burks, A. Grishin, D. Stablein and H. A. Sampson (2014). The natural history of egg allergy in an observational cohort. J Allergy Clin Immunol. 133(2): 492–499. doi: 410.1016/j.jaci.2013.1012.1041.

Sopo, S. M., V. Giorgio, I. Dello Iacono, E. Novembre, F. Mori and R. Onesimo (2012). A multicentre retrospective study of 66 Italian children with food protein-induced enterocolitis syndrome: different management for different phenotypes. Clin Exp Allergy. 42(8): 1257–1265. doi: 1210.1111/j.1365-2222.2012.04027.x.

Spergel, J. M., T. F. Brown-Whitehorn, A. Cianferoni, M. Shuker, M. L. Wang, R. Verma and C. A. Liacouras (2012). Identification of causative foods in children with eosinophilic esophagitis treated with an elimination diet. J Allergy Clin Immunol. 130(2): 461–467. e465. doi: 410.1016/j.jaci.2012.1005.1021. Epub 2012 Jun. 1027.

Staden, U., C. Rolinck-Werninghaus, F. Brewe, U. Wahn, B. Niggemann and K. Beyer (2007). Specific oral tolerance induction in food allergy in children: efficacy and clinical patterns of reaction. Allergy. 62(11): 1261–1269.

Suzuki, M., H. Fujii, H. Fujigaki, S. Shinoda, K. Takahashi, K. Saito, H. Wada, M. Kimoto, N. Kondo and M. Seishima (2010). Lipocalin-type prostaglandin D synthase and egg white cystatin react with IgE antibodies from children with egg allergy. Allergol Int. 59(2): 175–183. doi: 110.2332/allergolint.2309-OA-0121. Epub 2010 Feb. 2325.

Tan, J. W., D. E. Campbell, P. J. Turner, A. Kakakios, M. Wong, S. Mehr and P. Joshi (2013). Baked egg food challenges—clinical utility of skin test to baked egg and ovomucoid in children with egg allergy. Clin Exp Allergy. 43(10): 1189–1195. doi: 1110.1111/cea.12153.

Tan, J. W. and P. Joshi (2014). Egg allergy: an update. J Paediatr Child Health. 50(1): 11–15. doi: 10.1111/jpc.12408. Epub 12013 Oct. 12418.

Turner, P. J., M. H. Gowland, V. Sharma, D. Ierodiakonou, N. Harper, T. Garcez, R. Pumphrey and R. J. Boyle (2015). Increase in anaphylaxis-related hospitalizations but no increase in fatalities: an analysis of United Kingdom national anaphylaxis data, 1992–2012. J Allergy Clin Immunol. 135(4): 956–963.e951. doi: 910.1016/j.jaci.2014.1010.1021. Epub 2014 Nov. 1025.

Turner, P. J., S. Mehr, P. Joshi, J. Tan, M. Wong, A. Kakakios and D. E. Campbell (2013). Safety of food challenges to extensively heated egg in egg-allergic children: a prospective cohort study. Pediatr Allergy Immunol. 24(5): 450–455. doi: 410.1111/pai.12093. Epub 12013 Jun. 12016.

Turner, P. J., J. Southern, N. J. Andrews, E. Miller and M. Erlewyn-Lajeunesse (2015). Safety of live attenuated influenza vaccine in atopic children with egg allergy. J Allergy Clin Immunol. 136(2): 376–381. doi: 310.1016/j.jaci.2014.1012.1925. Epub 2015 Feb. 1013.

Urisu, A., H. Ando, Y. Morita, E. Wada, T. Yasaki, K. Yamada, K. Komada, S. Torii, M. Goto and T. Wakamatsu (1997). Allergenic activity of heated and ovomucoid-depleted egg white. J Allergy Clin Immunol. 100(2): 171–176.

Walsh, B. J., D. J. Hill, P. Macoun, D. Cairns and M. E. Howden (2005). Detection of four distinct groups of hen egg allergens binding IgE in the sera of children with egg allergy. Allergol Immunopathol (Madr). 33(4): 183–191.

Xepapadaki, P., A. Fiocchi, L. Grabenhenrich, G. Roberts, K. E. Grimshaw, A. Fiandor, J. Ignacio Larco, S. Sigurdardottir, M. Clausen, L. N. Papadopoulos, L. Dahdah, A. Mackie, A. B. Sprikkelman, A. A. Schoemaker, R. Dubakiene, I. Butiene, M. L. Kowalski, K. Zeman, S. Gavrili, T. Keil and K. Beyer (2015). Incidence and natural history of hen's egg allergy in the first 2 years of life—the EuroPrevall birth cohort study. Allergy. 29(10): 12801.

Yadav, A. and R. Naidu (2015). Clinical manifestation and sensitization of allergic children from Malaysia. Asia Pac Allergy. 5(2): 78–83. doi: 10.5415/apallergy.2015.5415.5412.5478. Epub 2015 Apr. 5429.

5

Fish Allergy

Annette Kuehn* and Karthik Arumugam

CONTENTS

Department of Infection and Immunity, Luxembourg Institute of Health, Luxembourg.
* Corresponding author: annette.kuehn@lih.lu

5.1 INTRODUCTION

5.1.1 Fish, a Staple Food

Fish is included in the food pyramid as a basic food to be eaten regularily in a balanced diet. It is an important source of essential fatty acids, fat-soluble vitamins and polyunsaturated fatty acids (Gil and Gil 2015). Fat content of commercially available fish varies between lean (less than 1% such as cod, haddock, pollock), low-fat (1 to 5% such as halibut, plaice, sole), fatty (5 to 10% such as salmon, tuna, redfish) and highly fatty fish species (more than 10% such as mackerel, herring, eel) (Rehbein and Oehlenschläger 2009). The content of valuable polyunsaturated fatty acids (PUFA) is especially high in fish species with high fat content such as herring with 2.3 g PFUA/100 g muscle.

The global demand for fish and fish products is increasing steadily. This is reflected by a tremendous growth of the world fishery production since the early seventies. The Food and Agriculture Organization (FAO) of the United Nations calculated a global mean of 19.2 kg fish consumption per capita in 2012 representing 17% of the worldwide animal protein intake (www.fao.org). However, there are notable differences between the fish consumption across continents and regions, fish is very common in the diet of developed countries (Figure 5.1A). More than 40% of the global fish supply is aquaculture-based and its share is growing. Among the thousands of known fish species, about 800 fishes are used for food and food production. In 2012, anchoveta, Alaska pollock, skipjack tuna, sardinella spp. and Atlantic herring represented the global top five of commonly consumed marine fish species according to their capture rates (Figure 5.1B) (www.fao.org). The market shares of fish species vary across countries according to the availability of the fishes and the regional eating habits. In the European Union, Atlantic herring, Atlantic mackerel, European sprat, sand eels and Atlantic cod are the top five species caught while pollock, menhaden, cod, salmon and sole are the highest volume species in North America.

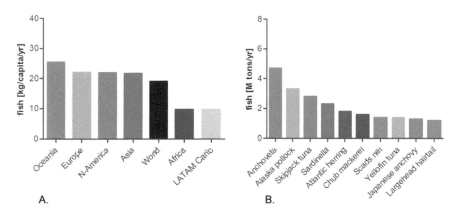

Figure 5.1 The global fish market (2012). **A.** The fish consumption varied in different continents/ territories by intake of fish in kg per capita per year. **B.** The most important fishes were ranked by their capture rates in million tons per year. LATAM: Latin America; M: million; yr: year.

5.1.2 Adverse Reactions to Fish: Intoxication and Allergy

Depending on the immunological mechanism, adverse reactions to fish can be grouped into IgE- and non-IgE-mediated reactions.

Fish allergy-like symptoms can occur as pharmacological reactions to histamine, a biogenic amine. These reactions have been described especially for members of the family *Scombridae* (e.g., mackerels, tunas, bonitos) (McLauchlin et al. 2006). Scombroid fishes have a naturally high content of the amino acid histidine while the normal concentration is below 0.1 mg/100 g of fresh muscle tissue (Rehbein and Oehlenschläger 2009). The US Food and Drug Administration (FDA) have considered a histamine content of more than 50 mg/100 g of fish toxic (www.fda.gov). Inadequate storage accompanied by bacterial contamination causes enzymatic decarboxylation of histidine to histamine (Attaran and Probst 2002). Binding of histamine to H1/H2-receptors in the cell membrane triggers the clinical manifestations of fish poisoning. Symptoms occur minutes to hours after ingestion, mostly as allergy-like reactions such as skin rash, nausea, abdominal pain and diarrhea. Histamine intolerance imitated a fish allergy (Thewes et al. 1999). Scombroid poisoning is a global health problem representing up to 40% of all seafood-related adverse reactions in the United States and Europe (Gould et al. 2013).

Non-IgE-mediated reactions to fish consumption appear to be less common but their clinical relevance is less well understood (Zapatero et al. 2005, Fernandes et al. 2012). Symptoms which are assumed to be T-cell-mediated develop only with a longer delay (> 4 hrs). Fish is considered a cause of food protein-induced enterocolitis (FPIES), a non-IgE-mediated reaction, which is commonly observed in children but only rarely reported in adults (Miceli Sopo et al. 2015).

Clinical symptoms of FPIES affect mostly the gastrointestinal tract and can include profound vomiting and diarrhea (Leonard and Nowak-Węgrzyn 2015). The epidemiology of FPIES is not well understood mainly because of the lack of common definitions but also because of missing availability of population-based studies (Berin 2015). However, the incidence of fish-induced FPIES might be higher in regions with important fish consumption than in others.

Most fish adverse reactions, defined as genuine fish allergy, are attributed to a Th2-driven IgE antibody response to fish allergens (Kuehn et al. 2011b, Sharp and Lopata 2014). During the development of this food allergy, the first step is the sensitization which develops often in childhood upon introduction of fish in the diet. Once the patient is re-exposed to this food, the allergic reaction starts minutes after ingestion affecting various organs such the skin (rash, urticarial, erythema), the gastrointestinal apparatus (epigastric pain, cramping, vomiting, diarrhea), the nervous (light-headedness, headache) and the respiratory system (bronchoconstriction, respiratory distress) and even severe anaphylaxis (circulatory collapse, breathing arrest) (Helbling et al. 1999, Bock et al. 2001, Pascual et al. 2008). A few milligrams of fish are sufficient to initiate the allergic cascade immediately after intake. Recent research suggests an eliciting dose predicted to provoke a reaction in 10% of individuals (ED_{10}) of 27.3 mg fish (Ballmer-Weber et al. 2014). However, the threshold dose defined as NOAEL (no observed adverse effect levels) seems to be much lower as intake of a few micrograms of fish can trigger allergic reactions. Eliciting doses and symptoms of fish allergy may also depend on the route of exposure. While digestion allows fish allergens to interact with mast cells of the oral and intestinal mucosa, allergenic proteins can trigger fish allergy symptoms also via inhalation or skin contact. Inhalation and skin contact have been mainly reported in the occupational context leading to symptoms such as skin rash, allergic rhinitis or asthma (Jeebhay and Cartier 2010, Jeebhay and Lopata 2012). Handling of fish seems to rarely cause severe adverse events such as anaphylaxis but affected workers may change their occupation because of intolerable allergic symptoms (Dickel et al. 2014). Eliciting doses have not yet been established in the occupational environment, although allergen exposure has been quantified in fish-processing factories previously (Lopata et al. 2005).

Fish is of global relevance in the human diet. This is in line with the fact that fish has been defined as one of the "big-8" of most important food allergens which are responsible for eliciting more than 90% of any type I hypersensitivity to food (Figure 5.2). Various epidemiology studies have established the IgE-mediated fish allergy prevalence ranging from 0.1% to more than 2% (Sharp and Lopata 2014). These prevalence rates vary between studies depending on study design such as different patient cohorts or different definitions

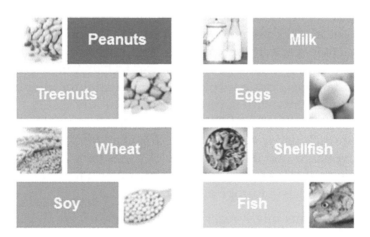

Figure 5.2 The most common triggers of food allergies from plant and animal origin have been referred to as the "big-8".

of a positive fish allergy diagnosis. Fish allergy is assumed to be more common in adults than in children, possibly because of its chronic nature but also because it may develop with late onset in adulthood. Reported prevalence of fish allergy ranges between 2.2% (self-reported), 0.6% (skin reactivity), and 0.1% (specific IgE to fish extracts or food challenge) (Rona et al. 2007, Nwaru et al. 2014). Studies showed that up to 8% of the workers developed allergic asthma during their occupation in fish-processing industries, most probably because of the high exposure rates with fish allergens in the work environment (Jeebhay et al. 2008). Prevalence estimates of 0.2–0.5% of the general population affected by this type of food allergy still need to be confirmed by comprehensive studies with a generally accepted research design.

5.1.3 IgE-mediated Fish Allergy: Clinical Phenotypes

An important clinical feature of fish allergy is the reaction upon ingestion of various fish species (Kuehn et al. 2014a). This cross-reactivity has been explained as a specific IgE-recognition of conserved antigen epitopes from distinct fishes manifesting as allergies to the corresponding species (see paragraph *Parvalbumins*) (Swoboda et al. 2002). Many clinical studies have described these fish poly-sensitized patients (Sten et al. 2004, Van Do et al. 2005, Perez-Gordo et al. 2011, Kuehn et al. 2013). Clinical cross-reactivity seems to be high especially in fish species which are closely related. However, the fact that fishes belong to separate families does not rule out the potential for allergic cross-reactivity. Patients with severe symptoms to any tested fish have been described. In general, fish-allergic patients react to fish species which are included in the local diet (Zinn et al. 1997, Lim et al. 2008). As eating habits and the availability of fishes vary across countries and regions, species with high allergenic potential might not be the same for instance in Europe and Asia.

Most patients might react to many fishes, others develop only symptoms to specific or single fishes only (Asero et al. 1999, Ebo et al. 2010, Kuehn et al. 2011a, Kuehn et al. 2014b, Calderon-Rodriguez et al. 2016). The global prevalence of these different phenotypes is not well understood as studies of appropriate design are missing. As many studies have previously focused on patients with reactivity to multiple fishes it may appear as if these patients constitute the vast majority of the fish-allergic population. However, patients with reactions to specific fishes may be underdiagnosed because of the lack of specific *in vitro* tests. According to the present state of knowledge, diagnostic food challenges would have to be performed to determine the tolerance of different fishes, which is difficult to realize in an epidemiological study. Clinical reactivity to single fishes has been mostly reported in case reports such as for sole, swordfish and cod (Kelso et al. 1996, Asero et al. 1999, Kuehn et al. 2014b). Allergy to several but specific fishes was described such as for tuna/marlin, pangasius/tilapia, salmon/trout and cod/perch (Kondo et al. 2006, Ebo et al. 2010, Kuehn et al. 2011a, Peñas et al. 2014, Kuehn et al. 2014b). More recently, a study with 62 fish-allergic patients suggested a phenotype distribution of about 60% poly-sensitized, about 30% oligo-sensitized and about 10% mono-sensitized fish-allergic patients (Kuehn et al. 2013).

However, selective tolerance to specific fishes may have different backgrounds: (i) Some fishes have naturally low allergen contents compared to others, which is why tuna is mostly tolerated by fish-allergic patients (Lim et al. 2005, Griesmeier et al. 2010a, Kuehn et al. 2010, Kobayashi et al. 2016). (ii) Some fish products might be reduced in allergen content as allergens are degraded by harsh food processing conditions such as for canned or pickled fish (Bernhisel-Broadbent et al. 1992, Sletten et al. 2010). (iii) Some patients might have IgE-antibodies to fish allergens recognizing species-specific B cell epitopes only as shown previously for monosensitivity to salmonid fishes and cod (Kuehn et al. 2011a, Vázquez-Cortés et al. 2012, Pérez-Gordo et al. 2012, Peñas et al. 2014, Kuehn et al. 2014b). Specific IgE to further species-specific allergens might also play a role such as reported for sole (Pérez-Gordo et al. 2010).

While there are still open questions around the *in vivo* digestion and resorption of fish allergens in sensitized patients, the biomolecular knowledge on these allergens is more advanced.

Molecular and allergenic characteristics of fish proteins will be set into context in the further paragraphs (see paragraph *Fish allergens*).

5.1.4 Fish Allergy Diagnosis and Therapy

A decision scheme has been proposed for the diagnostic procedure of fish allergy in order to optimize the outcome of individual tests and minimize the risk for the patient (Poulsen et al. 2016). As for many other food allergies, the clinical diagnosis of fish allergy includes four main steps: documenting the medical history, skin testing and specific IgE-detection in patient sera adding food provocation tests in selected cases (Poulsen 2015).

Diligent documentation of the clinical history of a patient is a key success driver in allergy diagnosis. A recent guideline of the European Academy of Allergy and Clinical Immunology (EAACI) taskforce on food allergy demonstrated the value of the systematic inquiry with key questions together with appropriate actions for the clinical diagnosis (Muraro et al. 2014a).

Skin testing of fish-allergic patients is performed usually with commercial extracts from a limited range of fishes such as cod, salmon, tuna, sole (Poulsen et al. 2016). The number of diagnostic fish extracts varies across countries. Besides testing the symptome-eliciting fish as suggested by the patient's medical record, further species such as native fish are commonly used in prick-to-prick analysis (Pitsios et al. 2010, Kuehn et al. 2013). Some studies reported higher diagnostic value with fresh fish, others promoted the use of heated or even processed food (Chikazawa et al. 2015). However, as closely related fishes might contain similar allergens, it has been recommended to test fishes from different families (Kuehn et al. 2016).

Only two fish allergens from carp and cod are available for diagnostic analysis (Cyp c 1, Gad c 1; see paragraph *Parvalbumins*) but the quantification of specific IgE in patient sera still relies mostly on fish extracts. As commonly consumed food fishes belong to restricted taxonomic orders, it might be advisable to test specific IgE to selected fishes such as cod extract for *Gadiformes*, salmon extract for *Salmoniformes* and mackerel extract for *Perciformes* (Kuehn et al.

2016, Poulsen et al. 2016). However, a careful selection is advisable to account for the local market and eating habits of the individual patient. The level of the fish-specific IgE-titer does not seem to necessarily correlate with the severity of the symptoms (Sampson and Ho 1997). Patients with very low specific IgE might experience anaphylactic reations which is confirmed by the low correlation reported between extract-specific IgE and the clinical reactivity of fish-allergic patients (Agabriel et al. 2010, Kuehn et al. 2013).

The gold standard for a final verification of fish allergy is still the double-blind, placebo-controlled food challenge (DBPCFC), which is performed only in selected cases because of ethical and financial reasons. In order to minimize the risk of severe reactions during such testing, a specific decision scheme should be followed (Muraro et al. 2014a). A negative DBPCFC result should be confirmed in an open food challenge.

The Precision Medicine in Allergy and Asthma (PRACTALL) program published a consensus paper of the American Academy of Allergy, Asthma & Immunology (AAAA) and EAACI including important practical recommendations for diagnostic food challenges (such as foods, timings, doses, recipes) for the diagnosis of fish allergy (Sampson et al. 2012).

At present, there is no specific immunotherapy available for the treatment of fish allergy. The desensibilization with cooked fish has been described in a case (Patriarca et al. 2007). However, the safety of this approach might be very low because of the thermostability of the main fish allergens. The elimination diet is an important step in the management of fish-allergic patients (Muraro et al. 2014b). After strict elimination, even the development of oral tolerance to fish has been described (Solensky 2003, Pite et al. 2012). At the same time there is a risk of patients becoming re-sensitized after re-consuming of fish (De Frutos et al. 2003). In such cases, tight clinical monitoring and care to the patient would be required which is often not feasible.

In order to help the food-allergic consumer to avoid the suspecting allergy triggers, specific food labeling regulations have been implemented in many countries within the US, Europe, Latin

America, Asia and Australia (Taylor and Baumert 2015). While allergen labeling rules vary across regions, fish is widely recognised as a potent allergenic food and needs to be declared as an ingredient on each product. This rule applies not only to packaged foods but also to other food such meals prepared in restaurants or canteens. Presently, US and European countries apply zero tolerance rules for allergenic fish which mandates declaration of fish content of any percentage. In order to avoid the "may contain"-indication on many products, efforts are undertaken to establish threshold levels governing the labeling of allergenic food components (Taylor et al. 2014). However, further research will be required to specify these guidelines for fish.

5.2 FISH ALLERGENS

5.2.1 Parvalbumins

Fish parvalbumin belongs to the class I food allergens which act as sensitizing proteins via the gastro-intestinal passage and elicit systemic reactions of allergy upon ingestion (Lorenz et al. 2015). Furthermore, these allergens have been characterized as low-molecular weight proteins of high stability towards food processing and enzymatic digestion (Griesmeier et al. 2010b). Parvalbumins are the major allergens of fish muscle where these 10–12 kDa-allergens occur in the cytoplasm (Bugajska-Schretter et al. 2000, Lim et al. 2008). According to their structure, they are members of the "EF hand"-protein family, which are characterized by structural motifs for specific divalent ion-binding (Figure 5.3) (Radauer et al. 2008). "EF-hand"-motifs are composed of a peptide helix/peptide loop/peptide helix-sequence. Fish parvalbumins have three EF-hand-motifs but only two sites are functional and bind calcium or magnesium ions. Parvalbumins have the physiological role of calcium-buffer proteins involved in the muscle relaxation process (Schwaller 2009). The ion binding is important for the stability of the molecule (Bugajska-Schretter et al. 2000). Notable differences in the structure stability of calcium-bound parvalbumin and the corresponding apo-form have been reported.

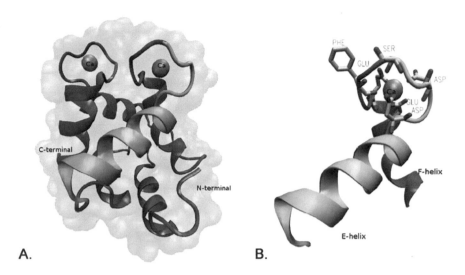

A.   B.

Figure 5.3 Fish parvalbumins have two active EF-hand motifs binding calcium ions (orange balls) which is coordinated especially by acidic amino acid residues.

These muscle proteins have been identified as important allergens in many fishes. Currently, parvalbumin homologs from twelve fishes, carp, cod, salmon, trout, mackerel, herring, pilchard, barramundi, tuna, swordfish, whiff and redfish, have been approved for their official allergen names by the allergen nomenclature sub-committee of the World Health Organization (WHO)/International Union of Immunological Societies (IUIS) (Table 5.1) (www.allergen.org). Fish parvalbumins share modest protein identities in their primary structure ranging from 65% to 75% but the three-dimensional structures are highly conserved. Comparative analysis of IgE binding to parvalbumin in the presence and absence of calcium showed that important epitopes are located in the ion-binding regions of the allergens (Figure 5.4A) (Bugajska-Schretter et al. 1998). These regions, EF-hand-motifs, are very similar in parvalbumins of many fishes and considered the reason for the high level of clinical cross-reactivity of poly-sensitized patients. Epitope analysis on different fish parvalbumins revealed that specific IgE-recognition sites might be also located in other regions of the molecule (Figure 5.4B) (Yoshida et al. 2008, Pérez-Gordo et al. 2012, Pérez-Gordo et al. 2013). Parvalbumins from different fishes have become available

Table 5.1 Twenty fish allergen names have been approved by the WHO/IUIS allergen nomenclature subcommittee. Most parvalbumins occur as several isoallergens or variants in the same fish.

Taxonomic order	Fish species	Allergen identity	Allergen name	Variant no.
Cichliformes	Nile tilapia (*Oreochromis niloticus*)	Tropomyosin	Ore m 4	–
Clupeiformes	Herring (*Clupea harengus*)	Parvalbumin	Clu h 1	3
	Pilchard (*Sardinops sagax*)	Parvalbumin	Sar sa 1	–
Cypriniformes	Carp (*Cyprinus carpio*)	Parvalbumin	Cyp c 1	2
Gadiformes	Atlantic cod (*Gadus morhua*)	Parvalbumin	Gad m 1	4
		Enolase	Gad m 2	–
		Aldolase	Gad m 3	–
	Baltic cod (*Gadus callarias*)	Parvalbumin	Gad c 1	–
Perciformes	Barramundi (*Lates calcarifer*)	Parvalbumin	Lat c 1	2
	Swordfish (*Xiphias gladius*)	Parvalbumin	Xip g 1	–
	Tuna (*Thunnus albacares*)	Parvalbumin	Thu a 1	–
		Enolase	Thu a 2	–
		Aldolase	Thu a 3	–
Pleuronectiformes	Megrim (*Lepidorhombus whiffiagonis*)	Parvalbumin	Lep w 1	–
Salmoniformes	Chum salmon (*Oncorhynchus keta*)	Vitellogenin*	Onc k 5	–
	Rainbow trout (*Oncorhynchus mykiss*)	Parvalbumin	Onc m 1	2
	Salmon (*Salmo salar*)	Parvalbumin	Sal s 1	–
		Enolase	Sal s 2	–
		Aldolase	Sal s 3	–
Scorpaeniformes	Redfish (*Sebastes marinus*)	Parvalbumin	Seb m 1	–

* allergen from fish roe (all others are fish muscle allergens); –, no variant(s) in www.allergen.org

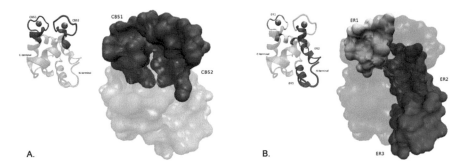

Figure 5.4 Fish parvalbumin epitopes in poly-sensitized patients. **A.** IgE-binding sites are located in the calcium-binding regions. **B.** Epitopes reported in different studies are spread over the molecule.

as recombinant allergens in research but only two molecules, carp and cod parvalbumin, have been added to IgE test kits in routine diagnosis. An ongoing European Union (EU)-project, the FAST (Food Allergy Specific ImmunoTherapy) study, aims at developing a specific immunotherapy for fish allergy based on the use of a hypoallergenic carp parvalbumin (Douladiris et al. 2015, Zuidmeer-Jongejan et al. 2015).

5.2.2 Fish Gelatin

Fish gelatin has been introduced to the market as a valid alternative to mammalian gelatins. It is a multifunctional ingredient commonly used by the cosmetic, pharmaceutical, medical and food industry (Boran and Regenstein 2010). Fish skin and scales from many fish species (mostly a mix of different fishes) are the main source for the production of fish gelatin. Collagen, a structural protein composed of a triple helix with two identical chains (α1) and one slightly different α2-chain, is extracted from these fish materials, purified and treated by alkaline or acid hydrolysis to derive fish gelatin.

The allergenic potency of fish gelatin has been addressed in only a few studies on fish-allergic patients. Several studies showed specific IgE-recognition or skin reactivity to this biopolymer (Sakaguchi et al. 2000, André et al. 2003). Food challenges have been only rarely performed, a clinical reaction at a cumulative dose of 7.6 g fish gelatin was described in one article (Hansen et al. 2004). Another case report identified fish gelatin as a trigger of life-threatening anaphylaxis (Kuehn et al. 2009). A cross-reactivity to bovine and porcine homologs was ruled out. The state of knowledge concerning the allergenicity and cross-reactivity among fish gelatins is still limited. Most studies do not specify origin nor ingredient fish species of the gelatin product in question. While up to 20% of the fish-allergic patients may have specific IgE to fish gelatin, the clinical consequences have not yet been addressed (Kuehn et al. 2013). Recombinant collagen is available from some sources such as the human homolog produced in yeast, but cDNA from fish collagen has been only cloned without recombinant expression (Olsen et al. 2003, Yu et al. 2014).

Based on the production process, there is a certain risk of fish gelatin to be contaminated with fish allergens. The low levels of fish allergens that can be detected in fish gelatin have been assumed to not trigger allergic reactions in fish-sensitized patients so far (Weber et al. 2009). The food allergen labeling in the US mandates the specific declaration of fish gelatin on food products but fish gelatin as well as fish oil are currently exempted from labeling in Europe, which may change with more *in vitro* and *in vivo* testings for fish gelatin-sensitization and cross-reactivity (Taylor and Hefle 2001, Pieretti et al. 2009). However, a hurdle for such research is the absence of fish gelatin for diagnostic testing purposes (IgE, skin).

5.2.3 Enolases and Aldolases

Enolases and aldolases are key enzymes of the catabolic glycolysis present in all tissues. Aldolase or 40 kDa-fructose-bisphosphate aldolase (EC 4.1.2.13) splits fructose 1,6-bisphosphate into triose phosphates dihydroxyacetone phosphate and glyceraldehyde 3-phosphate (4th step of glycolysis) (Garfinkel and Garfinkel 1985). Enolase or 50 kDa-phosphopyruvate hydratase (EC 4.2.1.11) is a metalloenzyme (Mg^{2+}-ions per molecule) catalysing the conversion of 2-phosphoglycerate to phosphoenolpyruvate (9th step of glycolysis). Both enzymes belong to the structural family of so-called "TIM barrel"-proteins (Kuehn et al. 2016). Eponym for this family is the triosephosphate isomerase (TIM), which was characterized as the first protein by a common structure of eight alpha-helices alternating with eight beta-strands. Despite of the structural homology within this family, there is a lack of substantial sequence identity between TIM barrel-proteins.

Enolases and aldolases have been identified as fish allergens in cod, salmon and tuna (Kuehn et al. 2013). These allergens have been approved by the WHO/IUIS-allergen nomenclature subcommittee as Gad m 2/Gad m 3 for cod, Sal s 2/Sal s 3 for salmon and Thu a 2/Thu a 3 for tuna (Table 5.1) (www.allergen.org). They can be found in notable levels in fish muscle but they seem to have a lower stability towards physical influences by food processing than parvalbumins. Still, specific IgE-binding to these enzymes was shown in patients with moderate to severe symptoms of fish allergy. *In vitro* cross-reactivity has been demonstrated for the homologs from cod, salmon and tuna, although IgE cross-reactivity was varible in magnitude with clear tendency indicating cod allergens as the most potent inhibitor. These fish allergens have not yet been produced as recombinant proteins as only partial allergen sequences have been characterized. Despite the fact that no comprehensive sequence comparison has been performed for fish enolases and aldolases substantiating the presence of common IgE epitopes in general, it can be assumed that cross-reactivity to homologs from other fishes, beyond cod, salmon and tuna, might occur.

5.2.4 Other Fish Allergens

Many studies have been perfomed over the past decades in order to address the identification of allergenic fish proteins. For most of the following proteins, the relevance as food or inhaled allergens still needs to be investigated in future studies.

The database of the WHO/IUIS-allergen nomenclature subcommittee comprises Onk k 5, vitellogenin as an allergen from roe (Table 5.1) (www.allergen.org). This allergen was described from chum salmon, *Oncorhynchus keta*, as a high molecular roe protein whose subunits seemed to be recognized by patients' specific IgE-antibodies (Shimizu et al. 2014). While fish vitellogenins have been reported as cross-reactive allergens, no cross-reactivity was found to homologs from hen's eggs. The prevalence of roe allergy is estimated to be very low, past studies relied on single case reports or small patient cohorts (Perez-Gordo et al. 2008, Shimizu et al. 2009). Allergens found in fish meat and roe are different explaining why fish-allergic patients do not react to fish eggs in general (Mäkinen-Kiljunen et al. 2003).

The WHO/IUIS-allergen database contains also Ore m 4, a 33 kDa-tropomyosin from *Oreochromis mossambicus* (Mozambique tilapia) (Table 5.1) (www.allergen.org). This fish muscle allergen has been described in a few patients so far (Liu et al. 2013). Tropomyosin is an alpha-helical, linear structure protein from the cell cytoskeleton. Serological cross-reactivity has been shown for Ore m 4 and the homologue main allergen from shrimp but the clinical significance of this *in vitro* assay has not yet been resolved. Furthermore, most of the patients in the study on Ore m 4 had been diagnosed for inflammatory bowel disease, an autoimmune disease involving auto-antibodies against the human tropomyosin isoform TM-5. Future investigations will have to address the link between allergy and this autoimmune disease.

More allergens have been described without being approved by the WHO/IUIS-allergen nomenclature database. Others such as aldehyde-phosphate dehydrogenase and creatine kinase have been proposed as food allergens (Das Dores 2002, Rosmilah et al. 2013). Glyceraldehyde-3-phosphate dehydrogenase was suggested as an occupational allergen in fish-processing workers (Van der Ventel et al. 2011). As the data on these allergens are still limited, they will not be further discussed in the present review.

5.3 TRANSLATIONAL ASPECTS: FROM BENCH TO BEDSIDE

Variations in clinical reactions to different fish species are a well-documented phenomenon reported in clinical studies (Bahna 2004, Mourad and Bahna 2015). At the forefront of fish allergy research, studies aim to identify the cause for this clinical observation as well as to develop an improved *in vitro* diagnostic procedure using single allergens for specific IgE-detection (defined as "component-resolved diagnosis"). Although the knowledge has advanced over the past decades, researchers are still working on closing remaining gaps.

5.3.1 Allergen Contents in Food

Parvalbumins have been described as pan-allergens in fish (Kuehn 2014a, Sharp and Lopata 2014). The content of parvalbumin has been quantified in different fish species using antibody-based immunoassays (Kuehn et al. 2010, Sletten et al. 2010, Lee et al. 2011, Shibahara et al. 2013, Saptarshi et al. 2014). These studies have reported a sensitivity issue for the parvalbumin quantification. While detection antibodies work sensitively with their respective target immunogen, their sensitivity decreases noticeably for the binding to another homolog. Using enzyme-linked immunosorbent assay (ELISA) and a set of specifically targeted anti-parvalbumin antibodies, parvalbumin levels were found to range from less than 0.05 mg for tuna, from 0.3 to 0.7 mg for mackerel, from 1 to 2.5 mg for salmon, trout and cod to more than 2.5 mg per gram raw muscle for carp, herring and redfish (Figure 5.5) (Kuehn et al. 2010). The very low allergen titer in tuna correlates with the clinical finding that this fish is often tolerated by fish-allergic patients (Lim et al. 2005). The content of intact parvalbumin was found to be up to 60% lower in processed samples when compared to native fish which

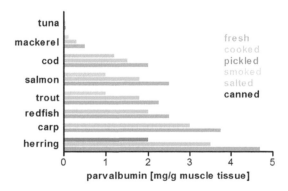

Figure 5.5 The content of native fish parvalbumin varies considerably in different fish species and fish products.

111

may suggest a lowered allergenic potential, depending on the IgE sensitization profile of the individual patient. This is in line with the clinical observation that canned fish is typically better tolerated by fish-allergic patient than steamed fish (Bernhisel-Broadbent 1992). However, knowledge on allergen contents in fish is still incomplete (Lee et al. 2012). Not only parvalbumins but also other fish allergens would have to be quantified in large number of commonly consumed fishes covering the global market. As reliable antibody-based assays seem not to be feasible for such a scope, other analytical methods will have to be developed to master this challenge.

However, parvalbumin levels do not explain the variable allergenicity of fishes with similar allergen titers such as salmon and cod. Thus, also the understanding of the effects of food processing and food matrices on the allergenicity of the fish products will have to improve (Somkuti et al. 2012, Shibahara et al. 2013). Food processing includes physical, enzymatic or chemical reactions (e.g., protein-protein crosslinking, amyloid formation, Maillard products) which may modify fish allergen epitopes leading to changed IgE-binding properties (de Jongh et al. 2011, Martínez et al. 2015). Food matrix effects (e.g., high protein, lipid or polysaccharids) can result in a modified *in vivo* digestion and absorption in the gastrointestinal system resulting in variable interactions with the immune system (Sletten et al. 2010). Further analytical challenges will have to be solved in order to simulate the fish allergen fate at the molecular level under physiological conditions.

5.3.2 Fish Allergens

The improvement of IgE diagnosis in fish allergy remains a key objective of ongoing research. The established predictive decision point (20 kU_A/L, codfish extract, >95%) seems to be of low value for the individual and in addition, the fish-extract based specific IgE fail to predict clinical reactivity to specific fishes (Sampson and Ho 1997, Schulkes et al. 2014).

Beta-parvalbumins have been described as the pan-allergens with broad IgE cross-reactivity among homologs from different fishes (Kuehn et al. 2014a). Parvalbumins from carp and cod display the most important IgE epitopes which are relevant in *in vitro* cross-reactivity (Ma et al. 2008). Parvalbumin-positive patients with clinical poly-sensitization to many fishes might be diagnosed efficiently using the carp or cod allergen (Table 5.2). This parvalbumin-based assay is already commercially available and might offer improved sensitivity in IgE-testing (Agabriel et al. 2010).

While carp and cod parvalbumins are useful in IgE-diagnosis of fish poly-sensitized patients, their diagnostic value seemed to be more limited in parvalbumin-positive patients with allergy to single fishes only (Table 5.2). Patients with clinical monosensitivity to salmonid fishes (salmons, trout) have specific IgE antibodies to salmon parvalbumin only (Kuehn et al. 2011a, Vázquez-Cortés et al. 2012, Pérez-Gordo et al. 2012, Peñas et al. 2014). Another clinical case with allergy to pangasius has been related to IgE-binding to single parvalbumins (Raith et al. 2014). Thus, a subgroup of fish-allergic patients have specific IgE to parvalbumins from single fish species only. This needs to be accounted for during sera analysis. Currently, these parvalbumins are not available for IgE diagnosis of fish allergy.

The prevalence of fish-allergic patients with specific IgE to parvalbumins has never been analyzed in an extensive study with a

Table 5.2 Fish allergens important for the application in improved IgE-based diagnosis of fish allergy.

Allergen	Clinical reactivity	Clinical phenotype(s)	Recombinant protein	Available for diagnosis
Parvalbumins	Muscle	Poly-/mono-sensitization	Yes	Cyp c 1, Gad c 1*
Enolases	Muscle	Poly-/oligo-/mono-sensitization	No	No
Aldolases	Muscle	Poly-/oligo-/mono-sensitization	No	No
Fish gelatin	Muscle/skin	Poly-/oligo-/mono-sensitization	No	No
Tropomyosin	Muscle	tbd	No	No
Vitellogenin	Roe	tbd	No	No

* other parvalbumins such as salmon Sal s 1 are missing; tbd: to be determined

representative number of patients of poly-, oligo- and monosensitized clinical phenotypes. Many studies claim that more than 90% of the fish-allergic patients have IgE-reactivity to the main allergen but most likely, the prevalence is lower ranging at about 70% as proposed more recently (Kuehn et al. 2013). Most of the parvalbumin-negative patients seemed to have specific IgE to fish enolase, aldolase or gelatin. In a small case series, even clincial monosensitivity to cod was correlated to specific IgE-binding to cod enolase and aldolase (Kuehn et al. 2014b). Fish enolases, aldolases and gelatin are not available for IgE-analysis of fish-allergic patients.

To summarize, the component-resolved approach in fish allergy diagnosis using a broad array of allergens, parvalbumins but also other fish allergens, seems to be a promising approach to improve the specificity and sensitivity of the current diagnosis which is based on non-standardized fish extracts.

ACKNOWLEDGEMENT

We thank Christiane Hilger, PhD for critical review of the manuscript.

Keywords: Fish allergy; fish allergen; parvalbumin; Enolase; Aldolase; food allergy

REFERENCES

Agabriel, C., P. Robert, P. Bongrand, J. Sarles and J. Vitte (2010). Fish allergy: in Cyp c1 we trust. Allergy. 65: 1483–4.

Asero, R., G. Mistrello, D. Roncarolo, M. Casarini and P. Falagiani (1999). True monosensitivity to a tropical sole. Allergy. 54: 1228–9.

Attaran, R. R. and F. Probst (2002). Histamine fish poisoning: a common but frequently misdiagnosed condition. Emerg Med J. 2002: 474–5.

Bahna, S. L. (2004). You can have fish allergy and eat it too! J Allergy Clin Immunol. 114: 125–6.

Ballmer-Weber, B. K., M. Fernandez-Rivas, K. Beyer, M. Defernez, M. Sperrin, A. R. Mackie et al. (2014). How much is too much? Threshold dose distributions for 5 food allergens. J Allergy Clin Immunol. 135: 964–71.

Berin, M. C. (2015). Immunopathophysiology of food protein-induced enterocolitis syndrome. J Allergy Clin Immunol. 135: 1108–13.

Bernhisel-Broadbent, J., D. Strause and H. A. Sampson (1992). Fish hypersensitivity. II: Clinical relevance of altered fish allergenicity caused by various preparation methods. J Allergy Clin Immunol. 90: 622–9.

Bock, S. A., A. Muñoz-Furlong and H. A. Sampson (2001). Fatalities due to anaphylactic reactions to foods. J Allergy Clin Immunol. 107: 191–3.

Boran, G. and J. M. Regenstein (2010). Fish gelatin. Adv Food Nutr Res. 60: 119–43.

Bugajska-Schretter, A., L. Elfman, T. Fuchs, S. Kapiotis, H. Rumpold, R. Valenta et al. (1998). Parvalbumin, a cross-reactive fish allergen, contains IgE-binding epitopes sensitive to periodate treatment and Ca^{2+} depletion. J Allergy Clin Immunol. 101: 67–74.

Bugajska-Schretter, A., M. Grote, L. Vangelista, P. Valent, W. R. Sperr, H. Rumpold et al. (2000). Purification, biochemical, and immunological characterisation of a major food allergen: different immunoglobulin E recognition of the apo- and calcium-bound forms of carp parvalbumin. Gut. 46: 661–9.

Calderon-Rodriguez, S., F. Pineda, R. Perez and C. Muñoz (2016). Tolerability to dogfish in children with fish allergy. Allergol Immunopathol (Madr). 44(2): 167–9. doi: 10.1016/j.aller.2015.05.007.

Chikazawa, S., T. Hashimoto, Y. Kobayashi and T. Satoh (2015). Fish-collagen allergy: a pitfall of the prick-to-prick test with raw fish. Br J Dermatol. doi: 10.1111/bjd.13962.

Das Dores, S., C. Chopin, A. Romano, A. V. Galland-Irmouli, D. Quaratino, C. Pascual et al. (2002). IgE-binding and cross-reactivity of a new 41 kDa allergen of codfish. Allergy. 57: 84–7.

De Frutos, C., L. Zapatero, A. Rodríguez, R. Barranco, E. Alonso and M. I. Martínez (2003). Re-sensitization to fish after a temporary tolerance. Allergy. 58: 1067–8.

de Jongh, H. H., S. L. Taylor and S. J. Koppelman (2011). Controlling the aggregation propensity and thereby digestibility of allergens by Maillardation as illustrated for cod fish parvalbumin. J Biosci Bioeng. 111: 204–11.

Dickel, H., T. Bruckner, P. Altmeyer and B. Künzlberger (2014). Seafood allergy in cooks: a case series and review of the literature. J Dtsch Dermatol Ges. 12: 891–902.

Douladiris, N., B. Linhart, I. Swoboda, A. Gstöttner, E. Vassilopoulou, F. Stolz et al. (2015). *In vivo* allergenic activity of a hypoallergenic mutant of the major fish allergen Cyp c 1 evaluated by means of skin testing. J Allergy Clin Immunol. doi: 10.1016/j.jaci.2015.01.015.

Ebo, D. G., A. Kuehn, C. H. Bridts, C. Hilger, F. Hentges and W. J. Stevens (2010). Monosensitivity to pangasius and tilapia caused by allergens other than parvalbumin. J Investig Allergol Clin Immunol. 20: 84–8.

Fernandes, B. N., R. J. Boyle, C. Gore, A. Simpson and A. Custovic (2012). Food protein-induced enterocolitis syndrome can occur in adults. J Allergy Clin Immunol. 130: 1199–200.

Garfinkel, L. and D. Garfinkel (1985). Magnesium regulation of the glycolytic pathway and the enzymes involved. Magnesium. 4: 60–72.

Gil, A. and F. Gil (2015). Fish, a Mediterranean source of n-3 PUFA: benefits do not justify limiting consumption. Br J Nutr. 113: 58–67.

Gould, L. H., K. A. Walsh, A. R. Vieira, K. Herman, I. T. Williams, A. J. Hall et al. (2013). Surveillance for foodborne disease outbreaks—United States, 1998–2008. MMWR Surveill Summ. 62: 1–34.

Griesmeier, U., S. Vázquez-Cortés, M. Bublin, C. Radauer, Y. Ma, P. Briza et al. (2010a). Expression levels of parvalbumins determine allergenicity of fish species. Allergy. 65: 191–8.

Griesmeier, U., M. Bublin, C. Radauer, S. Vázquez-Cortés, Y. Ma, M. Fernández-Rivas and H. Breiteneder (2010b). Physicochemical properties and thermal stability of Lep w 1, the major allergen of whiff. Mol Nutr Food Res. 54: 861–9.

Hansen, T. K., L. K. Poulsen, P. Stahl Skov, S. L. Hefle, J. J. Hlywka, S. L. Taylor et al. (2004). A randomized, double-blinded, placebo-controlled oral challenge study to evaluate the allergenicity of commercial, food-grade fish gelatin. Food Chem Toxicol. 42: 2037–44.

Helbling, A., R. Haydel Jr, M. L. McCants, J. J. Musmand, J. El-Dahr and S. B. Lehrer (1999). Fish allergy: is cross-reactivity among fish species relevant? Double-blind placebo-controlled food challenge studies of fish allergic adults. Ann Allergy Asthma Immunol. 83: 517–23.

Jeebhay, M. F. and A. Cartier. (2010). Seafood workers and respiratory disease: an update. Curr Opin Allergy Clin Immunol. 10: 104–13.

Jeebhay, M. F. and A. L. Lopata (2012). Occupational allergies in seafood-processing workers. Adv Food Nutr Res. 66: 47–73.

Jeebhay, M. F., T. G. Robins, M. E. Miller, E. Bateman, M. Smuts, R. Baatjies et al. (2008). Occupational allergy and asthma among salt water fish processing workers. Am J Ind Med. 51: 899–910.

Kelso, J. M., R. T. Jones and J. W. Yunginger (1996). Monospecific allergy to swordfish. Ann Allergy Asthma Immunol. 77: 227–8.

Kondo, Y., R. Komatsubara, Y. Nakajima, T. Yasuda, M. Kakami, I. Tsuge et al. (2006). Parvalbumin is not responsible for cross-reactivity between tuna and marlin: A case report. J Allergy Clin Immunol. 118: 1382–3.

Kuehn, A., C. Hilger and F. Hentges (2009). Anaphylaxis provoked by ingestion of marshmallows containing fish gelatin. J Allergy Clin Immunol. 123: 708–9.

Kuehn, A., T. Scheuermann, C. Hilger and F. Hentges (2010). Important variations in parvalbumin content in common fish species: a factor possibly contributing to variable allergenicity. Int Arch Allergy Immunol. 153: 359–66.

Kuehn, A., E. Hutt-Kempf, C. Hilger and F. Hentges (2011a). Clinical monosensitivity to salmonid fish linked to specific IgE-epitopes on salmon and trout beta-parvalbumins. Allergy. 66: 299–301.

Kuehn, A., C. Hilger and F. Hentges (2011b). Fish allergy—A comprehensive review of recent developments in fish allergy diagnosis and fish allergen detection. *In*: P. M. Rodgers (ed.). Food Allergies: Symptoms, Diagnosis, and Treatment. Nova Science Publishers: New York, USA.

Kuehn, A., C. Hilger, C. Lehners-Weber, F. Codreanu-Morel, M. Morisset, C. Metz-Favre et al. (2013). Identification of enolases and aldolases as important fish allergens in cod, salmon and tuna: component resolved diagnosis using parvalbumin and the new allergens. Clin Exp Allergy. 43: 811–22.

Kuehn, A., I. Swoboda, K. Arumugam, C. Hilger and F. Hentges (2014a). Fish allergens at a glance: variable allergenicity of parvalbumins, the major fish allergens. Front Immunol. 22: 179.

Kuehn, A., J. Fischer, C. Hilger, C. Sparla, T. Biedermann and F. Hentges (2014b). Correlation of clinical monosensitivity to cod with specific IgE to enolase and aldolase. Ann Allergy Asthma Immunol. 113: 670–671.

Kuehn, A., C. Radauer, I. Swoboda and J. Kleine-Tebbe (2016). Aktueller Stand und Perspektiven der Extrakt-basierten und der molekularen Diagnostik bei Fischallergie; Springer International Publishing: Cham, Switzerland, 2016.

Lee, P. W., J. A. Nordlee, S. J. Koppelman, J. L. Baumert and S. L. Taylor (2011). Evaluation and comparison of the species-specificity of 3 antiparvalbumin IgG antibodies. J Agric Food Chem. 59: 12309–16.

Lee, P. W., J. A. Nordlee, S. J. Koppelman, J. L. Baumert and S. L. Taylor (2012). Measuring parvalbumin levels in fish muscle tissue: relevance of muscle locations and storage conditions. Food Chem. 135: 502–7.

Leonard, S. A. and A. Nowak-Węgrzyn (2015). Food protein-induced enterocolitis syndrome. Pediatr Clin North Am. 62: 1463–77.

Lim, D. L., K. H. Neo, D. L. Goh, L. P. Shek and B. W. Lee (2005). Missing parvalbumin: implications in diagnostic testing for tuna allergy. J Allergy Clin Immunol. 115: 874–5.

Lim, D. L., K. H. Neo, F. C. Yi, K. Y. Chua, D. L. Goh, L. P. Shek et al. (2008). Parvalbumin—the major tropical fish allergen. Pediatr Allergy Immunol. 19: 399–407.

Liu, R., A. L. Holck, E. Yang, C. Liu and W. Xue (2013). Tropomyosin from tilapia (Oreochromis mossambicus) as an allergen. Clin Exp Allergy. 43: 365–77.

Lopata, A. L., M. F. Jeebhay, G. Reese, J. Fernandes, I. Swoboda, T. G. Robins et al. (2005). Detection of fish antigens aerosolized during fish processing using newly developed immunoassays. Int Arch Allergy Immunol. 138: 21–8.

Lorenz, A. R., S. Scheurer and S. Vieths (2015). Food allergens: Molecular and immunological aspects, allergen databases and cross-reactivity. *In*: Ebisawa et al. (eds.). Food Allergy: Molecular Basis and Clinical Practice. Karger Press, Basel, Switzerland.

Ma, Y., U. Griesmeier, M. Susani, C. Radauer, P. Briza, A. Erler et al. (2008). Comparison of natural and recombinant forms of the major fish allergen parvalbumin from cod and carp. Mol Nutr Food Res. 52: 196–207.

Mäkinen-Kiljunen, S., R. Kiistala and E. Varjonen (2003). Severe reactions from roe without concomitant fish allergy. Ann Allergy Asthma Immunol. 91: 413–6.

Martínez, J., R. Sánchez, M. Castellanos, A. M. Fernández-Escamilla, S. Vázquez-Cortés, M. Fernández-Rivas et al. (2015). Fish β-parvalbumin acquires allergenic properties by amyloid assembly. Swiss Med Wkly. doi: 10.4414/smw.2015.

McLauchlin, J., C. I., Little, K. A. Grant and V. Mithani (2006). Scombrotoxic fish poisoning. J Public Health (Oxf). 28: 61–2.

Miceli Sopo, S., S. Monaco, L. Badina, S. Barni, G. Longo, E. Novembre et al. (2015). Food protein-induced enterocolitis syndrome caused by fish and/or shellfish in Italy. Pediatr Allergy Immunol. doi: 10.1111/pai.12461.

Mourad, A. A. and S. L. Bahna (2015). Fish-allergic patients may be able to eat fish. Expert Rev Clin Immunol. 11: 419–30.

Muraro, A., T. Werfel, K. Hoffmann-Sommergruber, G. Roberts, K. Beyer, C. Bindslev-Jensen et al. (2014a). EAACI food allergy and anaphylaxis guidelines: diagnosis and management of food allergy. Allergy. 69: 1008–25.

Muraro, A., I. Agache, A. Clark, A. Sheikh, G. Roberts, C. A. Akdis et al. (2014b). EAACI food allergy and anaphylaxis guidelines: managing patients with food allergy in the community. Allergy. 69: 1046–57.

Nwaru, B., L. Hickstein, S. S. Panesar, G. Roberts, A. Muraro and A. Sheikh (2014). Prevalence of common food allergies in Europe: a systematic review and meta-analysis. Allergy. 69: 992–1007.

Olsen, D., C. Yang, M. Bodo, R. Chang, S. Leigh, J. Baez et al. (2003). Recombinant collagen and gelatin for drug delivery. Adv Drug Deliv Rev. 55: 1547–67.

Pascual, C. Y., M. Reche, A. Fiandor, T. Valbuena, T. Cuevas and M. M. Esteban (2008). Fish allergy in childhood. Pediatr Allergy Immunol. 19: 573–9.

Patriarca, G., E. Nucera, E. Pollastrini, C. Roncallo, T. De Pasquale, C. Lombardo et al. (2007). Oral specific desensitization in food-allergic children. Dig Dis Sci. 52: 1662–72.

Pérez-Gordo, M., S. Sanchez-Garcia, B. Cases, C. Pastor, F. Vivanco and J. Cuesta-Herranz (2008). Identification of vitellogenin as an allergen in Beluga caviar allergy. Allergy. 63: 479–80.

Pérez-Gordo, M., C. Pastor Vargas, B. Cases, H. M. De Las, A. Sanz, F. Vivanco et al. (2010). New allergen involved in a case of allergy to Solea solea, common sole. Ann Allergy Asthma Immunol. 104: 352–3.

Pérez-Gordo, M., J. Cuesta-Herranz, A. S. Maroto, B. Cases, M. D. Ibáñez, F. Vivanco et al. (2011). Identification of sole parvalbumin as a major allergen: study of cross-reactivity between parvalbumins in a Spanish fish-allergic population. Clin Exp Allergy. 41: 750–8.

Pérez-Gordo, M., J. Lin, L. Bardina, C. Pastor-Vargas, B. Cases, F. Vivanco et al. (2012). Epitope mapping of Atlantic salmon major allergen by peptide microarray immunoassay. Int Arch Allergy Immunol. 157: 31–40.

Pérez-Gordo, M., C. Pastor-Vargas, J. Lin, L. Bardina, B. Cases, M. D. Ibáñez et al. (2013). Epitope mapping of the major allergen from Atlantic cod in Spanish population reveals different IgE-binding patterns. Mol Nutr Food Res. 57: 1283–90.

Pieretti, M. M., D. Chung, R. Pacenza, T. Slotkin and S. H. Sicherer (2009). Audit of manufactured products: use of allergen advisory labels and identification of labeling ambiguities. J Allergy Clin Immunol. 124: 337–41.

Pitsios, C., A. Dimitriou, E. C. Stefanaki and K. Kontou-Fili (2010). Anaphylaxis during skin testing with food allergens in children. Eur J Pediatr. 169: 613–5.

Poulsen, L. K. (2015). Hints for diagnosis. *In*: Ebisawa et al. (eds.). Food Allergy: Molecular Basis and Clinical Practice. Karger Press, Basel, Switzerland.

Poulsen, L. K., M. Morisset and A. Kuehn (2016). Allergens from fish. *In*: Matricardi et al. (eds.). Molecular Allergology—A User's guide. EAACI Taskforce on 'Molecular Allergology' on www.eaaci.org.

Radauer, C., M. Bublin, S. Wagner, A. Mari and H. Breiteneder (2008). Allergens are distributed into few protein families and possess a restricted number of biochemical functions. J Allergy Clin Immunol. 121: 847–52.

Raith, M., C. Klug, G. Sesztak-Greinecker, N. Balic, M. Focke, B. Linhart et al. (2014). Unusual sensitization to parvalbumins from certain fish species. Ann Allergy Asthma Immunol. 113: 571–572.

Rehbein, H. and J. Oehlenschläger (2009). Fishery products: Quality, safety and authenticity; Wiley-Blackwell: West Sussex, United Kingdom.

Rosmilah, M., M. Shahnaz, J. Meinir, A. Masita, A. Noormalin and M. Jamaluddin (2013). Identification of parvalbumin and two new thermolabile major allergens of Thunnus tonggol using a proteomics approach. Int Arch Allergy Immunol. 162: 299–309.

Rona, R. J., T. Keil, C. Summers, D. Gislason, L. Zuidmeer, E. Sodergren et al. (2007). The prevalence of food allergy: a meta-analysis. J Allergy Clin Immunol. 120: 638–46.

Sakaguchi, M., M. Toda, T. Ebihara, S. Irie, H. Hori, A. Imai et al. (2000). IgE antibody to fish gelatin (type I collagen) in patients with fish allergy. J Allergy Clin Immunol. 106: 579–84.

Sampson, H. A., R. Gerth van Wijk, C. Bindslev-Jensen, S. Sicherer, S. S. Teuber, A. W. Burks et al. (2012). Standardizing double-blind, placebo-controlled oral food challenges: American Academy of Allergy, Asthma & Immunology-European Academy of Allergy and Clinical Immunology PRACTALL consensus report. J Allergy Clin Immunol. 130: 1260–74.

Sampson, H. A. and D. G. Ho (1997). Relationship between food-specific IgE concentrations and the risk of positive food challenges in children and adolescents. J Allergy Clin Immunol. 100: 444–51.

Saptarshi, S. R., M. F. Sharp, S. D. Kamath and A. L. Lopata (2014). Antibody reactivity to the major fish allergen parvalbumin is determined by isoforms and impact of thermal processing. Food Chem. 148: 321–8.

Schulkes, K. J., R. J. Klemans, L. Knigge, M. de Bruin-Weller, C. A. Bruijnzeel-Koomen, A. Marknell deWitt et al. (2014). Specific IgE to fish extracts does not predict allergy to specific species within an adult fish allergic population. Clin Transl Allergy. 4: 27.

Schwaller, B. (2009). The continuing disappearance of "pure" Ca^{2+} buffers. Cell Mol Life Sci. 66: 275–300.

Sharp, M. F. and A. L. Lopata (2014). Fish allergy: in review. Clin Rev Allergy Immunol. 46: 258–71.

Shibahara, Y., Y. Uesaka, J. Wang, S. Yamada and K. Shiomi (2013). A sensitive enzyme-linked immunosorbent assay for the determination of fish protein in processed foods. Food Chem. 136: 675–81.

Shimizu, Y., A. Nakamura, H. Kishimura, A. Hara, K. Watanabe and H. Saeki (2009). Major allergen and its IgE cross-reactivity among salmonid fish roe allergy. J Agric Food Chem. 57: 2314–9.

Sletten, G., T. Van Do, H. Lindvik, E. Egaas and E. Florvaag (2010). Effects of industrial processing on the immunogenicity of commonly ingested fish species. Int Arch Allergy Immunol. 151: 223–36.

Solensky, R. (2003). Resolution of fish allergy: a case report. Ann Allergy Asthma Immunol. 91: 411–2.

Somkuti, J., M. Bublin, H. Breiteneder and L. Smeller (2012). Pressure-temperature stability, Ca^{2+} binding, and pressure-temperature phase diagram of cod parvalbumin: Gad m 1. Biochemistry. 51: 5903–11.

Sten, E., T. K. Hansen, P. Stahl Skov, S. B. Andersen, A. Torp, U. Bindslev-Jensen et al. (2004). Cross-reactivity to eel, eelpout and ocean pout in codfish-allergic patients. Allergy. 59: 1173–80.

Swoboda, I., A. Bugajska-Schretter, P. Verdino, W. Keller, W. R. Sperr, P. Valent et al. (2002). Recombinant carp parvalbumin, the major cross-reactive fish allergen: a tool for diagnosis and therapy of fish allergy. J Immunol. 168: 4576–84.

Taylor, S. L., J. L. Baumert, A. G. Kruizinga, B. C. Remington, R. W. Crevel, S. Brooke-Taylor et al. (2014). Establishment of reference doses for residues of allergenic foods: report of the VITAL expert panel. Food Chem Toxicol. 63: 9–17.

Taylor, S. L. and J. L. Baumert (2015). Worldwide food allergy labeling and detection of allergens in processed foods. Chem Immunol Allergy. 101: 227–34.

Taylor, S. L. and S. L. Hefle (2001). Ingredient and labeling issues associated with allergenic foods. Allergy. 67: 64–9.

Thewes, M., J. Rakoski and J. Ring (1999). Histamine intolerance imitated a fish allergy. Acta Derm Venereol. 79: 89.

Vázquez-Cortés, S., B. Nuñez-Acevedo, L. Jimeno-Nogales, A. Ledesma and M. Fernández-Rivas (2012). Selective allergy to the Salmonidae fish family: a selective parvalbumin epitope? Ann Allergy Asthma Immunol. 108: 62–3.

van der Ventel, M. L., N. E. Nieuwenhuizen, F. Kirstein, C. Hikuam, M. F. Jeebhay I. Swoboda et al. (2011). Differential responses to natural and recombinant allergens in a murine model of fish allergy. Mol Immunol. 48: 637–46.

Van Do, T., S. Elsayed, E. Florvaag, I. Hordvik and C. Endresen (2005). Allergy to fish parvalbumins: studies on the cross-reactivity of allergens from 9 commonly consumed fish. J Allergy Clin Immunol. 116: 1314–20.

Weber, P., H. Steinhart and A. Paschke (2009). Competitive indirect ELISA for the determination of parvalbumins from various fish species in food grade fish gelatins and isinglass with PARV-19 anti-parvalbumin antibodies. J Agric Food Chem. 57: 11328–34.

Yoshida, S., A. Ichimura and K. Shiomi (2008). Elucidation of a major IgE epitope of Pacific mackerel parvalbumin. Food Chem. 111: 857–61.

Yu, E. M., B. H. Liu, G. J. Wang, D. G. Yu, J. Xie, Y. Xia et al. (2014). Molecular cloning of type I collagen cDNA and nutritional regulation of type I collagen mRNA expression in grass carp. J Anim Physiol Anim Nutr (Berl). 98: 755–65.

Zapatero Remón, L., E. Alonso Lebrero, E. Martín Fernández and M. I. Martínez Molero (2005). Food-protein-induced enterocolitis syndrome caused by fish. Allergol Immunopathol (Madr). 33: 312–6.

Zinn, C., A. Lopata, M. Visser and P. C. Potter (1997). The spectrum of allergy to South African bony fish (Teleosti). Evaluation by double-blind, placebo-controlled challenge. S Afr Med J. 87: 146–52.

Zuidmeer-Jongejan, L., H. Huber, I. Swoboda, N. Rigby, S. A. Versteeg, B. M. Jensen et al. (2015). Development of a Hypoallergenic recombinant Parvalbumin for first-in-man subcutaneous immunotherapy of fish allergy. Int Arch Allergy Immunol. 166: 41–51.

6

Recent Advances in Diagnosis and Management of Shellfish Allergy

Sandip D. Kamath,[1] *Roni Nugraha*[1,2] *and*
Andreas L. Lopata[1,]*

CONTENTS

[1] Department of Molecular and Cell Biology, Australian Institute of Tropical Health and Medicine, James Cook University, Townsville, Australia.
[2] Department of Aquatic Product Technology, Bogor Agricultural University, Bogor, Indonesia.
* Corresponding author: andreas.lopata@jcu.edu.au

6.1 INTRODUCTION

Shellfish allergy has been an increasing health concern over the last decade. Currently, over 2% of the world population is affected by food allergy to shellfish. However, in some parts of Asia, as much as 5–11% of children and young adults have been shown to be sensitised to shellfish. The prevalence and distribution of shellfish allergy seems to be partly dependent on the dietary habits, consumption and availability of various edible species. Shellfish is considered one of the Big Eight Food groups which also include milk, egg, peanut, tree nuts, fish, soy and wheat. These food groups are responsible for more than 90% of all food allergic cases. Shellfish allergy similar to peanuts, tree nuts and fish allergy, is persistent and continues into adulthood. Clinical symptoms of shellfish allergy can range from mild to medium reactions such as oral allergy syndrome, urticaria and angioedema to the life threatening anaphylaxis. Shellfish is one of the highly implicated food group in terms of food-related emergency department visits in the United States alone (Sampson 2000). Due to the changing dietary habits in recent times, there has been a spurt in production and consumption of various seafood products.

Several shellfish allergens have been characterised, but the muscle protein tropomyosin is considered the major allergen. Allergenic proteins from shellfish have been implicated not only in food allergy, but also in inhalational allergy, which affects the seafood processing industry. Most of the identified shellfish allergens have been identified to be highly stable to heat- and food-processing methods.

Current diagnostic methods for shellfish allergy includes *in vivo* methods such as skin prick testing and double blind placebo controlled food challenge, as well as *in vivo* methods including serum IgE quantification tests to whole shellfish extract or purified allergens. The major challenge faced by current diagnostic techniques is the occurrence of false-negative results often as a result of testing against select species that are not consumed in other parts of the world.

The following chapter will present a brief overview of the classification of shellfish species, prevalence of allergies to shellfish, a brief description of currently identified allergenic proteins and its implication in shellfish sensitisation and current diagnostic methods for shellfish allergy.

6.2 CLASSIFICATION OF SHELLFISH

Shellfish is a general term used to identify commonly consumed invertebrate species belonging to phylum arthropoda and mollusca. Shellfish species can be broadly classified into crustaceans and molluscs (Figure 6.1). Crustacean species that are commonly consumed are almost all decapods and include over 200 different species worldwide. In broad terms, these include prawns, shrimps, crabs, lobsters and crayfish.

The phylum Mollusca is the second richest group in the animal kingdom after arthropods. This phylum is divided into three classes including—Gastropoda, Bivalvia, and Cephalopoda (Telford and Budd 2011). Abalone, snail, limpet and whelk are the major food sources from the gastropod group; while mussels and clams are the major bivalves; and squid, cuttlefish and octopus are primary species of cephalopods (Haszprunar and Wanninger 2012) (see Table 6.1).

6.3 PREVALENCE OF SHELLFISH ALLERGY

Increased incidences of reactivity to shellfish have been reported over the past decade, probably due to increased consumption of shellfish. The prevalence and distribution of shellfish allergy is dependent on

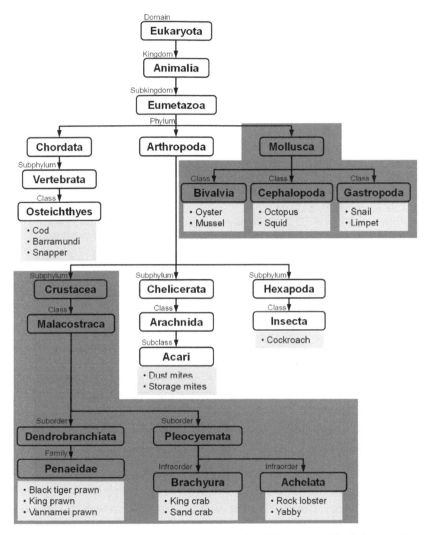

Figure 6.1 Classification of commonly consumed shellfish species (shaded in grey).

geographical regions (availability of shellfish) and changing dietary habits. Several risk factors are proposed to influence food allergy or sensitisation, including sex, ethnicity, genetics, atopy, vitamin D insufficiency, increased hygiene and the timing and route of exposure to foods (Sicherer and Sampson 2006, Sicherer and Sampson 2014).

Table 6.1 Commonly consumed crustacean and mollusc species.

Class		Common name	Scientific name
Crustacean	Prawn	Black tiger prawn Vannamei prawn Banana prawn Northern shrimp	*Penaeus monodon* *Litopenaeus vannamei* *Fenneropenaeus mergueinsis* *Crangon crangon*
	Crab	Snow crab King crab Edible crab Blueswimmer crab	*Chionoecetes opilio* *Paralithodes camtschaticus* *Cancer pagurus* *Portunus pelagicus*
	Lobster	Rock lobster Crayfish Langoustine	*Jasus lalandii* *Cherax quadricarinatus* *Nephrops norvegicus*
Mollusc	Bivalves	Pacific oyster Eastern oyster Blue mussel Green mussel	*Crassostrea gigas* *Crassostrea virginica* *Mytilus edulis* *Perna viridis*
	Gastropods	Abalone Snail Limpet	*Haliotus discus* *Helix aspersa* *Cellana exarata*
	Cephalopods	Squid Octopus Cuttlefish	*Loligo vulgaris* *Amphioctopus fangsiao* *Sepia officinalis*

Allergy to crustaceans and molluscs is more common in adults than in children. In general, childhood food allergies to milk, egg, wheat and soy usually resolve with age, and allergies to shellfish, nuts and fish persist into adulthood.

A telephone survey conducted in the US including 14,648 participants demonstrated that adults seem to be more affected by shellfish allergy than children, with a prevalence of 2% and 0.3%, respectively. Of the identified participants with shellfish allergy, 38% and 49% have perceived allergies to crustaceans and molluscs, respectively with only 14% reacting to both shellfish groups (Sicherer et al. 2004) (Table 6.2). In a randomised cross-sectional survey administered in US households, involving a total of 38,480 children, the prevalence of shellfish allergy was found to be 1.3% (Lau et al. 2012). A study in Spain involving 355 children established that 6.8% of patients reacted to crustaceans by skin prick test (Crespo et al. 1995). A study from South Africa with perceived adverse reactions

Table 6.2 Prevalence of shellfish allergy from different international studies.

Region	Population	Method of survey	Implicated shellfish	Prevalence	Age (years)	Reference
Taiwan	30018	Expert decision	Mollusc	1.3%	All	(Wu et al. 2012)
	1010 (asthma patients)	Serum IgE	Clam Squid Oyster Scallop Abalone Lobster Pacific squid Octopus	4.75% 2.28% 9.90% 24.85% 25.05% 17.33% 6.83% 7.52%	6–8	(Wan and Chiu 2011)
United States	34480	Parent report, convincing history	Shellfish	1.3%	0–18	(Lau et al. 2012)
	14948	Telephone surveys	Shellfish	2%	All	(Sicherer et al. 2004)
	9184 (Urban minority)	Retrospective review	Shellfish	0.5% 1.5%	0–5 6–21	(Taylor-Black and Wang 2012)
	512	Prospective study	Shellfish	1.2%	0.3–1.5	(Fleischer et al. 2012)
	40104	Parent report, convincing history	Shellfish	1.4%	0–18	(Gupta et al. 2011)
	14948	Self-report	Mollusc	0.4%	All	(Sicherer et al. 2004)
	20747 (Urban minority)	Retrospective review	Shellfish	0.4%	23–81	(Agarwal and Wang 2014)
	20686	Self-report	Shellfish	1.76%	All	(McGowan and Keet 2013)

Table 6.2 contd. ...

...Table 6.2 contd.

Region	Population	Method of survey	Implicated shellfish	Prevalence	Age (years)	Reference
Thailand	452	Parent report, convincing history	Mollusc	0.2%	3–7	(Lao-araya and Trakultivakorn 2012)
Australia	2848	Skin Prick test	Shrimp	0.9%	0.9–1.5	(Osborne et al. 2011)
Singapore	11318	Convincing history	Shellfish	3.4%	4–16	(Shek et al. 2010)
Philippines	11158	Convincing history	Shellfish	5.12%	14–16	
Canada	9667	Convincing history	Shellfish	1.42%	All	(Ben-Shoshan et al. 2010)
	14904 (Vulnerable population)	Self-report	Shellfish	New Canadian = 1.3%, Immigrant = 1.5% Born in Canada = 1.8%	All	(Soller et al. 2015)
Denmark	843	Convincing history (Food challenge)	Shrimp Octopus	2.0% (0.2%) 0.4% (0.1%)	±22	(Osterballe et al. 2009)
China	7047 (Allergic patients)	Serum IgE	Crab Shrimp	2.28% 2.27%	All	(Sun et al. 2014)
Korea	16749	Cross-sectional	Crustacean	0.5%	0–6	(Park et al. 2014)
Chile	488	Cross-sectional	Shellfish	0.7%	5, 10 and 15	(Hoyos-Bachiloglu et al. 2014)
Europe	17366	Cross-sectional followed by ImmunoCAP	Shrimp	4.79%	20–54	(Burney et al. 2014)
Greece	3673	Internet survey	Seafood/ shellfish	5.9%	18–74	(Kalogeromitros et al. 2013)

to seafood confirmed the sensitisation to prawn and rock lobster. Of the 131 positive reactions by ImmunoCAP, 50% reacted to four crustacean species (Zinn et al. 1997, Lopata et al. 2010). A recent study from Australia involving 167 children with definited reaction to seafood established anaphylactic reaction in nearly one fifth of the patient cohort. Moreover, 25% of all seafood allergic children elicited reactivity to shrimps (Turner et al. 2011).

However, recent studies have demonstrated higher prevalence in south East Asian countries where seafood is part of the staple diet. A study conducted in Singapore based on a structured written questionnaire, involving 25,692 school children, demonstrated a prevalence of 1.2% in children aged 4–6 years and 5.2% in children aged 14–16 years (Shek et al. 2010). A Hong Kong based study involving 3677 Chinese pre-school children aged 2–7 years incorporating parent reported and doctor diagnosed adverse food reactions, revealed shellfish to be the leading cause of allergy among 15.8% children (Leung et al. 2009).

The prevalence of food allergy to molluscs is not precisely known. Previous studies on mollusc allergy were based on only questionnaire-based surveys or retrospective reviews of allergic patient clinical data, which are likely to generate an overestimated prevalence data set (Woods et al. 2002, Taylor 2008). The prevalence of mollusc allergy ranges between 0.15% and 1.3% (Sicherer et al. 2004, Rance et al. 2005, Osterballe et al. 2009, Lao-araya and Trakultivakorn 2012, Wu et al. 2012). The variation in reported prevalence could be explained by two reasons; age of survey participants and geographical differences. The lack of accurate epidemiological tests, for example double-blind placebo-controlled oral food challenge (DBPCFC), for studying specific food allergy to molluscs has resulted in insufficient prevalence data.

6.4 CLINICAL MANIFESTATIONS AND ROUTES OF EXPOSURE

Gastrointestinal uptake is the major route of sensitisation to seafood allergens including crustaceans and molluscs. Clinical manifestation of shellfish allergy can range from mild symptoms such as hives (urticarial), swelling (angioedema), gut reactions (vomiting and diarrhoea) and oral allergy syndrome to asthma and systemic anaphylactic reactions (Table 6.3).

Exposure and allergic symptoms can occur via multiple routes such as ingestion, inhalation and even contact in some cases. While most cases reported for allergic reactions to shellfish are via ingestion, sensitisation and reactivity can occur by inhaling cooking fumes, which generates bioaerosols containing air-borne allergens.

Occupational sensitisation has also been found to trigger allergic reaction against crustaceans and molluscs especially among seafood-processing workers. Workers exposed to daily doses of particulate allergens in the industrial setting, may be at a high risk of developing respiratory conditions eventually leading to ingestion-induced food allergy (Jeebhay et al. 2001, Lopata and Jeebhay 2013). The prevalence of occupational asthma caused by mollusc species alone may be as high as 23% (Jeebhay et al. 2001).

Shellfish has been deemed the most common trigger in adults and regarded as persistent with a high risk of anaphylaxis (Sicherer 2011, Lau et al. 2012). In a study conducted by Lau et al. in the US, nearly half of the children with shellfish allergy had a history of severe reactions (Lau et al. 2012). Similarly, shellfish is the leading cause of food-induced anaphylaxis in South-East Asia, Hong Kong and Taiwan (Yi et al. 2002, Ben-Shoshan et al. 2010, Shek et al. 2010, Sicherer 2011). Shrimps have been shown to be the most common trigger of allergic reactions of all consumed crustacean species.

Contact urticaria and eczematous contact dermatitis are two major allergy manifestations in skin. A recent study demonstrated that one third of a cohort of chefs and culinary trainees, developed allergy to molluscs. Most of these individuals, showed symptoms of dermatitis very early in their career as chefs (Dickel et al. 2014). Allergic reactions to shellfish occur within minutes of exposure. However, several cases have been reported for delayed type reactions frequently to mollusc species such as oyster, abalone and squid. In addition exercise-induced anaphylaxis has also been observed (Teo et al. 2009).

Table 6.3 Manifestation of clinical symptoms on exposure to various shellfish species.

Route of exposure	Symptoms	Species	Reference
Ingestion	Face wheal, erythema and dyspnea	Freshwater clam	(Zhang et al. 2012)
	Dyspnea, urticaria, nausea, and stomachache	Abalone	(Masuda et al. 2012)
	Chronic bronchitis and highblood pressure	King Broderip clam	(Rodriguez-Del Rio et al. 2009)
	Urticaria and anaphylaxis	Scallops	(Masuda et al. 2008)
	Pruritus and facial angioedema	Razor clam *Ensis macha*	(Jiménez et al. 2005)
	Anaphylaxis	Limpet	(Azofra and Lombardero 2003)
	Angioedema and urticaria	*Octopus vulgaris*	(Damiani et al. 2010)
	Urticaria and facial angioedema	Limpet *Patella vulgata*	(Gutierrez-Fernandez et al. 2009)
	Edema, widespread wheals with pruritus, sneezes, rhinorrhea, nasal itching, cough, chest tightness, and dyspnea	razor shell (*Ensis macha*)	(Martin-Garcia et al. 2007)
	Anaphylaxis	Snail	(Peroni et al. 2000)
	Oral swelling and pain	Scallop	(Zhang et al. 2006)
	Pruritus, urticaria, dyspnea, and dizziness	Shrimp	(Vital et al. 2004)
Occupational	Asthma, rhinitis and conjunctivitis	Squid	(Wiszniewska et al. 2013)
	Asthma	Octopus	(Rosado et al. 2009)
	Asthma and urticaria	Queen scallop (*Chlamys opercularis*) and king scallop (*Pectin maximus*)	(Barraclough et al. 2006)
	Asthma	Freshwater shrimp *Gammarus*	(Baur et al. 2000)
	Asthma	Lobster and Shrimp	(Lemiére et al. 1996)
Skin	Dermatitis	Baby squid *Loligo vulgaris*	(Garcia-Abujeta et al. 1997)
	Erythema, oedema, itchin and burning	Squid *Loligo japonica*	(Valsecchi et al. 1996)
	Dermatitis	Pearl oyster	(Nakamori et al. 1996)
	Eczema	Squid	(Goday Bujan et al. 1991)
	Anaphylactic	Shrimp	(Steensma 2003)

6.5 SHELLFISH ALLERGENS

Over the past decade shellfish allergens, particularly from crustaceans have been identified and characterised. Most of these allergens belong to six proteins families (Table 6.3). Crustacean and mollusc allergens are characterised by low molecular weight, high water solubility, high heat stability, and acidic isoelectric point. Most of the known characterised allergens are found in the edible portions of the shellfish. Having a primary role in muscle movement and metabolism, most of the allergenic proteins are found in abundant concentrations in the edible portions. Currently, 34 allergens have been identified and characterised in detail from various crustacean and mollusc species and registered with the IUIS Allergen Nomenclature. However many more have been identified in the Allfam and Allergome databases, however not completely molecular or immunological characterised (Radauer et al. 2008). The relevant major and minor allergens are described below in detail and depicted in Figure 6.2.

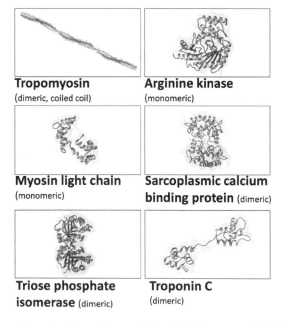

Tropomyosin (dimeric, coiled coil)

Arginine kinase (monomeric)

Myosin light chain (monomeric)

Sarcoplasmic calcium binding protein (dimeric)

Triose phosphate isomerase (dimeric)

Troponin C (dimeric)

Figure 6.2 Three dimensional homology models of six characterised shellfish allergens.

6.5.1 Tropomyosin

Tropomyosin is the major allergenic protein across all edible crustacean and mollusc species. More than 60% of shellfish allergic patients are sensitized and reactive to TM, often leading to severe systemic reactions. Tropomyosin-specific IgE is frequently used to predict clinical outcomes of shrimp allergy with a positive predictive value of 0.72 (Gámez et al. 2011, Pascal et al. 2015). Tropomyosin plays a primary role in muscle contraction and regulation. For this reason, the primary structure of tropomyosin is highly conserved across various invertebrate species. This is probably the main reason for IgE-mediated allergenic cross-reactivity across various shellfish species and even groups. Interestingly, even though crustacean and mollusc tropomyosin are allergenic, they share an amino acid sequence identity of only 55–70%. Tropomyosins are generally low molecular weight (33–38 kDa) and are highly stable to heat-treatment, retaining allergenicity even after cooking. However, some studies have demonstrated modulation of IgE recognition to tropomyosin due to heat induced Maillard reaction with sugar residues, which may occur in some shellfish species (Nakamura et al. 2005, Nakamura et al. 2006).

6.5.2 Arginine Kinase

Arginine kinase, which was first characterised in the Indian meal-moth, has been identified currently in over seven crustacean and molluscs species according to the IUIS allergen nomenclature, has a molecular weight of 40–42 kDa and is unstable to acid or alkali treatment. Arginine kinase catalysed the transfer of the high-energy phosphoryl group from ATP to arginine, thus yielding ADP and N-phosphoarginine (Yu et al. 2003). IgE sensitisation to arginine kinase has been demonstrated in 21–50% of adults and 67% children (Yang et al. 2010, Kamath et al. 2014). However, the frequency of clinical reactivity to arginine kinase has not been investigated in detail. Crustacean arginine kinase along with TM has been implicated in inhalational exposure and sensitisation among crab processing workers (Abdel Rahman et al. 2011). Similar to tropomyosin, arginine kinase may cause immunological cross-reactivity between crustaceans and molluscs. Although the similarity between octopus arginine kinase and shrimp arginine kinase is less than 54%, their three dimensional structure are highly similar and share identical amino acid sequence in several regions.

6.5.3 Myosin Light Chain

The myosin light chain group of proteins belong to the EF-Hand domain family of allergenic proteins. It is mainly found in smooth muscles in complex with myosin heavy chain motor domains.

Myosin light chains have a molecular weight between 17–20 kDa and well characterised in four crustacean species. IgE binding has been demonstrated to heat treated myosin light chain. Currently, there is a lack of data on cross-reactivity of MLC among crustaceans, molluscs or other invertebrate species.

6.5.4 Sarcoplasmic Calcium Binding Protein

Sarcoplasmic calcium binding proteins are also members of the EF hand calcium binding protein family, incorporating the helix-loop-helix motif in the primary amino acid sequence.

SCBP is ubiquitously expressed throughout the organism, but more abundant in the abdominal muscle (Gao et al. 2006). In molluscs it is found in a tissue-specific manner (Hermann and Cox 1995). It has a molecular weight of approximately 20 kDa, an isoelectric point of 5 and can elicit IgE binding even after heat treatment (Kamath et al. 2014). SCP along with myosin light chain has been demonstrated to have higher IgE recognition in children compared to adults.

6.5.5 Troponin C

Troponin C has been characterised not only in shrimps but also as a cockroach allergen (Bla g 6 and Per a 6). Similar to SCBP and MLC, Troponin C is an EF hand calcium binding protein. Troponin C has not yet been characterised in mollusc species and IgE binding frequency is not as high as TM, AK or SCBP.

6.5.6 Triose Phosphate Isomerase

This allergen has been characterised in shrimps, crayfish and cockroach with an approximate molecular weight of 28 kDa. Its clinical and immunological relevance and cross-reactivity are not yet well understood.

6.5.7 Paramyosin

Paramyosin has also been reported as a major allergen in abalone (*Haliotis discus discus*) and a cross-reactive allergen in some molluscs (Suzuki et al. 2011). However unlike tropomyosin which is heat stable, paramyosin is readily degraded after heat treatment and less IgE-reactive (Suzuki et al. 2011, Suzuki et al. 2014). The allergenicity of paramyosin in invertebrate is quite well known especially in house dust mites (Der p 11) (Tsai et al. 2005) and the fish parasite *Anisakis simplex* (Perez-Perez et al. 2000) and *Anisakis pegreffii* (Quiazon et al. 2013). Paramyosin is a major component of myofibril protein in some species, in which the content of paramyosin can reach 50% of the total myofibril proteins (Ehara et al. 2004).

6.6 CLINICAL AND IMMUNOLOGICAL CROSS REACTIVITY

One of the important features of major shellfish allergens is the phenomena of IgE antibody cross-reactivity. Cross-reactivity occurs between allergenic proteins derived from different sources because of the similar IgE binding regions shared by the molecules (Bonds et al. 2008). Tropomyosin is the major pan-allergen in invertebrate organisms (Reese et al. 1999) and conserved regions of IgE-binding epitope of tropomyosin are shared between crustaceans and molluscs. It is known that tropomyosin has linear IgE epitopes and is of great importance in determining the degree of cross reactivity between different shellfish species. Furthermore, the expressions of different forms of allergens from one species, for example in bivalve (Fujinoki et al. 2006), increases the complexity of cross reaction in shellfish allergy. A simple amino acid sequence alignment and comparison of amino acid sequences of IgE binding epitopes may be able to predict the level of IgE cross-reactivity. However, an in-depth investigation into the conservation or relevance of specific IgE epitopes among various tropomyosins through experimental analysis is required for confirming clinical cross-reactivity among shellfish allergic patients. Clinical and immunological cross-reactivity among various crustacean and mollusc species poses a challenge in pin pointing the primary sensitizing food source.

Tropomyosin is highly conserved among various crustacean species such as prawn, crabs and lobsters with amino acid identities reaching 95–100%. (Table 6.4) For this reason, IgE cross-reactivity is very frequent among crustacean species (Ayuso et al. 2002, Zhang et al. 2006, Motoyama et al. 2007, Nakano et al. 2008, Abramovitch et al. 2013).

Hypersensitivity cross-reaction within mollusc species is often found in allergic individuals. A study in 2006 (Motoyama et al. 2006) determined IgE cross reactivity on 10 species of cephalopod and found cross-reaction in all species tested. Similar results were shown for four species of gastropods (disc abalone, turban shell, whelk and Middendorf's buccinum) and seven species of bivalves (bloody cockle, Japanese oyster, Japanese cockle, surf clam, horse clam, razor clam and short neck clam) (Emoto et al. 2009). Cross-reactivity has also been shown to play a role in occupational allergy to seafood where a seafood handler elicited asthma and contact urticarial to both shrimp and scallops (Goetz and Whisman 2000).

IgE cross-sensitisation to tropomyosin maybe responsible for reactivity to shellfish and other invertebrates such as house dust-mites and cockroaches (Figure 6.3). It was demonstrated that IgE antibodies against mite tropomyosin (Der p 10) reacted very strongly to shrimp tropomyosin, although tropomyosin is present in very low concentrations in house dust mites (Arlian et al. 2009). More interestingly, reactivity to shrimp has been demonstrated in subjects

Table 6.4 Biochemical and immunological details of currently known shellfish allergens.

Allergen	Molecular weight	Heat stability	Routes of exposure	IgE sensitisation	Protein function
Tropomyosin	34–38 kDa	Highly heat stable and IgE reactive	Ingestion Inhalation	• Pen a 1, 51% (n–45), (Gámez et al. 2011) Lit v 1, 61% (n–19) (Ayuso et al. 2010) Pen m 1, 62% (n–16) (Kamath et al. 2014)	Coiled-coil protein that binds to actin and regulates interaction of troponin and myosin
Arginine kinase	40–45 kDa	Labile but can elicit IgE binding	Ingestion Inhalation	• Pen m 2, 50% (n–16) (Kamath et al. 2014) Lit v 2, 21% (n–19) (Ayuso et al. 2010)	A kinase that catalyzes reversible transfer of phosphoryl group from ATP to arginine
Myosin light chain	17–20 kDa	Stable	Ingestion	• Pen m 3, 31% (n–15) (Kamath et al. 2014) Lit v 3, 31% (n–19) (Ayuso et al. 2010)	Regulatory function in smooth muscle contraction when phosphorylated my MLC kinase
Sarcoplasmic calcium binding protein	20–25 kDa	Stable	Ingestion	• Pen m 4, 19% (n–16) (Kamath et al. 2014) Lit v 4, 21% (n–19) (Ayuso et al. 2010)	Binds to cytosolic calcium (Ca2+) and acts as a calcium buffer regulating calcium based signalling
Troponin C	20–21 kDa	Unknown	Ingestion	Cra c 6, 29% (n–31) (Bauermeister et al. 2011)	Regulates interaction of actin and myosin during muscle contraction on binding to calcium

Table 6.4 contd. ...

...Table 6.4 contd.

Triose-Phosphate isomerase	28 kDa	Labile	Ingestion Inhalation	• Pen m 8, 19% (n–16) (Kamath et al. 2014) • Cra c 8, 23% (n–31) (Bauermeister et al. 2011)	Key enzyme in glycolysis; catalyses conversion of dihydroxyacetone phosphate to glyceraldehyde 3-phosphate
Paramyosin	100 kDa	Labile	Ingestion	• 84% (Suzuki et al. 2011)	Core component of invertebrate thick filaments, responsible for the "catch" mechanism

Table 6.5 Amino acid sequence identity (%) comparison of the major allergen tropomyosin from various crustacean and mollusc species.

	Common names	No.	\multicolumn{8}{Crustacean}								Mollusc					
			1	2	3	4	5	6	7	8	9	10	11	12	13	14
Crustacean	Black tiger prawn	1	100													
	King prawn	2	95	100												
	Vannamei prawn	3	100	95	100											
	Banana prawn	4	100	95	100	100										
	Blue swimmer crab	5	97	94	97	97	100									
	Snow crab	6	89	88	89	89	88	100								
	Slipper lobster	7	97	95	97	97	99	88	100							
	Rock lobster	8	95	100	95	95	94	88	95	100						
Mollusc	Green mussel	9	55	57	55	55	55	54	55	57	100					
	Blue mussel	10	57	58	57	57	56	56	56	58	94	100				
	Oyster	11	61	62	61	61	61	59	61	62	78	78	100			
	Sea snail	12	60	61	60	60	60	58	60	61	68	72	73	100		
	Octopus	13	63	64	63	63	63	63	63	64	69	70	76	76	100	
	Squid	14	62	63	62	62	62	62	63	63	70	71	75	76	91	100

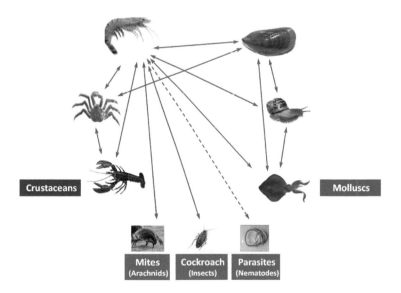

Crustaceans   Molluscs

Mites (Arachnids)   Cockroach (Insects)   Parasites (Nematodes)

Figure 6.3 Schematic representation of immunological cross-reactivity among crustaceans, molluscs and other invertebrate species based on the major allergen tropomyosin.

with house dust-mite allergy, who have never been exposed to shrimps (Fernandes et al. 2003).

Most of the clinical studies on cross-reactivity have been conducted using tropomyosin as the major pan-allergen. However, other shellfish allergens may play a role in immunological cross-sensitisation. A recent study has shown that allergens other than tropomyosin, such as arginine kinase might also be responsible for seafood-mite cross-reactivity (Marinho et al. 2006, Gamez et al. 2014). Hemocyanin has been demonstrated to play a role in seafood-mite sensitivity as well as being characterised as a cockroach allergen (Giuffrida et al. 2014, Khurana et al. 2014).

6.7 ALLERGY DIAGNOSIS AND MANAGEMENT

Diagnostic methods for establishing a true seafood allergy include various *in vivo* and *in vitro* tests to demonstrate the presence of specific IgE antibodies (Sastre 2010). Due to the possible unavailability of the exact species preparation for performing skin prick test (SPT) and blood IgE assays, positive and negative test results should be supported by a convincing clinical history of the patient and/or oral challenge where possible. An accurate evaluation of shellfish allergy using the best *in vivo* and *in vitro* tests will result in a less restricted dietary curtailment than is currently recommended.

Specific questions, interpretation of sensitisation tests (e.g., SPT, IgE) and optional food challenges help to establish the diagnosis of shellfish allergy. The following work-up described in Figure 6.4 might facilitate better diagnosis.

Case history: A precise and detailed history is very important to gain information regarding the seafood species under suspicion, nature of the symptoms and the atopic status of the patient. In addition, the identification of the implicated seafood species using specific diagnostic procedures is of importance, particularly if mislabelling of a seafood product is a possibility.

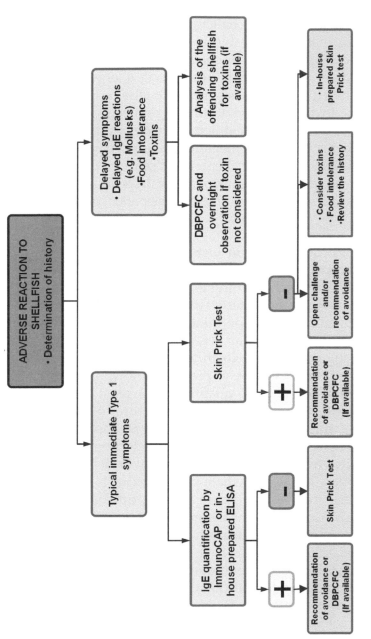

Figure 6.4 Suggested seafood allergy diagnostic strategy.

Skin prick test (SPT): The use of commercial skin prick tests is often performed in addition to food challenge or serum IgE quantification. If specific extracts are not available, so called in-house prepared SPT extracts can be utilised, if they are confirmed to be safe for testing (no toxins/histamine) and contain the appropriate allergens (Jeebhay et al. 2008). Despite the drawbacks of possible false positive/negative results obtained with skin prick testing, if performed properly and with the appropriate shellfish extracts, it is a quick and sensitive test (Wu and Williams 2004). False negative results can be attributed to low concentrations of specific allergens, absence or degradation of heat-sensitive proteins, and the use of shellfish species in commercial preparations, which are not consumed by the patient or not available in the region.

IgE-Testing: IgE quantification tests are available for a variety of crustacean and mollusc species as well as for cross-reactive invertebrate species such as anisakis, dust mites and cockroaches. There are several commercial tests available to quantify specific IgE antibodies; however, the most prominent system is the ImmunoCAP (Thermo Scientific), which has been used as a model system to demonstrate the gaps and needs in the context of seafood allergy diagnosis. The ImmunoCAP test (previous known as CAP-RAST) is an *in vitro* diagnostic test to quantify IgE against whole protein extract of specific species or to purified natural or recombinant allergens (Lopata et al. 1997, Gill et al. 2009, Sastre 2010). The accuracy of this assay is dependent on the selection of the correct seafood species and is restricted to the panel of commercially available species. The use of whole protein extracts however may lead to false negative results due to low IgE titres or low limit of detection of IgE to low abundant allergens. IgE quantification tests for component allergens are available only for shrimp tropomyosin (rPen a 1), and other minor allergens have not yet been incorporated in *in vitro* testing. However, component resolved diagnostics are available in microarray format (ISAC chip, Thermo Scientific) for prawn tropomyosin (nPen m 1), arginine kinase (nPen m 2) and sarcoplasmic calcium binding protein (rPen m 4).

Food challenge or double blind placebo controlled food challenge (DBPCFC) are considered the gold standard for diagnosis of food allergy. These tests are performed to confirm clinical reactivity to crustacean and mollusc species. Such provocation tests are not performed routinely because of increased risk and costs, and are performed for investigating individual cases. Food challenges does not distinguish between allergic (IgE mediated) and non-allergic hypersensitivity involving different antibody types, cellular immune mechanisms and reactions based on intolerance or toxins. However, performing oral food challenges can improve the quality of life, particularly when the results are favourable (van der Velde et al. 2012).

6.8 FOOD PROCESSING AND EFFECT ON ALLERGENS

Foods may undergo various processing stages for the purpose of preservation from microbes, modification to suit the end use (such as texture, taste or colour) and the improvement of digestibility. These processes can significantly alter the physicochemical and structural properties of the allergenic proteins thereby increasing or attenuating their allergenicity (Mills et al. 2009, Lepski and Brockmeyer 2013). Moreover, food processing can modulate the digestibility of allergenic proteins, which may subsequently affect its presentation to intestinal immune cells (de Jongh et al. 2011).

Food processing methods include mechanical processes, separation, biochemical processes, thermal processes, high-pressure treatment, electric field treatment, and irradiation. Broadly, these methods can be categorised into thermal and non-thermal processes. Food processing methods may enhance, reduce or eliminate the allergenic potential of food. This is in turn affected by the factors such as heat, pH, moisture, pressure, concentration of proteins, etc. In general, the conditions of processing, allergen composition, food matrix, allergen protein structure, presence of linear or conformational epitopes on the allergen and inherent stability can all influence the immune responses to modified or natural allergens.

However, the effect of food processing on the immunological reactivity or digestibility is not yet entirely predictable, though some general rule exists (Davis et al. 2001, Lepski and Brockmeyer 2013).

Several commercially important shellfish species undergo food-processing, such as preservation and sterilisation, prior to consumption. In the majority of the cases, thermal processing techniques are used for this purpose. However, the effects of heat processing on the IgE reactivity and structural stability of shellfish allergens have not been investigated in detail, with only a few studies conducted in the past ten years.

A study conducted in 2010 by Liu et al. analysed the effects of boiling shrimps on the IgE binding of the major allergen tropomyosin. It was demonstrated that the IgE binding to tropomyosin was enhanced after heat treatment (Liu et al. 2010, Kamath et al. 2013, Kamath et al. 2014). In another study, three different processing methods; boiling, ultrasound and high-pressure steaming were used prior to simulated digestion of the allergenic protein. It was demonstrated that ultrasound and high pressure processing promoted the digestion of tropomyosin, subsequently reducing the IgE binding properties *in vitro*. However, the processing methods employed in this study were based on lab scale equipment and did not represent industrial grade food processing. Enhanced IgE binding was observed to the Maillard products of tropomyosin from scallop after heat treatment. In contrast, another study in 2006 demonstrated reduced IgE binding to the Maillard products of squid tropomyosin. In both these studies, tropomyosin was analysed from the crude mollusc extracts (Nakamura et al. 2005, Nakamura et al. 2006).

Recent advances in the field of proteomics and functional cellular assays have made it possible to analyse the effects of thermal processing of shellfish on the IgE reactivity of allergens in more detail. Identification of the whole allergen repertoire in the different shellfish species and analysis of the effects of thermal processing on the stability and IgE reactivity using isolated allergens is essential in the development of improved allergen diagnostic approaches.

6.9 CONCLUSIONS

Approximately 2% of the world population is affected by shellfish allergy, which includes the crustacean and mollusc groups. The allergenic proteins present in the shellfish group have variable primary structures and often present a challenge in allergen detection and diagnosis. The distinction of crustacean from mollusc is important from clinical point of view, as molecular cross-reactivity, particular between crustaceans, seems to be determined by the close relationship to insects and mites. Currently at least seven different shellfish allergens have been identified, mostly from crustaceans, however only three recombinant allergens are available for IgE-based routine diagnostic, including tropomyosin, arginine kinase and sarcoplasmic Ca-binding protein. Other allergens include myosin light-chain, troponin C, Triose phosphate isomerase and actin.

ACKNOWLEDGEMENT

SK is a research fellow of the Australian Institute of Tropical Health and Medicine. AL is holder of an Australian Research Council Future Fellowship.

Keywords: Shellfish allergy; shellfish allergens; prawn allergy; tropomyosin; shellfish diagnosis; food allergy

REFERENCES

Abdel Rahman, A. M., S. D. Kamath, A. L. Lopata, J. J. Robinson and R. J. Helleur (2011). Biomolecular characterization of allergenic proteins in snow crab (*Chionoecetes opilio*) and de novo sequencing of the second allergen arginine kinase using tandem mass spectrometry. Journal of Proteomics. 74(2): 231–241.

Abramovitch, J. B., S. Kamath, N. Varese, C. Zubrinich, A. L. Lopata, R. E. O'Hehir and J. M. Rolland (2013). IgE reactivity of blue swimmer crab *Portunus pelagicus* tropomyosin, Por p 1, and other allergens; cross-reactivity with black tiger prawn and effects of heating. PLoS ONE. 8(6): e67487.

Agarwal, S. and J. Wang (2014). Prevalence and characteristics of food allergy in urban minority adults. Annals of Allergy, Asthma & Immunology. 112(5): 476–478.

145

Arlian, L., M. Morgan, D. Vyszenski-Moher and D. Sharra (2009). Cross-reactivity between storage and dust mites and between mites and shrimp. Experimental and Applied Acarology. 47(2): 159–172.

Ayuso, R., G. Reese, S. Leong-Kee, M. Plante and S. B. Lehrer (2002). Molecular basis of arthropod cross-reactivity: IgE-binding cross-reactive epitopes of shrimp, house dust mite and cockroach tropomyosins. International Archives of Allergy and Immunology. 129(1): 38–48.

Ayuso, R., S. Sanchez-Garcia, J. Lin, Z. Fu, M. D. Ibanez, T. Carrillo, C. Blanco, M. Goldis, L. Bardina, J. Sastre and H. A. Sampson (2010). Greater epitope recognition of shrimp allergens by children than by adults suggests that shrimp sensitization decreases with age. Journal of Allergy and Clinical Immunology. 125(6): 1286–1293 e1283.

Azofra, J. and M. Lombardero (2003). Limpet anaphylaxis: cross-reactivity between limpet and house-dust mite Dermatophagoides pteronyssinus. Allergy. 58(2): 146–149.

Barraclough, R. M., J. Walker, N. Hamilton, D. Fishwick and A. D. Curran (2006). Sensitization to king scallop (Pectin maximus) and queen scallop (Chlamys opercularis) proteins. Occupational Medicine. 56(1): 63–66.

Bauermeister, K., A. Wangorsch, L. P. Garoffo, A. Reuter, A. Conti, S. L. Taylor, J. Lidholm, A. M. Dewitt, E. Enrique, S. Vieths, T. Holzhauser, B. Ballmer-Weber and G. Reese (2011). Generation of a comprehensive panel of crustacean allergens from the North Sea Shrimp Crangon crangon. Mol Immunol. 48(15-16): 1983–1992.

Baur, X., H. Huber and Z. Chen (2000). Asthma to Gammarus shrimp. Allergy. 55(1): 96–97.

Ben-Shoshan, M., D. W. Harrington, L. Soller, J. Fragapane, L. Joseph, Y. St Pierre, S. B. Godefroy, S. J. Elliot and A. E. Clarke (2010). A population-based study on peanut, tree nut, fish, shellfish, and sesame allergy prevalence in Canada. Journal of Allergy and Clinical Immunology. 125(6): 1327–1335.

Bonds, R. S., R. Midoro-Horiuti, T. Fau-Goldblum and R. Goldblum (2008). A structural basis for food allergy: the role of cross-reactivity. Current Opinion in Allergy and Clinical Immunology. 8(1528-4050): 5.

Burney, P. G. J., J. Potts, I. Kummeling, E. N. C. Mills, M. Clausen, R. Dubakiene, L. Barreales, C. Fernandez-Perez, M. Fernandez-Rivas, T. M. Le, A. C. Knulst, M. L. Kowalski, J. Lidholm, B. K. Ballmer-Weber, C. Braun-Fahlander, T. Mustakov, T. Kralimarkova, T. Popov, A. Sakellariou, N. G. Papadopoulos, S. A. Versteeg, L. Zuidmeer, J. H. Akkerdaas, K. Hoffmann-Sommergruber and R. van Ree (2014). The prevalence and distribution of food sensitization in European adults. Allergy. 69(3): 365–371.

Crespo, J. F., C. Pascual, A. W. Burks, R. M. Helm and M. M. Esteban (1995). Frequency of food allergy in a pediatric population from Spain. Pediatric Allergy and Immunology. 6(1): 39–43.

Damiani, E., M. G. Aloia Am Fau - Priore, S. Priore Mg Fau - Nardulli, E. Nardulli S Fau - Nettis, A. Nettis E Fau - Ferrannini and A. Ferrannini (2010). Adverse reaction after ingestion of raw and boiled Octopus vulgaris. Allergy. (1398–9995).

Davis, P. J., C. M. Smales and D. C. James (2001). How can thermal processing modify the antigenicity of proteins? Allergy. 56 Suppl. 67: 56–60.

de Jongh, H. H., S. L. Taylor and S. J. Koppelman (2011). Controlling the aggregation propensity and thereby digestibility of allergens by Maillardation as illustrated for cod fish parvalbumin. J Biosci Bioeng. 111(2): 204–211.

Dickel, H., T. Bruckner, P. Altmeyer and B. Künzlberger (2014). Seafood allergy in cooks: a case series and review of the literature. JDDG: Journal der Deutschen Dermatologischen Gesellschaft. 12(10): 891–901.

Ehara, T., K. Nakagawa, T. Tamiya, S. F. Noguchi and T. Tsuchiya (2004). Effect of paramyosin on invertebrate natural actomyosin gel formation. Fisheries Science. 70(2): 306–313.

Emoto, A., S. Ishizaki and K. Shiomi (2009). Tropomyosins in gastropods and bivalves: Identification as major allergens and amino acid sequence features. Food Chemistry. 114(2): 634–641.

Fernandes, J., A. Reshef, L. Patton, R. Ayuso, G. Reese and S. B. Lehrer (2003). Immunoglobulin E antibody reactivity to the major shrimp allergen, tropomyosin, in unexposed Orthodox Jews. Clinical and Experimental Allergy. 33(7): 956–961.

Fleischer, D. M., T. T. Perry, D. Atkins, R. A. Wood, A. W. Burks, S. M. Jones, A. K. Henning, D. Stablein, H. A. Sampson and S. H. Sicherer (2012). Allergic reactions to foods in preschool-aged children in a prospective observational food allergy study. Pediatrics. 130(1): e25–e32.

Fujinoki, M., M. Ueda, T. Inoue, N. Yasukawa, R. Inoue and T. Ishimoda-Takagi (2006). Heterogeneity and tissue specificity of tropomyosin isoforms from four species of bivalves. Comparative Biochemistry and Physiology Part B: Biochemistry and Molecular Biology. 143(4): 500–506.

Gámez, C., S. Sánchez-García, M. D. Ibáñez, R. López, E. Aguado, E. López, B. Sastre, J. Sastre and V. del Pozo (2011). Tropomyosin IgE-positive results are a good predictor of shrimp allergy. Allergy. 66(10): 1375–1383.

Gamez, C., M. P. Zafra, M. Boquete, V. Sanz, C. Mazzeo, M. D. Ibanez, S. Sanchez-Garcia, J. Sastre and V. Del Pozo (2014). New shrimp IgE-binding proteins involved in mite-seafood cross-reactivity. Molecular Nutrition and Food Research.

Gao, Y., C. M. Gillen and M. G. Wheatly (2006). Molecular characterization of the sarcoplasmic calcium-binding protein (SCP) from crayfish *Procambarus clarkii*. Comp Biochem Physiol B Biochem Mol Biol. 144(4): 478–487.

Garcia-Abujeta, J. L., F. Rodriguez, E. Maquiera, I. Picans, L. Fernandez, I. Sanchez, D. Martin-Gil and J. Jerez (1997). Occupational protein contact dermatitis in a fishmonger. Contact Dermatitis. 36(3): 163.

Gill, B. V., T. R. Rice, A. Cartier, D. Gautrin, B. Neis, L. Horth-Susin, M. Jong, M. Swanson and S. B. Lehrer (2009). Identification of crab proteins that elicit IgE reactivity in snow crab-processing workers. Journal of Allergy and Clinical Immunology. 124(5): 1055–1061.

Giuffrida, M. G., D. Villalta, G. Mistrello, S. Amato and R. Asero (2014). Shrimp allergy beyond Tropomyosin in Italy: clinical relevance of Arginine Kinase, Sarcoplasmic calcium binding protein and Hemocyanin. European Annals of Allergy and Clinical Immunology. 46(5): 172–177.

Goday Bujan, J., A. Aguirre and N. Gil Ibarra (1991). Allergic contact dermatitis from squid (Loligo opalescens). Contact Dermatitis. 24(4): 307.

Goetz, D. W. and B. A. Whisman (2000). Occupational asthma in a seafood restaurant worker: cross-reactivity of shrimp and scallops. Annals of Allergy Asthma & Immunology. 85(6): 461–466.

Gupta, R. S., E. E. Springston, M. R. Warrier, B. Smith, R. Kumar, J. Pongracic and J. L. Holl (2011). The Prevalence, Severity, and Distribution of Childhood Food Allergy in the United States. Pediatrics.

Gutierrez-Fernandez, D., F. -V. Ms, B. B. Zavala, A. Foncubierta-Fernandez, J. Lucas-Velarde and A. Leon-Jimenez (2009). Urticaria-angioedema due to limpet ingestion. Journal of Investigational Allergology and Clinical Immunology. 19(1): 77–79.

Haszprunar, G. and A. Wanninger (2012). Molluscs. Current Biology. 22(13): R510–R514.

Hermann, A. and J. A. Cox (1995). Sarcoplasmic calcium-binding protein. Comp Biochem Physiol B Biochem Mol Biol. 111(3): 337–345.

Hoyos-Bachiloglu, R., D. Ivanovic-Zuvic, J. Álvarez, K. Linn, N. Thöne, M. de los Ángeles Paul and A. Borzutzky (2014). Prevalence of parent-reported immediate hypersensitivity food allergy in Chilean school-aged children. Allergologia et Immunopathologia. 42(6): 527–532.

Jeebhay, M. F., T. G. Robins, S. B. Lehrer and A. L. Lopata (2001). Occupational seafood allergy: a review. Occupational and Environmental Medicine. 58(9): 553–562.

Jeebhay, M. F., T. G. Robins, M. E. Miller, E. Bateman, M. Smuts, R. Baatjies and A. L. Lopata (2008). Occupational allergy and asthma among salt water fish processing workers. Am J Ind Med. 51(12): 899–910.

Jiménez, M., F. Pineda, I. Sánchez, I. Orozco and C. Senent (2005). Allergy due to Ensis macha. Allergy. 60(8): 1090–1091.

Kalogeromitros, D., M. P. Makris, C. Chliva, T. N. Sergentanis, M. K. Church, M. Maurer and T. Psaltopoulou (2013). An internet survey on self-reported food allergy in Greece: clinical aspects and lack of appropriate medical consultation. Journal of the European Academy of Dermatology and Venereology. 27(5): 558–564.

Kamath, S. D., A. M. Abdel Rahman, T. Komoda and A. L. Lopata (2013). Impact of heat processing on the detection of the major shellfish allergen tropomyosin in crustaceans and molluscs using specific monoclonal antibodies. Food Chemistry. 141(4): 4031–4039.

Kamath, S. D., A. M. Rahman, A. Voskamp, T. Komoda, J. M. Rolland, R. E. O'Hehir and A. L. Lopata (2014). Effect of heat processing on antibody reactivity to allergen variants and fragments of black tiger prawn: A comprehensive allergenomic approach. Molecular Nutrition & Food Research. 58(5): 1144–1155.

Khurana, T., M. Collison, F. T. Chew and J. E. Slater (2014). Bla g 3: a novel allergen of German cockroach identified using cockroach-specific avian single-chain variable fragment antibody. Annals of Allergy, Asthma & Immunology. 112(2): 140–145. e141.

Lao-araya, M. and M. Trakultivakorn (2012). Prevalence of food allergy among preschool children in northern Thailand. Pediatrics International. 54(2): 238–243.

Lau, C. H., E. E. Springston, B. Smith, J. Pongracic, J. L. Holl and R. S. Gupta (2012). Parent report of childhood shellfish allergy in the United States. Allergy and Asthma Proceedings. 33(6): 474–480.

Lemiére, C., A. Desjardins, S. Lehrer and J. L. Malo (1996). Occupational asthma to lobster and shrimp. Allergy. 51(4): 272–273.

Lepski, S. and J. Brockmeyer (2013). Impact of dietary factors and food processing on food allergy. Molecular Nutrition & Food Research. 57(1): 145–152.

Leung, T. F., E. Yung, Y. S. Wong, C. W. Lam and G. W. Wong (2009). Parent-reported adverse food reactions in Hong Kong Chinese pre-schoolers: epidemiology, clinical spectrum and risk factors. Pediatric Allergy and Immunology. 20(4): 339–346.

Liu, G. M., H. Cheng, J. B. Nesbit, W. J. Su, M. J. Cao and S. J. Maleki (2010). Effects of boiling on the IgE-binding properties of tropomyosin of shrimp (*Litopenaeus vannamei*). Journal of Food Science. 75(1): T1–T5.

Lopata, A. L. and M. F. Jeebhay (2013). Airborne seafood allergens as a cause of occupational allergy and asthma. Current Allergy and Asthma Reports 13(3): 288–297.

Lopata, A. L., R. E. O'Hehir and S. B. Lehrer (2010). Shellfish allergy. Clinical and Experimental Allergy. 40(6): 850–858.

Lopata, A. L., C. Zinn and P. C. Potter (1997). Characteristics of hypersensitivity reactions and identification of a unique 49 kd IgE-binding protein (Hal-m-1) in abalone (Haliotis midae). Journal of Allergy and Clinical Immunology. 100(5): 642–648.

Marinho, S., M. Morais-Almeida, A. Gaspar, C. Santa-Marta, G. Pires, I. Postigo, J. Guisantes, J. Martinez and J. Rosado-Pinto (2006). Barnacle allergy: allergen characterization and cross-reactivity with mites. Journal of Investigational Allergology and Clinical Immunology. 16(2): 117–122.

Martin-Garcia, C., J. Carnes, R. Blanco, J. C. Martinez-Alonso, A. Callejo-Melgosa, A. Frades and T. Colino (2007). Selective hypersensitivity to boiled razor shell. Journal of Investigational Allergology and Clinical Immunology. 17(4): 271–273.

Masuda, K., N. Katoh, K. Fukuba, K. Shimakura and S. Kishimoto (2008). Case of anaphylaxis due to fish and shellfish. J Dermatol. 35(3): 181–182.

Masuda, K., S. Tashima, N. Katoh and K. Shimakura (2012). Anaphylaxis to abalone that was diagnosed by prick test of abalone extracts and immunoblotting for serum immunoglobulin E. International Journal of Dermatology. 51(3): 359–360.

McGowan, E. C. and C. A. Keet (2013). Prevalence of self-reported food allergy in the National Health and Nutrition Examination Survey (NHANES) 2007–2010. Journal of Allergy and Clinical Immunology. 132(5): 1216–1219.e1215.

Mills, E. N. C., A. I. Sancho, N. M. Rigby, J. A. Jenkins and A. R. Mackie (2009). Impact of food processing on the structural and allergenic properties of food allergens. Molecular Nutrition & Food Research. 53(8): 963–969.

Motoyama, K., S. Ishizaki, Y. Nagashima and K. Shiomi (2006). Cephalopod tropomyosins: identification as major allergens and molecular cloning. Food and Chemical Toxicology. 44(12): 1997–2002.

Motoyama, K., Y. Suma, S. Ishizaki, Y. Nagashima and K. Shiomi (2007). Molecular cloning of tropomyosins identified as allergens in six species of crustaceans. Journal of Agricultural and Food Chemistry. 55(3): 985–991.

Nakamori, M., I. Matsuo and M. Ohkido (1996). Coexistence of contact urticaria and contact dermatitis due to pearl oysters in an atopic dermatitis patient. Contact Dermatitis. 34(6): 438.

Nakamura, A., F. Sasaki, K. Watanabe, T. Ojima, D. H. Ahn and H. Saeki (2006). Changes in allergenicity and digestibility of squid tropomyosin during the maillard reaction with ribose. Journal of Agricultural and Food Chemistry. 54(25): 9529–9534.

Nakamura, A., K. Watanabe, T. Ojima, D. H. Ahn and H. Saeki (2005). Effect of Maillard reaction on allergenicity of scallop tropomyosin. Journal of Agricultural and Food Chemistry. 53(19): 7559–7564.

Nakano, S., T. Yoshinuma and T. Yamada (2008). Reactivity of shrimp allergy-related IgE antibodies to krill tropomyosin. International Archives of Allergy and Immunology. 145(3): 175–181.

Osborne, N. J., J. J. Koplin, P. E. Martin, L. C. Gurrin, A. J. Lowe, M. C. Matheson, A. -L. Ponsonby, M. Wake, M. L. K. Tang, S. C. Dharmage and K. J. Allen (2011). Prevalence of challenge-proven IgE-mediated food allergy using population-based sampling and predetermined challenge criteria in infants. Journal of Allergy and Clinical Immunology. 127(3): 668–676. e662.

Osterballe, M., C. G. Mortz, T. K. Hansen, K. E. Andersen and C. Bindslev-Jensen (2009). The prevalence of food hypersensitivity in young adults. Pediatric Allergy and Immunology. 20(7): 686–692.

Park, M., D. Kim, K. Ahn, J. Kim and Y. Han (2014). Prevalence of immediate-type food allergy in early childhood in seoul. Allergy Asthma and Immunology Research. 6(2): 131–136.

Pascal, M., G. Grishina, A. C. Yang, S. Sanchez-Garcia, J. Lin, D. Towle, M. D. Ibanez, J. Sastre, H. A. Sampson and R. Ayuso (2015). Molecular diagnosis of shrimp allergy: Efficiency of several allergens to predict clinical reactivity. Journal of Allergy and Clinical Immunology—In Practice. 3(4): 521–529 e510.

Perez-Perez, J., E. Fernandez-Caldas, F. Maranon, J. Sastre, M. L. Bernal, J. Rodriguez and C. A. Bedate (2000). Molecular cloning of paramyosin, a new allergen of Anisakis simplex. International Archives of Allergy and Immunology. 123(2): 120–129.

Peroni, D. G., G. L. Piacentini, A. Bodini and A. L. Boner (2000). Snail anaphylaxis during house dust mite immunotherapy. Pediatric Allergy and Immunology. 11(4): 260–261.

Quiazon, K. M. A., K. Zenke and T. Yoshinaga (2013). Molecular characterization and comparison of four Anisakis allergens between Anisakis simplex sensu stricto and Anisakis pegreffii from Japan. Molecular and Biochemical Parasitology. 190(1): 23–26.

Radauer, C., M. Bublin, S. Wagner, A. Mari and H. Breiteneder (2008). Allergens are distributed into few protein families and possess a restricted number of biochemical functions. Journal of Allergy and Clinical Immunology. 121(4): 847–852 e847.

Rance, F., X. Grandmottet and H. Grandjean (2005). Prevalence and main characteristics of schoolchildren diagnosed with food allergies in France. Clinical and Experimental Allergy. 35(2): 167–172.

Reese, G., R. Ayuso and S. B. Lehrer (1999). Tropomyosin: An invertebrate pan–allergen. International Archives of Allergy and Immunology. 119(4): 247–258.

Rodriguez-Del Rio, P., J. Sanchez-Lopez, T. Robledo Echarren, C. Martinez-Cocera and M. Fernandez-Rivas (2009). Selective allergy to Venus antiqua clam. Allergy. 64(5): 815.

Rosado, A., M. A. Tejedor, C. Benito, R. Cárdenas and E. González-Mancebo (2009). Occupational asthma caused by octopus particles. Allergy. 64(7): 1101–1102.

Sampson, H. A. (2000). Food anaphylaxis. Br Med Bull. 56(4): 925–935.

Sastre, J. (2010). Molecular diagnosis in allergy. Clinical and Experimental Allergy. 40(10): 1442–1460.

Shek, L. P. -C., E. A. Cabrera-Morales, S. E. Soh, I. Gerez, P. Z. Ng, F. C. Yi, S. Ma and B. W. Lee (2010). A population-based questionnaire survey on the prevalence of peanut, tree nut, and shellfish allergy in 2 Asian populations. Journal of Allergy and Clinical Immunology. 126(2): 324–331. e327.

Sicherer, S. H. (2011). Epidemiology of food allergy. Journal of Allergy and Clinical Immunology. 127(3): 594–602.

Sicherer, S. H., A. Muñoz-Furlong and H. A. Sampson (2004). Prevalence of seafood allergy in the United States determined by a random telephone survey. Journal of Allergy and Clinical Immunology. 114(1): 159–165.

Sicherer, S. H. and H. A. Sampson (2006). Food allergy. Journal of Allergy and Clinical Immunology. 117(2): S470–S475.

Sicherer, S. H. and H. A. Sampson (2014). Food allergy: Epidemiology, pathogenesis, diagnosis, and treatment. Journal of Allergy and Clinical Immunology. 133(2): 291–307; quiz 308.

Soller, L., M. Ben-Shoshan, D. W. Harrington, M. Knoll, J. Fragapane, L. Joseph, Y. St Pierre, S. La Vieille, K. Wilson, S. J. Elliott and A. E. Clarke (2015). Prevalence and predictors of food allergy in Canada: A focus on vulnerable populations. Journal of Allergy and Clinical Immunology. In Practice. 3(1): 42–49.

Steensma, D. P. (2003). The kiss of death: A severe allergic reaction to a shellfish induced by a good-night kiss. Mayo Clinic Proceedings. 78(2): 221–222.

Sun, B. -q., P. -y. Zheng, X. -w. Zhang, H. -m. Huang, D. -h. Chen and G. -q. Zeng (2014). Prevalence of allergen sensitization among patients with allergic diseases in Guangzhou, Southern China: a four-year observational study. Multidisciplinary Respiratory Medicine. 9(1): 2.

Suzuki, M., Y. Kobayashi, Y. Hiraki, H. Nakata and K. Shiomi (2011). Paramyosin of the disc abalone Haliotis discus discus: Identification as a new allergen and cross-reactivity with tropomyosin. Food Chemistry. 124(3): 921–926.

Suzuki, M., K. Shimizu, Y. Kobayashi, S. Ishizaki and K. Shiomi (2014). Paramyosin from the Disc Abalone Haliotis Discus Discus. Journal of Food Biochemistry. 38(4): 444–451.

Taylor-Black, S. and J. Wang (2012). The prevalence and characteristics of food allergy in urban minority children. Annals of Allergy, Asthma & Immunology. 109(6): 431–437.

Taylor, S. L. (2008). Molluscan shellfish allergy. Advances in Food and Nutrition Research. 54(1043-4526): 139–177.

Telford, Maximilian J. and Graham E. Budd (2011). Invertebrate evolution: Bringing order to the Molluscan chaos. Current Biology. 21(23): R964–R966.

Teo, S. L., I. F. Gerez, E. Y. Ang and L. P. Shek (2009). Food-dependent exercise-induced anaphylaxis—a review of 5 cases. Annals of the Academy of Medicine Singapore. 38(10): 905–909.

Tsai, L. C., H. J. Peng, C. S. Lee, P. L. Chao, R. B. Tang, J. J. Tsai, H. D. Shen, M. W. Hung and S. H. Han (2005). Molecular cloning and characterization of full-length cDNAs encoding a novel high-molecular-weight Dermatophagoides pteronyssinus mite allergen, Der p 11. Allergy. 60(7): 927–937.

Turner, P., I. Ng, A. Kemp and D. Campbell (2011). Seafood allergy in children: a descriptive study. Annals of Allergy, Asthma, and Immunology. 106(6): 494–501.

Valsecchi, R., B. Pansera, A. Reseghetti, P. Leghissa, R. Cortinovis and L. Cologni (1996). Contact urticaria from Loligo japonica. Contact Dermatitis. 35(6): 367–368.

van der Velde, J. L., B. M. Flokstra-de Blok, H. de Groot, J. N. Oude-Elberink, M. Kerkhof, E. J. Duiverman and A. E. Dubois (2012). Food allergy-related quality of life after double-blind, placebo-controlled food challenges in adults, adolescents, and children. Journal of Allergy and Clinical Immunology. 130(5): 1136–1143 e1132.

Vital, C. J., C. Baggett, S. Kamboj and P. Kumar (2004). Shrimp ingestion dependent exercise induced anaphylaxis. Journal of Allergy and Clinical Immunology. 113(2, Supplement): S242–S243.

Wan, K. -S. and W. -H. Chiu (2011). Food hypersensitivity in primary school children in Taiwan: relationship with asthma. Food and Agricultural Immunology. 23(3): 247–254.

Wiszniewska, M., D. Tymoszuk, A. Pas-Wyroślak, E. Nowakowska-Świrta, D. Chomiczewska-Skóra, C. Pałczyński and J. Walusiak-Skorupa (2013). Occupational allergy to squid (Loligo vulgaris). Occupational Medicine. 63(4): 298–300.

Woods, R. K., R. M. Stoney, J. Raven, E. H. Walters, M. Abramson and F. C. Thien (2002). Reported adverse food reactions overestimate true food allergy in the community. European Journal of Clinical Nutrition. 56(1): 31–36.

Wu, A. Y. and G. A. Williams (2004). Clinical characteristics and pattern of skin test reactivities in shellfish allergy patients in Hong Kong. Allergy and Asthma Proceedings. 25(4): 237–242.

Wu, T. C., T. C. Tsai, C. F. Huang, F. Y. Chang, C. C. Lin, I. F. Huang, C. H. Chu, B. H. Lau, L. Wu, H. J. Peng and R. B. Tang (2012). Prevalence of food allergy in Taiwan: A questionnaire-based survey. Intern Med J. 42(12): 1310–1315.

Yang, A. C., L. K. Arruda, A. B. R. Santos, M. C. R. Barbosa, M. D. Chapman, C. E. S. Galvao, J. Kalil and F. F. Morato-Castro (2010). Measurement of IgE antibodies to shrimp tropomyosin is superior to skin prick testing with commercial extract and measurement of IgE to shrimp for predicting clinically relevant allergic reactions after shrimp ingestion. Journal of Allergy and Clinical Immunology. 125(4): 872–878.

Yi, F. C., N. Cheong, L. P. Shek, D. Y. Wang, K. Y. Chua and B. W. Lee (2002). Identification of shared and unique immunoglobulin E epitopes of the highly conserved tropomyosins in Blomia tropicalis and Dermatophagoides pteronyssinus. Clinical and Experimental Allergy. 32(8): 1203–1210.

Yu, C. J., Y. F. Lin, B. L. Chiang and L. P. Chow (2003). Proteomics and immunological analysis of a novel shrimp allergen, Pen m 2. Journal of Immunology. 170(1): 445–453.

Zhang, Y., E. Matsuo H Fau - Morita and E. Morita (2006). Cross-reactivity among shrimp, crab and scallops in a patient with a seafood allergy. Journal of Dermatology. (0385–2407).

Zhang, Y., T. Wang, S. Gao and E. Morita (2012). Novel allergen from the freshwater clam and the related allergy. Journal of Dermatology. 39(7): 672–674.

Zinn, C., A. Lopata, M. Visser and P. C. Potter (1997). The spectrum of allergy to South African bony fish (Teleosti). Evaluation by double-blind, placebo-controlled challenge. South African Medical Journal. 87(2): 146–152.

7

Anisakis, Allergy and the Globalization of Food

Fiona J. Baird,[1,]* *Yasuyuki Morishima*[2] *and Hiromu Sugiyama*[2]

CONTENTS

[1] School of Medical Sciences, Griffith University, Gold Coast, Australia.
[2] Laboratory of Helminthology, Department of Parasitology, National Institute of Infectious Diseases, Tokyo, Japan.
* Corresponding author: f.baird@griffith.edu.au

7.1 INTRODUCTION

Over the last fifty years, mass transportation has allowed citizens to explore the world more than ever before. This freedom of travel not only applies to humans but also to our food supply. Food security is a global concern and has led to local supplies being supplemented with foreign imports to ensure that all people have access to sufficient food for their dietary needs (Pinstrup-Andersen 2009) but also allows seasonal products to be available year round. This globalized supply chain is essential for food security; however, food safety of these supplies can sometimes be neglected. The World Health Organization (WHO) selected Food Safety as their focus for World Health Day 2015 due to the increasing complexity of the stakeholders involved in the globalized food chain (Fukuda 2015). The WHO and the United Nations Food and Agriculture Organization (FAO) are at the forefront of galvanizing all food chain stakeholders to improve food safety across all borders as a step towards improving health on a global scale.

Food-borne illnesses are estimated to kill more than 2 million people annually and impedes economies by removing healthy workers from the workforce and in developing countries, causing a cycle of malnutrition and illness (World Health Organization 2014a, World Health Organization 2014b). The most common foodborne pathogens are *Salmonella*, *Campylobacter* and enterohaemorrhagic *Escherichia coli* which are well known by the general public due to local health education initiatives (World Health Organization 2014a,b). Other food-borne pathogens such as viruses and parasites are often not considered in the local context due to high local food standards and regulations; however the global supply chain allows yearlong availability of seasonal produce sourced from around the world. Therefore, other pathogens such as the nematode *Anisakis*, need to be considered when a patient presents with gastrointestinal symptoms and particularly if they subsequently develop a food allergy. This chapter will explore how *Anisakis* species and its associated illness, anisakiasis, should be considered by clinicians outside of endemic areas and how food globalization have contributed to this situation.

7.2 THE PARASITE

Anisakiasis is a parasitic disease caused by inadvertent ingestion of larval nematodes, mainly belonging to the genera *Anisakis* and *Pseudoterranova*, found in raw or improperly cooked seafood. The first confirmed cases of anisakiasis in humans were recorded in the Netherlands in 1960 (van Thiel et al. 1960). Five years later, two cases were reported in Japan (Asami et al. 1965). Since the first reports of human anisakiasis, until 1996, 31,575 cases have been reported across 27 countries, including Japan (Takahashi et al. 1998). With the increased popularity of eating undercooked or raw fish dishes, the annual number of anisakiasis cases is expected to increase.

Anisakiasis is not restricted to humans. In 1967, the first reported case of anisakiasis in a dog occurred in Japan (Kitayama et al. 1967). Anisakid nematodes can also infect cats, and thus, anisakiasis is currently regarded as one of the most important zoonotic parasites around the world. However, the larvae do not mature into adults in humans or in those animals, because the definitive hosts are marine mammals such as whales and sea lions. The third-stage larvae of anisakid nematodes are commonly found in marine fish and squid, the ingestion of which is the primary pathway by which humans become infected. Berland (1961) classified the larvae of *Anisakis* into two types, Types I and II, based on morphological characteristics such as shapes of the ventriculus (oblong for Type I and square for Type II) and presence (Type I) or absence (Type II) of a mucron at the tip of the tail. Conversely, presence of intestinal cecum is a key characteristic used in the morphological identification of Pseudoterranova larvae. The intestinal cecum is the part of the intestine that projects anteriorly and is located behind the ventriculus; *Anisakis* larvae, both Types I and II, lack the intestinal cecum (Figure 7.1). Among these parasitic larvae, *Anisakis* Type I was recognized as the most frequently associated with human illness (Smith et al. 1978).

As *Anisakis* Type I is a morphotype and includes several species. Several isoenzyme studies coupled with molecular analyses have been conducted with the goal of identifying the different species within the genus (Mattiucci et al. 2006). For example, the species of *Anisakis* Type I larvae were genetically characterized using samples

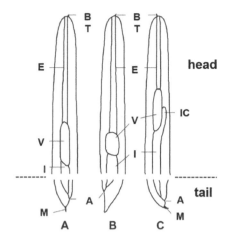

Figure 7.1 Comparative morphology of three larvae of parasitic nematodes. **(A).** *Anisakis* Type I; **(B).** *Anisakis* Type II; and **(C)**, *Pseudoterranova* sp. **(BT)** boring tooth; **(E)** oesophagus; **(V)** ventriculus; **(I)** intestine, **(IC)** intestinal cecum; and **(M)** mucron.

isolated from fish and anisakiasis patients in Japan. Although both *A. simplex* sensu stricto and *A. pegreffii* were identified in fish, almost all larvae (99%) isolated from patients were identified as *A. simplex* s.s. (Umehara et al. 2006, Umehara et al. 2007). This shows a striking discrepancy between the predominant species found in fish and that affecting patients. Suzuki et al. (Suzuki et al. 2010) explained that this discrepancy was due to the high penetration rate of *A. simplex* s.s. into the muscles of fish. Thus, *A. simplex* s.s. larvae are ingested by humans, while *A. pegreffii* larvae are usually removed when the internal organs are discarded. Because it is not possible to differentiate *Anisakis* Type I larvae by using morphology alone, molecular methods are indispensable for identifying each clinically and epidemiologically important species.

7.3 ANISAKIASIS: A COMMONLY OVERLOOKED INFECTION

7.3.1 Clinical Features

Anisakiasis is typically a self-limiting infection where the live nematode is killed by the immune system and can be misdiagnosed as viral or bacterial gastroenteritis (Baird et al. 2014), gastro-oesophageal reflux disease (Carrascosa et al. 2015) ileitis, diverticulitis (Baron et al. 2014) or appendicitis (World Health Organization 2008). The infective dose is one live nematode which is ingested whole and either is non-invasive where it remains in the gastrointestinal tract and is passed; or becomes invasive where it burrows and penetrates the intestinal wall and moves into other tissues and organs in the body (Food and Drug Administration 2012). The most usual symptoms associated with anisakiasis are: vomiting, diarrhoea, severe stomach and/or abdominal pain and nausea with occasional tingling in the throat or coughing within 24 hours of consuming raw or pickled fish (World Health Organization 2008, Food and Drug Administration 2012). There are four different types of anisakiasis—gastric, intestinal, ectopic and allergic (Ishikura et al. 1993, Kakizoe et al. 1995, Ito et al. 2007, Hochberg et al. 2010). Gastric is the most common and most resembles typical gastroenteritis with similar symptoms with nausea, vomiting, low grade fever with epigastric pain and rapid onset (Kakizoe et al. 1995). Intestinal anisakiasis has a longer incubation period of five to seven days post-ingestion and has intermittent or constant abdominal pain with development of peritonitis and/or ascites in some cases (Ishikura et al. 1993). When anisakids penetrate the stomach or intestinal wall and travel through the vicera, this can be known as ectopic, extra-gastrointestinal or intraperitoneal anisakiasis (Ito et al. 2007). Finally allergic anisakiasis is when the patient becomes sensitised to *Anisakis* allergens and suffers from associated symptoms such as urticaria, angioedema and/or anaphylaxis (Lopez-Serrano et al. 2000).

Whilst the infection is usually self-limiting, the resemblance to bacterial and viral gastroenteritis usually results in the patient being advised to rest and the infection resolves itself with the patient expelling the parasite. However, anisakiasis should be considered in cases where traditional interventions for gastroenteritis, such as antimicrobial treatments, have failed and the patient continues to present with symptoms that could fit multiple diagnoses. In 2013, the first documented case of anisakiasis was reported in China when a 56 year old male patient presented with vomiting, peripheral umbilicus and abdominal distension, and frequent mucous diarrhoea over a month long period (Qin et al. 2013). The patient was hospitalised and upon examination was given a primary diagnosis of colonic polyps then diagnosed as probable colon cancer. Abdominal and rectal examinations were normal and upon examination using an electronic gastroscope a worm was discovered in his stomach in addition to a duodenal and gastric ulcer, moderate to severe chronic superficial gastritis and mild sinus atrophy (Qin et al. 2013). Had the correct diagnosis not been made gastroscopically this patient would have been subjected to incorrect surgical intervention resulting in the removal of healthy tissue without resolving his illness. During 2013, the first case of anisakiasis presenting as a bowel obstruction in a child in Croatia (Juric et al. 2013) and the first anisakid detected in an endocervical adenocarcinoma in a German-American female from the United States of America (Ramanan et al. 2013) were both reported in the scientific literature.

7.3.2 Prevalence and Epidemiology

The country with the highest prevalence of anisakiasis is Japan. This is due to their national dishes of sushi and sashimi where the raw fish transmit the *Anisakis* worms to any person who ingests it. It is estimated that there are 2,000–3,000 cases of anisakiasis being reported annually (Yorimitsu et al. 2013). Ishikura (Ishikura 1995) analysed the incidence of anisakiasis in Japan, by tallying the cases that appeared as articles/abstracts in journals/proceedings for 5 years, from 1987 to 1991. Simultaneously, he conducted a customized questionnaire survey of anisakiasis cases among physicians. From the analysis and survey results, he concluded the number of patients in this period was 14,302. This finding indicated that at the time, the average annual number of anisakiasis cases in Japan was 2,860. Another study (Kawanaka et al. 2006) showed an estimated 2,000 or more anisakiasis cases annually, after the year 2000.

An examination of the literature published after 2001, using a similar method was conducted to investigate the number of anisakiasis cases in an eleven year period. The Japan Centra Revuo Medicina (Ichushi, a type of Index Medicus in Japan) was selected as the search engine, and articles (in English as well as in Japanese) registered in the Ichushi from 2001 to 2012 were searched using the keyword "anisakiasis". Original articles with this keyword were scrutinised and local symptoms and general conditions were ascertained. This search resulted in 270 articles/abstracts being identified, which contained 353 anisakiasis cases, indicating that the average annual case number was 27.2. Thus, it is apparent that literature search alone will not provide the accurate total number of recent anisakiasis cases. This is probably due to the under-reporting of gastric cases with severe upper abdominal pain due to invasion of anisakid larvae, as these cases are easily resolved by removal of the larvae by using a biopsy forceps under endoscopic examination. Reports of anisakiasis in the literature are usually of atypical cases where the clinical presentation made differential diagnosis difficult. Of the literature examined, 41 cases (11.6% of the 353 cases) reported anisakiasis with urticaria as one of the general symptoms. Within these 41 cases, 4 cases had dyspnoea and 1 with exanimation; these 5 cases were regarded as fulminant anisakiasis (Table 7.1) (Fukunaga et al. 2001, Kameyama et al. 2006, Hoshino et al. 2011, Iijima et al. 2012).

The ever-increasing diversity of provincial cuisine offered in new markets has also increased the number and distribution of consumers eating raw fish. Other regional dishes, such as ceviche (South America), salted or pickled herring (Holland and Nordic

Table 7.1 Fulminant anisakiasis cases reported between 2001 and 2013 in Japan.

Case			Diagnostic criteria	Major symptom	References
No.	Reported year	Prefecture			
1	2001	Hyogo	prick test and immunoblotting	dyspnoea	(Fukunaga et al. 2001)
2	2001	Hyogo	prick test and immunoblotting	dyspnoea	(Fukunaga et al. 2001)
3	2006	Aichi	prick test and immunoblotting	dyspnoea	(Kameyama et al. 2006)
4	2011	Fukushima	worm recovery from the stomach	dyspnoea	(Hoshino et al. 2011)
5	2012	Ibaraki	prick test, immunoblotting, and ELISA	exanimation	(Iijima et al. 2012)

countries), lomi-lomi (Hawaii), gravlax (Nordic countries) and boquerones (Spain), also appear to have contributed to an increase in anisakiasis in humans. It is recognised that *A. simplex* (s.s.) is the main causative agent of anisakidosis in Japan, as determined by polymerase chain reaction (PCR)-based restriction fragment length polymorphism (RFLP) analysis of the internal transcribed spacers 1 (ITS-1) and 2 (ITS-2) regions (Umehara et al. 2007) and its relatively high prevalence in chub mackerel and other fish species (Suzuki et al. 2010, Arizono et al. 2012, Jabbar et al. 2012b).

Other regions in Europe are only now being explored as potential anisakiasis endemic areas. For example marinated and/or raw fish is a culinary tradition in some regions of Italy, the actual incidence of anisakiasis is believed to be severely underestimated in these regions due to a high prevalence of parasitized fish in the Mediterranean region (Mattiucci et al. 2008). As the fish are highly parasitized, high consumption rates of these fish indicate a high risk of human infection (Mattiucci et al. 2011). In Italy, there are several reports of complicated anisakiasis cases in the various regions of Italy linked to *A. pegreffii*. This species of anisakid was identified using PCR-RFLP from a Southern Italian patient suffering from gastric anisakiasis (D'Amelio et al. 1999). In another case in Viterbo, Italy, *A. pegreffii* was molecularly identified and confirmed as the cause of granulomatous lesions in a patient suffering from anisakiasis (Mattiucci et al. 2011). The *A. pegreffii* nematode has also been associated with multiple cases of gastro-allergic anisakiasis in multiple locations across Italy (Mattiucci et al. 2013). Whilst anisakiasis is a common disease in Japan and potentially Italy as well, clinical cases have also been reported from other countries. In 2011, the first human anisakidosis case was reported in Adelaide, South Australia, after a patient had consumed raw, locally caught mackerel (Shamsi et al. 2011) and in 2013, the first case of anisakiasis in China (Qin et al. 2013).

7.3.3 Diagnosis and Treatment

Differential diagnosis is exceedingly difficult with anisakiasis as many other conditions have similar symptoms. Recently there has been reports of a mesenteric tumor diagnosis associated with chronic anisakiasis (Menendez Sanchez 2015), an adhesive intestinal abscess caused by extragastrointestinal anisakiasis (Takamizawa et al. 2015), and a case where a sub-mucosal tumour was found to be a rare case of asymptomatic colonic anisakiasis (Tamai et al. 2015). Computed tomographic imaging has been useful in expedient diagnosis in some emergency department presentations (Takabayashi et al. 2014). A common occurrence is the discovery of a worm in an ulcerative lesion during endoscopic examination or exploratory surgery (Hochberg et al. 2010, Baron et al. 2014). After surgical removal of the nematode, the most commonly utilised method of identification is morphology; however this usually requires a trained parasitologist or a pathologist with experience in parasitic zoonoses. Histopathological diagnosis is based on the morphological identification of the nematode in tissue sections using visible features such as an unpaired excretory gland (or renette cell), Y-shaped lateral epidermal cords, no apparent reproductive system, and a ventriculus (glandular oesophagus plus intestine); and an inflammatory eosinophil infiltration in the surrounding tissue accompanied with an increased neutrophil count (Baron et al. 2014). Serology using an enzyme-linked immunosorbent assay can also be used for identification of IgG and IgE against known *Anisakis* allergens which will confirm a recent anisakiasis infection (Moore et al. 2002).

Molecular methods are also being used to confirm identify of isolated nematodes from patients post-endoscopic or explorative surgery. PCR amplification of the internal transcribed spacers 1 and 2 (ITS-1, ITS-2) between rRNA genes and then restriction fragment length polymorphism (RFLP) digestion using HinfI (Umehara et al. 2006, Umehara et al. 2007) can be utilised to identify worms. This method identified *A. simplex* s.s. as the main causative agent of anisakiasis in Japan (Umehara et al. 2007). There are two distinct RFLP profiles for the two main *Anisakis* species found in human anisakiasis. Multiple cases have been published where PCR-RFLP has been used as a confirmative diagnosis of anisakiasis post-recovery: *A. pegreffii* in a paraffin-preserved granuloma isolated from an Italian man (Mattiucci et al. 2011), *A. pegreffii* in a 51 year old woman in southern Italy (D'Amelio et al. 1999); *A. pegreffii* in two Italian women (a 49 year old and a 59 year old) both suffering from gastritis (Fumarola et al. 2009); and *A. pegreffii* or *A. simplex* s.s. in multiple Japanese patients (Umehara et al. 2007). Upon surgical removal of the worm, the patient usually recovers quickly (Takabayashi et al. 2014); however use of albendazole as an alternative has been reported when the patient history is highly suggestive of anisakiasis and they are unable to have surgical or endoscopic interventions (Moore et al. 2002).

7.3.4 Allergy and Misdiagnosis of Fish Allergy Post-Infection

Allergic anisakiasis, like the other forms of anisakiasis, is underreported in both healthcare statistics and the scientific literature. All forms of anisakiasis can result in subsequent sensitization to the *Anisakis* nematode (Audicana et al. 2008). Traditionally for the last thirty years, radioallergosorbent test using a blood sample (now referred to as allergen-specific IgE) has been primarily used to confirm allergic sensitisation to anisakid allergens post-anisakiasis (Desowitz et al. 1985). However, most patients who experience gastroenteritis and/or allergic symptoms after consuming fish will recover from their infection, never see a doctor specifically about this illness and consequently self-diagnose themselves as having food poisoning. If there is a later repeat of the symptoms upon re-ingestion of fish, the majority of patients will self-diagnose themselves with a fish allergy (Mourad et al. 2015).

Most often upon serologically examination of allergen-specific IgE, this will be demonstrated to be an incorrect diagnosis (Sharp et al. 2014) meaning that the patient has subjected themselves to an unnecessary exclusion diet. The allergen-specific IgE test coupled with allergen-specific IgG tests are confirmatory of a prior anisakiasis infection/s; however high levels of IgE is not indicative of a reactive food allergy especially when total extracts are used as key allergens may be underrepresented in the extract (Larenas-Linnemann et al. 2008). Basophil activation tests (BAT) are also useful in determining if allergic immune responses will be induced in the presence of allergens and allergen-specific IgE in the patient's sera. Basophil activation tests have demonstrated to be helpful when the patient's other test results are inconclusive. For example when the allergen-specific IgE level is low but the patient is experiencing allergic symptoms when exposed to the allergen either through ingestion or through skin prick test (Pignatti et al. 2015). Skin prick test using purified allergens or direct food-source-prick to patient-skin-prick, and double-blind placebo-controlled food-challenges under supervised conditions are the most accurate tests available at confirming a true reactive food allergy (Hoffmann-Sommergruber et al. 2015).

The increase in patients with suffering from a food allergy has become a concerning social issue in Japan. While agricultural and livestock products such as eggs, milk, and soybeans are common causative agents of food allergy, seafood has become one of the commonest causes of food allergy in Japan. Pioneering studies such as Kasuya et al. (Kasuya et al. 1990) indicated that the allergens in seafood causing urticaria is not the seafood itself but the anisakid larvae that parasitizes seafood. These authors reported that 11 patients who developed urticaria after ingesting mackerel showed a positive reaction to *A. simplex* larval antigen, whereas none reacted to mackerel antigen. Thirteen allergens of *A. simplex* s.s. have been identified and characterized (Baird et al. 2014) with haemoglobin being the newest allergen identified (González-Fernández et al. 2015). As new recombinant allergens are used as immunodiagnostic tools, further anisakiasis cases with the primary symptom of urticaria have been diagnosed in Japan (Shigehira et al. 2010).

Sensitization to *Anisakis* spp. can be determined by the presence of *Anisakis* allergen-specific antibodies or immune-responsiveness via BAT using native or recombinant *Anisakis* antigens. Once this is determined further skin-prick testing can be conducted using commercial total extracts from *A. simplex* to determine reactivity to the allergens. However, in most healthcare systems around the world, these specialised appointments to conduct these gold standard tests can have long wait times due to the scarcity of allergologists (Rodero et al. 2004, Nieuwenhuizen et al. 2006). Sensitization can be found in many countries. For example, in Northern Morocco, no cases of anisakiasis have been reported for this population; however in one study a sero-prevalence of 5.1% was recorded, with the highest proportion of sensitized people being between the ages of 31–43 years (Abattouy et al. 2012). Another example is Brazil, where again there is no evidence of anisakiasis in humans, yet a sero-prevalence of 20.9% was recorded in a cohort of military personnel (*n* = 67) that regularly ingested fish suggesting a significant association between fish consumption and serology to *A. simplex* (Junior et al. 2013). Even though these people are sensitized to *Anisakis* allergens does not immediately classify them as having allergic anisakiasis as upon subsequent exposure they may not manifest an allergic reaction.

However examination of patients suffering from chronic allergic symptoms does support *Anisakis* as an underreported cause of allergic symptoms. In an Italian study, chronic urticaria patients of 65 years of age or older were tested for hypersensitivity to *A. simplex* allergens and it was found that 55% of these patients had anti-*A. simplex* serum antibodies with 35% also reactive to house dust mite (Ventura et al. 2013). An earlier study found (Lopez-Saez et al. 2003) that the prevalence of anti-*Anisakis* antibodies did not correlate with regular fish-ingestion, acute or chronic gastrointestinal disease and/or abdominal surgery. Nevertheless, a murine model have demonstrated that *Anisakis* proteins can generate urticaria and systemic hypersensitivity to *Anisakis* antigens in the absence of an acute infection (Nieuwenhuizen et al. 2009). In addition, inhalation of *Anisakis*-derived proteins can induce severe respiratory symptoms and sensitisation in mice (Kirstein et al. 2010). Moreover, in a study of fish-processing employees (Nieuwenhuizen et al. 2006), 8% exhibited *Anisakis*-specific IgE responses against an extract from *A. simplex* (s.s.) L3s. In this case, the severity of allergic symptoms related to the extent of fish consumption. The induction of allergic reactions in *Anisakis*-sensitised patients, and the role of live *Anisakis* larvae or *Anisakis* derived proteins in these allergic reactions are yet to be fully elucidated (Daschner et al. 2012). In the meantime, the allergens from *A. simplex* and *A. pegreffii* are medically important in human populations, in which there is high risk of exposure and requires detailed investigation by clinicians.

7.4 CLINICAL IMPLICATIONS OF TRAVELLING AND GLOBALIZATION OF FOOD PRODUCTS ON HEALTH

Travel has become more accessible with low cost airlines and the "experience the unknown" ideology is inspiring countless number of people to travel to new destinations like never before. In some cases the affordability has dropped so much that a person can purchase a return airline ticket for $1. This is particularly enticing for Australians who are geographically isolated from the other continents; however not all these destinations have the same food standards and regulations that Australia has. This increase in tourism to more diverse destinations has had unexpected consequences when travellers pick up exotic diseases that are endemic to the area and bring them home with them or have long-term effects which requires the local healthcare to manage. It is this increased burden on the local healthcare system that prompted the Smart Traveller campaign by the Australian government to be prepared when travelling abroad (http://www.smartraveller.gov.au/).

It is quite common for anecdotes of travel gone wrong to lead a news bulletin; however it is the ones that are dismissed that often has consequences long after the traveller has returned home. For example, during a seminar at a local hospital on Anisakiasis, a woman identified herself as having had a very similar infection upon travelling overseas which she had dismissed as a food allergy. Upon serological examination it was found that she had circulating *Anisakis*-specific IgE and no circulating IgE antibodies against the fish she had believed she had an allergy to (F.J.B., manuscript in preparation). This case is interesting as the woman had self-diagnosed herself as having a fish allergy which resulted in her changing her diet even though it was more likely a reaction to an Anisakid than the fish she ingested. This outcome has been documented and is a growing problem globally (Prester 2015).

Over the past year, Australia has had multiple food biosafety incidents which have resulted in persons falling ill after consumption of imported goods. The most publicised incident was that of the mixed berries from two leading Australian brands that were contaminated with hepatitis A and were imported from China (Food Standards

Australia and New Zealand 2015a, Food Standards Australia and New Zealand 2015b). During this outbreak, the Department of Agriculture and Food Standards Australia and New Zealand reiterated that routine testing of foods for viruses is of limited use due to: levels of active virus being below the detecting limit of most tests; unequal distribution of the virus throughout the food; and the incidence of false positive results due to inactive genomic material is too high for the wide spread application of the test (Food Standards Australia and New Zealand 2015d). A biosafety lapse which is specific to imported fish is that of histamine (scombroid) poisoning where the fish has been improperly stored and histamine accumulates in the tissues. In February 2015, there was an outbreak of scombroid poisoning in New South Wales, Australia where imported canned tuna from Thailand was implicated (Food Standards Australia and New Zealand 2015c). Even though all the Department of Agriculture examined 100% of the consignments that contain high risk species such as tuna and mackerel, these substandard canned tuna evaded detection. There is a real need to develop more sensitive rapid tests considering that all food should be examined under the Imported Food Control Act 1992 and the Imported Food Control Regulations 1993; however in the case of the tuna, they were missed during the inspection period.

Another strategy that is under review to improve food safety in Australia is country of origin food labelling. The major benefit of having labels of origin is that consumers can decide for themselves if they want to purchase that product based on their faith in that country's food standards. In Australia, the Joint Food Standards—Code Standard 1.2.11 required all foods origins to be labelled (Food Standards Australia New Zealand 2005, Wood et al. 2013). In 2015, this has undergone a major overhaul which as of 2016, will make the labels standardised across all food products and will denote if a product is only packaged in Australia (Department of Industry; Innovation; and Science 2015). Research into consumer choices based on labelling found that food safety is the largest driver of purchase rather than origin labels (Loureiro et al. 2007); however that may have changed over the last few years in light of different food recalls and outbreaks.

In 2002, in response to emerging biosafety issues such as bovine spongiform encephalopathy, the United States of America had passed a country of origin labelling law to improve their meat traceability (Loureiro et al. 2007, Dinopoulos et al. 2010). This law took over a decade to be implemented and when it was in 2014, Canada and Mexico appealed to the World Trade Organization (WTO) that the law was a Technical Barrier to Trade which discriminates against their imported products as the cost of tracking their movement would be prohibitive to their industry. The two countries are now seeking retaliatory actions valued at US$3.7 billion a year collectively (Tracy 2015). As a result of the WTO ruling against the United States of America, Sections 281 and 282 covering country of origin labelling requirements for beef, pork, and chicken of the Agricultural Marketing Act of 1946 was repealed (Country of Origin Labeling Amendments Act of 2015 (USA)). This outcome demonstrates the extreme complexity of the issue of food safety where government trade agreements can influence the information given to a consumer at the point of purchase.

7.5 CONCLUSIONS

Anisakiasis is an established infection in many regions of the world where fish is regularly consumed and this is reflected in how patients are treated when they present to a hospital displaying gastrointestinal symptoms. However as anisakiasis is often viewed as a self-limiting infection, allergic consequences are rarely investigated. Self-reporting of fish allergies may be concealing the true prevalence of *Anisakis*-sensitisation in these populations. Another aspect is the globalisation of food where in any one country you can purchase products from a wide variety of foreign countries with vastly different food standards to those of the developed world. As anisakiasis is emerging in more countries, health professionals will need to incorporate a few more questions into the workup of the presenting illness to eliminate anisakiasis. This is particularly important for recently returned travellers who may have been exposed whilst travelling overseas and for atypical gastroenteritis cases where the presenting clinical features are omnipresent.

Keywords: Anisakis allergy; infection and allergy; parasite allergy; Anisakis nematode; food allergy

REFERENCES

Country of Origin Labeling Amendments Act of 2015 (USA), H. R. 2393, 94th Congress.

Abattouy, N., A. Valero, J. Martin-Sanchez, M. C. Penalver and J. Lozano (2012). Sensitization to *Anisakis simplex* species in the population of northern Morocco. J Investig Allergol Clin Immunol. 22(7): 514–519.

Arizono, N., M. Yamada, T. Tegoshi and M. Yoshikawa (2012). *Anisakis simplex* sensu stricto and *Anisakis pegreffii*: biological characteristics and pathogenetic potential in human anisakiasis. Foodborne Pathog Dis. 9(6): 517–521.

Asami, K., T. Watanuki, H. Sakai, H. Imano and R. Okamoto (1965). Two cases of stomach granuloma caused by Anisakis-like larval nematodes in Japan. Am J Trop Med Hyg. 14: 119–123.

Audicana, M. T. and M. W. Kennedy (2008). *Anisakis simplex*: from obscure infectious worm to inducer of immune hypersensitivity. Clin Microbiol Rev. 21(2): 360–379.

Baird, F. J., R. B. Gasser, A. Jabbar and A. L. Lopata (2014). Foodborne anisakiasis and allergy. Mol Cell Probes. 28(4): 167–174.

Baron, L., G. Branca, C. Trombetta, E. Punzo, F. Quarto, G. Speciale and V. Barresi (2014). Intestinal anisakidosis: Histopathological findings and differential diagnosis. Pathol Res Pract. 210(11): 746–750.

Berland, B. (1961). Nematodes from some Norwegian marine fishes. Sarsia. 2: 1–50.

Carrascosa, M. F., J. C. Mones, J. R. Salcines-Caviedes and J. G. Roman (2015). A man with unsuspected marine eosinophilic gastritis. Lancet Infect Dis. 15(2): 248.

D'Amelio, S., K. D. Mathiopoulos, O. Brandonisio, G. Lucarelli, F. Doronzo and L. Paggi (1999). Diagnosis of a case of gastric anisakidosis by PCR-based restriction fragment length polymorphism analysis. Parassitologia. 41(4): 591–593.

Daschner, A., C. Cuéllar and M. Rodero (2012). The *Anisakis* allergy debate: does an evolutionary approach help? Trends in Parasitol. 28(1): 9–15.

Department of Industry; Innovation; and Science (2015). Proposed reforms to country of origin food labels—overview. Retrieved Sept. 28, 2015, from http://www.industry.gov.au/industry/IndustrySectors/FoodManufacturingIndustry/Pages/Proposed-reforms-to-country-of-origin-food-labels-overview.aspx.

Desowitz, R. S., R. B. Raybourne, H. Ishikura and M. M. Kliks (1985). The radioallergosorbent test (RAST) for the serological diagnosis of human anisakiasis. Trans R Soc Trop Med Hyg. 79(2): 256–259.

Dinopoulos, E., G. Livanis and C. West (2010). Country of Origin Labeling (C.O.O.L.): How cool is it? Int Rev Econ Finance. 19(4): 575–589.

Food and Drug Administration (2012). Bad Bug Book, Foodborne Pathogenic Microorganisms and Natural Toxins. Second Edition. USA.

Food Standards Australia and New Zealand (2015a, 14/02/2015). Nanna's brand frozen berries (Mixed Berries and Raspberries) Recall. Retrieved September 15, 2015, from http://www.foodstandards.gov.au/industry/foodrecalls/recalls/Pages/Nanna's-Mixed-Berries.aspx.

Food Standards Australia and New Zealand (2015b). Creative Gourmet Mixed Berries Recall. Retrieved September 15, 2015, from http://www.foodstandards.gov.au/industry/foodrecalls/recalls/Pages/Creative-Gourmet-Mixed-Berries.aspx.

Food Standards Australia and New Zealand (2015c, 26 Feb. 2015). Histamine (Scombroid) fish poisoning. Retrieved September 15, 2015, from http://www.foodstandards.gov.au/consumer/safety/Pages/Histamine-(Scombroid)-fish-poisoning.aspx.

Food Standards Australia and New Zealand (2015d, May 2015). FSANZ advice on hepatitis A and imported ready-to eat berries, from http://www.foodstandards.gov.au/consumer/safety/Pages/FSANZ-advice-on-hepatitis-A-and-imported-ready-to-eat-berries.aspx.

Food Standards Australia New Zealand (2005). Country of Origin Requirements (Australia only), Vol. 1.2.11. Canberra, Australia, Department of Health and Ageing.

Fukuda, K. (2015). Food safety in a globalized world. Bull World Health Organ. 93: 212.

Fukunaga, A., T. Chihara, S. Harada and T. I. Horikawa, M (2001). Two cases of anaphylaxis probably provoked by cross-reactivity among crustaceans, mollusks, Anisakis, roundworms, cockroaches, and house-dust mite. Skin Res. 43(67–73).

Fumarola, L., R. Monno, E. Ierardi, G. Rizzo, G. Giannelli, M. Lalle and E. Pozio (2009). Anisakis pegreffi etiological agent of gastric infections in two Italian women. Foodborne Pathog Dis. 6(9): 1157–1159.

González-Fernández, J., A. Daschner, N. E. Nieuwenhuizen, A. L. Lopata, C. D. Frutos, A. Valls and C. Cuéllar (2015). Haemoglobin, a new major allergen of Anisakis simplex. Int J Parasitol. 45(6): 399–407.

Hochberg, N. S., D. H. Hamer, J. M. Hughes and M. E. Wilson (2010). Anisakidosis: perils of the deep. Clin Infect Dis. 51(7): 806–812.

Hoffmann-Sommergruber, K., S. Pfeifer and M. Bublin (2015). Applications of molecular diagnostic testing in food allergy. Curr Allergy Asthma Rep. 15(9): 557.

Hoshino, T. and M. Narita (2011). *Anisakis simplex*-induced anaphylaxis. J Inf Chemother. 17: 544–546.

Iijima, S., T. Moriyama, H. Ichikawa, Y. Kobayashi and K. Shiomi (2012). A case of an allergic reaction due to *Anisakis simplex* possibly after the ingestion of squid–successful detection of four *A. simplex* allergens, Ani s 1, Ani s 2, Ani s 12 and troponin C-like protein. Jpn J Allergol. 61: 1104–1110.

Ishikura, H. (1995). Analysis of the incidence of anisakidosis in Japan. Jpn J Clin Exp Med. 72: 1152–1158.

Ishikura, H., K. Kikuchi, K. Nagasawa, T. Ooiwa, H. Takamiya, N. Sato and K. Sugane (1993). Anisakidae and anisakidosis. Prog Clin Parasitol. 3: 43–102.

Ito, Y., Y. Ikematsu, H. Yuzawa, Y. Nishiwaki, H. Kida, S. Waki, M. Uchimura, T. Ozawa, T. Iwaoka and T. Kanematsu (2007). Chronic gastric anisakiasis presenting as pneumoperitoneum. Asian J Surg. 30(1): 67–71.

Jabbar, A., A. Asnoussi, L. J. Norbury, A. Eisenbarth, S. Shamsi, R. B. Gasser, A. L. Lopata and I. Beveridge (2012b). Larval anisakid nematodes in teleost fishes from Lizard Island, northern Great Barrier Reef, Australia. Mar Freshwater Res. 63(12): 1283–1299.

Junior, I. F., M. A. Vericimo, L. R. Cardoso, S. C. Clemente, E. R. do Nascimento and G. A. Teixeira (2013). Cross-sectional study of serum reactivity to *Anisakis simplex* in healthy adults in Niteroi, Brazil. Acta Parasitol. 58(3): 399–404.

Juric, I., Z. Pogorelic, R. Despot and I. Mrklic (2013). Unusual cause of small intestine obstruction in a child: Small intestine anisakiasis: report of a case. Scott Med J. 58(1): e32–36.

Kakizoe, S., H. Kakizoe, K. Kakizoe, Y. Kakizoe, M. Maruta, T. Kakizoe and S. Kakizoe (1995). Endoscopic findings and clinical manifestation of gastric anisakiasis. Am J Gastroenterol. 90(5): 761–763.

Kameyama, R., A. Yagami, T. Yamakita, M. Nakagawa, K. Nagase, I. H. and M. K. (2006). A case of immediate allergy to Anisakis following saury intake. Jpn J Allergol. 55: 1429–1432.

Kasuya, S., H. Hamano and S. Izumi (1990). Mackerel-induced urticaria and Anisakis. Lancet. 335(8690): 665.

Kawanaka, M. and J. Araki (2006). Anisakidosis: status and prevention. Food Sanit Res. 56(6): 17–22.

Kirstein, F., W. G. Horsnell, N. Nieuwenhuizen, B. Ryffel, A. L. Lopata and F. Brombacher (2010). *Anisakis pegreffii*-induced airway hyperresponsiveness is mediated by gamma interferon in the absence of interleukin-4 receptor alpha responsiveness. Infect Immun. 78(9): 4077–4086.

Kitayama, H., M. Ohbayashi, H. Satoh and Y. Kitamura (1967). Studies on parasitic granuloma in the dog. Jpn J Parasitol. 16: 28–35.

Larenas-Linnemann, D. and L. S. Cox (2008). European allergen extract units and potency: review of available information. Ann Allergy Asthma Immunol. 100(2): 137–145.

Lopez-Saez, M. P., J. M. Zubeldia, M. Caloto, S. Olalde, R. Pelta, M. Rubio and M. L. Baeza (2003). Is *Anisakis simplex* responsible for chronic urticaria? Allergy Asthma Proc. 24(5): 339–345.

Lopez-Serrano, M. C., A. A. Gomez, A. Daschner, A. Moreno-Ancillo, J. M. de Parga, M. T. Caballero, P. Barranco and R. Cabanas (2000). Gastroallergic anisakiasis: findings in 22 patients. J Gastroenterol Hepatol. 15(5): 503–506.

Loureiro, M. L. and W. J. Umberger (2007). A choice experiment model for beef: What US consumer responses tell us about relative preferences for food safety, country-of-origin labeling and traceability. Food Policy. 32(4): 496–514.

Mattiucci, S., P. Fazii, A. De Rosa, M. Paoletti, A. S. Megna, A. Glielmo, M. De Angelis, A. Costa, C. Meucci, V. Calvaruso, I. Sorrentini, G. Palma, F. Bruschi and G. Nascetti (2013). Anisakiasis and gastroallergic reactions associated with *Anisakis pegreffii* infection, Italy. Emerg Infect Dis. 19(3): 496–499.

Mattiucci, S. and G. Nascetti (2006). Molecular systematics, phylogeny and ecology of anisakid nematodes of the genus Anisakis Dujardin, 1845: an update. Parasite. 13(2): 99–113.

Mattiucci, S. and G. Nascetti (2008). Chapter 2. Advances and Trends in the Molecular Systematics of Anisakid Nematodes, with Implications for their Evolutionary Ecology and Host—Parasite Co-evolutionary Processes. D. Rollinson and S. I. Hay (eds.). Advances in Parasitology, Academic Press. Volume 66: 47–148.

Mattiucci, S., M. Paoletti, F. Borrini, M. Palumbo, R. M. Palmieri, V. Gomes, A. Casati and G. Nascetti (2011). First molecular identification of the zoonotic parasite *Anisakis pegreffii* (Nematoda: Anisakidae) in a paraffin-embedded granuloma taken from a case of human intestinal anisakiasis in Italy. BMC Infect Dis. 11: 82.

Menendez Sanchez, P. (2015). Mesenteric tumor due to chronic anisakiasis. Rev Esp Enferm Dig. 107.

Moore, D. A., R. W. Girdwood and P. L. Chiodini (2002). Treatment of anisakiasis with albendazole. Lancet. 360(9326): 54.

Mourad, A. A. and S. L. Bahna (2015). Fish-allergic patients may be able to eat fish. Expert Rev Clin Immunol. 11(3): 419–430.

Nieuwenhuizen, N., D. R. Herbert, F. Brombacher and A. L. Lopata (2009). Differential requirements for interleukin (IL)-4 and IL-13 in protein contact dermatitis induced by *Anisakis*. Allergy. 64(9): 1309–1318.

Nieuwenhuizen, N., A. L. Lopata, M. F. Jeebhay, D. R. Herbert, T. G. Robins and F. Brombacher (2006). Exposure to the fish parasite *Anisakis* causes allergic airway hyperreactivity and dermatitis. J Allergy Clin Immunol. 117(5): 1098–1105.

Pignatti, P., M. R. Yacoub, C. Testoni, G. Pala, M. Corsetti, G. Colombo, A. Meriggi and G. Moscato (2015). Evaluation of basophil activation test in suspected food hypersensitivity. Cytometry B Clin Cytom.

Pinstrup-Andersen, P. (2009). Food security: definition and measurement. Food Security. 1(1): 5–7.

Prester, L. (2015). Seafood allergy, toxicity, and intolerance: A review. J Am Coll Nutr. 1–13.

Qin, Y., Y. Zhao, Y. Ren, L. Zheng, X. Dai, Y. Li, W. Mao and Y. Cui (2013). Anisakiasis in china: the first clinical case report. Foodborne Pathog Dis. 10(5): 472–474.

Ramanan, P., A. K. Blumberg, B. Mathison and B. S. Pritt (2013). Parametrial anisakidosis. J Clin Microbiol. 51(10): 3430–3434.

Rodero, M., C. Cuellar, T. Chivato, A. Jimenez, J. M. Mateos and R. Laguna (2004). Evaluation by the skin prick test of *Anisakis simplex* antigen purified by affinity chromatography in patients clinically diagnosed with Anisakis sensitization. J Helminthol. 78(2): 159–165.

Shamsi, S. and A. R. Butcher (2011). First report of human anisakidosis in Australia. Med J Aust. 194(4): 199–200.

Sharp, M. F. and A. L. Lopata (2014). Fish allergy: in review. Clin Rev Allergy Immunol. 46(3): 258–271.

Shigehira, Y., N. Inomata, R. Nakagawara, T. Okawa, H. N. Sawaki, K. Y. Kobayashi, K. Shiomi and Z. Ikezawa (2010). A case of an allergic reaction due to *Anisakis simplex* after the ingestion of salted fish guts made of sagittated calamari: allergen analysis with recombinant and purified *Anisakis simplex* allergens. Jpn J Allergol. 59: 55–60.

Smith, J. W. and R. Wootten (1978). Anisakis and Anisakiasis. R. M. W. H. R. Lumsden and J. R. Baker (eds.). Advances in Parasitology, Academic Press. Volume 16: 93–163.

Suzuki, J., R. Murata, M. Hosaka and J. Araki (2010). Risk factors for human *Anisakis* infection and association between the geographic origins of *Scomber japonicus* and anisakid nematodes. Int J Food Microbiol. 137(1): 88–93.

Takabayashi, T., T. Mochizuki, N. Otani, K. Nishiyama and S. Ishimatsu (2014). Anisakiasis presenting to the ED: clinical manifestations, time course, hematologic tests, computed tomographic findings, and treatment. Am J Emerg Med. 32(12): 1485–1489.

Takahashi, S., H. Ishikura and K. Kikuchi (1998). Anisakidosis: Global point of view. pp. 109–120. *In*: H. Ishikura, M. Aikawa, H. Itakura and K. Kikuchi (eds.). Host Response to International Parasitic Zoonoses. Springer Japan.

Takamizawa, Y. and Y. Kobayashi (2015). Adhesive intestinal obstruction caused by extragastrointestinal anisakiasis. Am J Trop Med Hyg. 92(4): 675–676.

Tamai, Y. and K. Kobayashi (2015). Asymptomatic colonic anisakiasis. Intern Med. 54(6): 675.

Tracy, T. (2015). House votes to remove country-of-orgin labels on meat sold in U.S. The Wall Street Journal.

Umehara, A., Y. Kawakami, J. Araki and A. Uchida (2007). Molecular identification of the etiological agent of the human anisakiasis in Japan. Parasitol Int. 56(3): 211–215.

Umehara, A., Y. Kawakami, T. Matsui, J. Araki and A. Uchida (2006). Molecular identification of Anisakis simplex sensu stricto and Anisakis pegreffii (Nematoda: Anisakidae) from fish and cetacean in Japanese waters. Parasitol Int. 55(4): 267–271.

van Thiel, P., F. C. Kuipers and R. T. Roskam (1960). A nematode parasitic to herring, causing acute abdominal syndromes in man. Trop Geogr Med. 12: 97–113.

Ventura, M. T., S. Napolitano, R. Menga, R. Cecere and R. Asero (2013). *Anisakis simplex* hypersensitivity is associated with chronic urticaria in endemic areas. Int Arch Allergy Immunol. 160(3): 297–300.

Wood, A., T. Tenbensel and J. Utter (2013). The divergence of country of origin labelling regulations between Australia and New Zealand. Food Policy. 43: 132–141.

World Health Organization (2008). Foodborne disease outbreaks: Guidelines for Investigation and Control, World Health Organization.

World Health Organization (2014a). Advancing food safety initiatives: Strategic plan for food safety including foodborne zoonoses 2013 2022. Geneva, Switzerland, World Health Organization, 32.

World Health Organization (2014b). Food Safety Fact Sheet #399.

Yorimitsu, N., A. Hiraoka, H. Utsunomiya, Y. Imai, H. Tatsukawa, N. Tazuya, H. Yamago, Y. Shimizu, S. Hidaka, T. Tanihira, A. Hasebe, Y. Miyamoto, T. Ninomiya, M. Abe, Y. Hiasa, B. Matsuura, M. Onji and K. Michitaka (2013). Colonic intussusception caused by anisakiasis: a case report and review of the literature. Intern Med. 52(2): 223–226.

8

Occupational Allergy and Asthma Associated with Inhalant Food Allergens

Mohamed F. Jeebhay[1],* and *Berit Bang*[2]

CONTENTS

[1] Division of Occupational Medicine and Centre for Environmental and Occupational Health Research, School of Public Health and Family Medicine, University of Cape Town, South Africa.
[2] Department of Occupational and Environmental Medicine, University Hospital of North Norway, Tromso, Norway.
* Corresponding author: Mohamed.Jeebhay@uct.ac.za

176

8.6   Preventive Approaches
8.7   Conclusion
References

8.1 INTRODUCTION—FOOD INDUSTRY AND HIGH RISK WORKING POPULATIONS

The food industry is one of the largest employers of workers exposed to numerous allergens that are capable of inducing immunological reactions resulting in allergic disease (Jeebhay 2002a, Cartier 2010, Sikora 2008). Such allergic reactions can occur at every level of the industry, from growing/harvesting of crops or animals, storage of grains, processing and cooking, conversion, preparation, preservation and packaging of food substances (Gill 2002). It is estimated that at least one third of the world's population is engaged in the agricultural sector, the figure increases to 40% in developing countries and 50% for the African population (FAO (year 2010). The International Labour Organisation estimates that the food industry comprises about 10% of the global working population.

The largest food-handling population is employed in the agricultural sector followed by the food manufacturing and processing industry that employs workers involved in a broad spectrum of occupations. These include sectors involved in processing of fruit, vegetables, meat, fish, oils and fats; dairy products; grain mill products, starches and starch products (e.g., sweets, chocolates, confectionery); prepared animal feeds; and beverages. Materials processed include both naturally occurring biological raw products (plant/vegetable, animal or microbial origin) as well as chemicals for food preservation, flavouring, packaging and labelling. Both these biological and chemical materials are known to contain sensitising agents capable of causing occupational allergies among high risk working populations (Jeebhay 2002b).

Workers considered to be at increased risk include farmers who grow and harvest crops; factory workers involved in food processing, storage and packing; as well as those involved in food preparation (chefs and waiters) and transport.

8.2 FOOD PROCESSING ACTIVITIES AND ALLERGEN SOURCES

In the occupational setting, hazardous constituents of food products enter the body either through inhalation or dermal contact resulting in adverse reactions on an irritant or allergic basis. Allergic diseases commonly encountered in the food industry include respiratory diseases such as occupational asthma, rhinitis, conjunctivitis and hypersensitivity pneumonitis, as well as skin disease such as contact dermatitis (Sikora 2008, Gill 2002).

Tables 8.1 and 8.2 outline common food sources (cereals, plants/vegetables/fruits/spices, seeds, herbal teas, mushrooms, farm products) as well additives (colorants, thickening agents, sulphites and enzymes) and food contaminants (mites and other insects, fungi, parasites) associated with food storage that are found in food processing industries. Most of these are biological agents containing high molecular weight (>10 kDa) proteins derived from plant or animal sources, that are both naturally occurring or synthetically derived, and which act as allergic respiratory sensitisers (James and Crespo 2007, Cartier 2010).

Various work processes are employed in the food industry that produce wet aerosols and dust particulates that are capable of being inhaled and causing allergic reactions. This is typically illustrated in the seafood industry in which processes such as cutting, scrubbing or cleaning, cooking or boiling, and drying are commonly used (Table 8.3) (Jeebhay 2001). Various immunological techniques have been developed to determine the allergen concentrations produced by these work processes in the various industrial sectors (Raulf 2014). For some dust particulate there is a strong linear correlation with airborne allergen concentrations as has been observed for flour dust measurements in the baking industries, whereas this has not been borne out for studies in the seafood processing industry due to the nature of the aerosolised particles (Baatjies 2010, Jeebhay 2005a). Other food processing activities such as storage, thermal denaturation, acidification and fermentation may destroy allergens, cause conformational changes or result in the formation of new

Table 8.1 Food allergens responsible for occupational asthma.

Agent	Occupational exposure
Cereals	
Wheat, rye, barley	Baker, pastry maker (Cartier 2010)
Gluten	Baker (Cartier 2010)
Corn	Making stock feed (Cartier 2010)
Rice	Rice miller (Sikora 2008, Cartier 2010)
Malt	Machine operator (Miedinger 2009)
Plants, vegetables, fruits, and spices	
Spinach	Baker (handling spinach) (Sikora 2008, Cartier 2010)
Asparagus	Harvesting asparagus (Sikora 2008, Cartier 2010)
Broccoli, cauliflower	Plant breeder, restaurant worker (Sikora 2008)
Artichokes	Warehouse (packaging artichokes) (Cartier 2010)
Bell peppers	Greenhouse worker (Cartier 2010)
Courgettes (zucchini)	Warehouse (packaging courgette) (Cartier 2010)
Carrots	Cook (handling and cutting raw carrots) (Sikora 2008)
Tomatoes (flower)	Greenhouse grower (Cartier 2010)
Raspberries	Chewing gum coating (Cartier 2010)
Peaches	Farmer, factory worker handling peaches (Cartier 2010)
Oranges (pollen and zest/flavido)	Farmer (de las Marinas 2013), orange peeling (Felix 2013)
Aniseed	Meat industry (handling spices) (Cartier 2010)
Saffron (pollen)	Saffron worker (Cartier 2010)
Hops	Baker (Cartier 2010), brewery chemist (Sikora 2008)
Soybeans	Dairy food product company, baker, animal food preparation (Sikora 2008, Cartier 2010)
Chicory	Factory producing inulin from chicory roots, chicory grower (Cartier 2010)
Coffee beans (raw and roasted)	Roasting green coffee beans (Cartier 2010)
Green beans	Handling green beans (Cartier 2010)
Cacao	Confectionery (Cartier 2010)
Anise	Anise liqueur factory (Cartier 2010)
Almonds	Almond-processing plant (Cartier 2010)
Olive oil	Olive mill worker (Cartier 2010)
Devil's tongue root (maiko)	Food processor (Cartier 2010)
Garlic, onion, chilli pepper	Sausage makers, garlic harvesters, spice factory, packing and handling garlic (Sikora 2008, Cartier 2010, van der Walt 2010)

Table 8.1 contd. ...

179

...Table 8.1 contd.

Agent	Occupational exposure
Plants, vegetables, fruits, and spices	*Plants, vegetables, fruits, and spices*
Aromatic herbs (rosemary, thyme, bay leaf, garlic)	Butcher (Cartier 2010), greenhouse worker (Sikora 2008)
Paprika, coriander, mace	Anise liqueur factory (Cartier 2010)
Seeds	
Red onion (Allium cepa) seeds	Seed-packing factory worker (Cartier 2010)
Sesame seeds	Miller (grounding waste bread for animal food), baker (Cartier 2010)
Fennel seeds	Sausage-manufacturing plant (Cartier 2010)
Lupine seeds	Agricultural research worker (Cartier 2010)
Buckwheat flour	Health food products, noodle maker, cook (Sikora 2008, Cartier 2010)
Herbal teas	
Tea	Green tea factory, tea packer (Cartier 2010)
Cinnamon	Worker processing cinnamon (Cartier 2010)
Chamomile	Tea-packing plant worker (Cartier 2010)
Sarsaparilla root	Herbal tea worker (Cartier 2010)
Mushrooms	
Boletus edulis (porcino or king bolete)	Pasta factory (Cartier 2010)
Saccharomyces cerevisiae	Mixing baker's yeast (Cartier 2010)
Mushroom powder	Food manufacturer (Cartier 2010)
Pleurotus cornucopiae	Mushroom grower (Cartier 2010)
Seafood (shellfish and fish)	
Crustaceans	
Snow crabs, Alaskan king crabs, dungeness crabs, tanner crabs, rock crabs	Crab-processing worker (Cartier 2010, Lopata and Jeebhay 2013)
Prawns, shrimp/shrimpmeal, clams	Prawn processor, food processor (lyophilized powder), fishmonger, seafood delivery (Cartier 2010, Lopata and Jeebhay 2013)
Lobster	Cook, fishmonger (Cartier 2010, Lopata and Jeebhay 2013)

Table 8.1 contd. ...

...Table 8.1 contd.

Agent	Occupational exposure
Mollusks	
Cuttlefish	Deep sea fisherman (Cartier 2010, Lopata and Jeebhay 2013)
Mussels	Mussels opener, cook (Cartier 2010, Lopata and Jeebhay 2013)
King and queen scallops	Processor (Cartier 2010, Lopata and Jeebhay 2013)
Abalone	Fisherman (Cartier 2010, Lopata and Jeebhay 2013)
Octopi and squid	Processor (Cartier 2010, Rosado 2009, Wiszniewska 2013, Lopata and Jeebhay 2013)
Fish	
Salmon, pilchard, anchovy, plaice, hake, tuna, trout, turbot, cod, swordfish, sole, pomfret, yellowfin, herring, fishmeal flour	Fish processor, fishmonger (Cartier 2010, Lopata and Jeebhay 2013)
Farm products	
Pork (raw)	Meat-processing plant (Cartier 2010), meat packer (Hilger 2010)
Beef (raw)	Cook (Cartier 2010)
Lamb (raw)	Cutting raw lamb meat (Cartier 2010)
Hogs	Pig farmer (Sikora 2008)
Cows	Dairy farmer (Sikora 2008)
Poultry (turkey, chicken)	Food-processing plant, poultry slaughterhouse (Cartier 2010)
Eggs	Confectionary worker, bakery, egg-processing plant (Cartier 2010)
Pheasants, quails, doves	Breeder (Sikora 2008)
Milk derivatives	
A-lactalbumin	Candy maker, baker (Cartier 2010)
Lactoserum	Cheese maker (Cartier 2010)
Casein	Delicatessen factory, milking sheep, candy maker (Cartier 2010)
Rennet	Cheese maker (Cartier 2010)
Bovine serum albumin powder	Laboratory worker (Choi 2009)
Bees, honey, pollens	Beekeeper, honey processor, cereal producer (Sikora 2008, Cartier 2010)

(Adapted from Cartier 2010 and Sikora 2008 with permission)

Table 8.2 Food additives and contaminants responsible for occupational asthma.

Agent	Occupational exposure
Food additives	
Colorants	
Carmine	Butcher (production of sausages) (Sikora 2008)
Chinese red rice (derived from *Monascus ruber*)	Delicatessen manufacturing plant (Cartier 2010)
Marigold flour (derived from *Tagetes erecta*)	Porter in animal fodder factory (Lluch-Perez 2009)
Bacterial enzymes	
Transglutaminase (*Bacillus subtilis*)	Superintendent involved in ingredient commercialisation for food industry (De Palma 2014)
Fungal enzymes	
A-amylase, cellulase, xylanase	Baker (Cartier 2010)
Glucoamylase	Baker (Cartier 2010)
Pectinase, glucanase	Fruit salad processing (Cartier 2010)
Papain, bromelain	Meat tenderizer (Sikora 2008)
Thickening agents	
Carob bean flour	Jam factory (Sikora 2008), ice cream maker (Cartier 2010)
Pectin	Candy maker, preparation of jam (Cartier 2010)
Konjac glucomannan	Food-manufacturing plant (Cartier 2010)
Vitamins (thiamine)	Castor oil Factory and dock workers (Cartier 2010)
Gluten	Manufacturing-enriched breakfast cereals (Cartier 2010)
Sodium metabisulfite	Biscuit maker (Cartier 2010)
Food contaminants	
Insects	
Poultry mites (*Ornithonyssus sylviarum*)	Poultry worker (Sikora 2008)
Grain storage mites (*Glycyphagus destructor*)	Grain worker (Sikora 2008)
Storage mite (*Tyrophagus putrescentiae*)	Van driver for dry cured ham (Rodriguez 2012)
Spider mites (*Tetranychus urticae*), Panonychus ulmi	Table grape (Jeebhay 2007), apple (Kim 1999), citrus farmers (Burches 1996)
Flour moth (*Ephestia kuehniella*)	Cereal stocker, baker (Cartier 2010)

Table 8.2 contd. ...

...*Table 8.2 contd.*

Agent	Occupational exposure
Champignon flies	Champignon cultivator (Cartier 2010)
Cockroaches (*Blattella* spp.)	Baker (Cartier 2010)
Granary weevils (*Sitophilus granarius*)	Baker (Cartier 2010)
Rice flour beetles (*Tribolium confusum*)	Baker (Sikora 2008)
Fungi	
Aspergillus niger	Brewer (contaminated malt) (Cartier 2010)
Chrysonilia (Neurospora) sitophila	Service operator of coffee dispenser (Cartier 2010)
Aspergillus, Alternaria spp.	Baker (Cartier 2010)
Verticillium alboatrum	Greenhouse tomato grower (Sikora 2008)
Penicillium nalgiovensis	Semi-industrial pork butcher (Talleu 2009)
Parasites	
Anisakis simplex	Fish-processing workers, frozen fish factory (Cartier 2010)
Plants	
Hoya (sea squirts)	Oysters handlers (Cartier 2010)
Others	
Soft red coral	Spiny lobster fisherman (Cartier 2010)

(Adapted from Cartier 2010 and Sikora 2008 with permission)

sensitising epitopes which may increase the allergenicity of the food protein (Lopata 2010a, van der Walt 2010).

8.3 EPIDEMIOLOGY AND RISK FACTORS

Various studies have demonstrated that between 10–25% of occupational allergic rhinitis or asthma reported to voluntary respiratory surveillance programmes are due to food and food products (Meredith and Nordman 1996). Esterhuizen et al. also reported that the food processing industry in South Africa has been one of the top three industries reporting workers with occupational asthma under the SORDSA voluntary surveillance programme (Esterhuizen 2002). The proportion of occupational asthma cases reported in food handlers was 14.4%. The majority of cases were due to flour and grain

Table 8.3 Common processing techniques employed for seafood groups that are sources of potential high risk exposure to seafood products.

Seafood category	Processing techniques	Sources of potential high-risk exposure to seafood product/s
Crustaceans		
Crabs, lobsters	cooking (boiling or steaming) "tailing" lobsters, "cracking", butchering and degilling crabs, manual picking of meat, cutting, grinding, mincing, scrubbing and washing, cooling, crab leg "blowing	inhalation of wet aerosols from lobster "tailing", crab "cracking", butchering and degilling, boiling, scrubbing and washing, spraying, cutting, grinding, mincing, crab leg blowing
Prawns, shrimps	heading, peeling, deveining, prawn "blowing" (water jets or compressed air)	prawn "blowing", cleaning processing lines/tanks with pressurised water
Molluscs		
Oysters, mussels, cuttlefish, scallops, octopi	washing, oyster "shucking", shellfish depuration, chopping, dicing, slicing	inhalation of wet aerosols from oyster "shucking", washing
Finfish		
Various species: Salmon, pilchard, anchovy, plaice, hake, tuna, trout, turbot, cod, swordfish, sole, pomfret, yellowfin, herring	heading, degutting, skinning, mincing, filleting, trimming, cooking (boiling or steaming), spice/batter application, frying, milling, bagging	inhalation of wet aerosols from fish heading, degutting, boiling
		inhalation of dry aerosols from fishmeal bagging
		cleaning floors, trays and machineries using pressurized water

(Updated and modified from Jeebhay 2001 with permission with references from Sikora 2008 and Shiryaeva 2014)

(80%), with baking and milling contributing almost half the cases (Figure 8.1). The common agents responsible for these cases were flour, grain/maize, onion and garlic (Figure 8.2).

Comprehensive data for the prevalence of occupational asthma in various food sectors are not available. However, in those food-related industries in which prevalence of occupational asthma is available, rates do not significantly differ from those found in non-food industries. For example, occupational asthma occurs in 3% to 10% of workers exposed to green coffee beans, 4% to 13% of bakers, 4% to 36% of shellfish and 2 to 8% of bony fish processors (Sikora 2008, Baatjies and Jeebhay 2013, Pacheco

2013). This is also observed in the South African industrial setting (Jeebhay 2012), although what is evident is that the prevalence of work-related asthma is higher in the plant (4–25%) as opposed to the animal handling or processing industry (4–12%) (Table 8.4). Although the differences in prevalence observed may be due to the use of varying definitions of occupational asthma, the allergenic potential of the specific proteins as well as the type of work process causing excessive exposure, do play a role.

Various epidemiological studies and case reports indicate that ocular-nasal symptoms and allergic rhinitis are commonly encountered in food exposed workers (Sikora 2008, Baatjies and Jeebhay 2013, Pacheco 2013). Frequently, this is the first indicator of underlying allergic disease and a large proportion of individuals with occupational asthma also report co-existing occupational rhinitis. Rhino-conjunctivitis may therefore precede or coincide with the onset of occupational asthma. The prevalence of occupational rhinitis associated with food proteins appears to be double the prevalence of occupational asthma in these settings.

Occupational allergic respiratory disease is commonly the result of an interaction between genetic, environmental and host

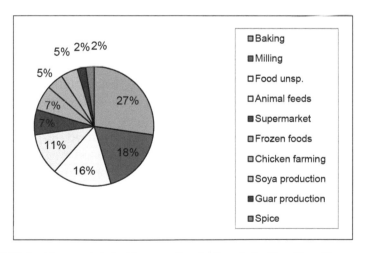

Figure 8.1 Industries associated with occupational asthma in food handlers: 44 cases reported to SORDSA (Reproduced with permission from Esterhuizen 2002).

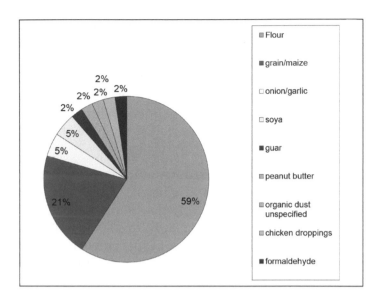

Figure 8.2 Agents causing occupational asthma in food handlers: 44 cases reported to SORDSA. (Reproduced with permission from Esterhuizen 2002)

factors giving rise to various allergic disease phenotypes. The most important environmental risk factors are exposure to the causative agent and elevated exposure to the sensitising agent. Some agents such as crustaceans (e.g., crab) and cereal flours (e.g., wheat, rye) appear to be more potent sensitizers than others in their food grouping. Studies in the seafood industry also indicate that exposure to raw seafood may be less sensitizing to individuals than cooked seafood during processing activities (Lopata and Jeebhay 2013). There is increasing evidence that the risk of sensitisation and occupational asthma is increased with higher exposures to food aerosols. These studies have been reported in workers exposed to flour (wheat, rye), fungal alpha-amylase, green coffee, castor bean, seafood (crab, prawn, salmon, pilchard and anchovy fish) (Nicholson 2005, Pacheco 2013, Baatjies et al. 2015). Other workplace organisational factors can mediate hazardous exposures and worker vulnerability especially agricultural workers due to their rural locations, being a migrant and seasonal workforce, divisions of labour along gender and racial

Table 8.4 Epidemiological studies of food industry workers in South Africa (1985–2015).

Type of workforce (author, yr published)	N	Outcome measure	Agent/s implicated	Prevalence/incidence (%)
Bakeries (supermarkets) (Baatjies 2009)	517	Occupational asthma	Cereal flour (wheat, rye) and fungal alpha-amylase	13%
Grain mill A (Jeebhay 2000, 2005b)	111	Work-related grain dust allergic asthma*	Wheat, storage pests (mealworm, cockroach, storage mites)	17%
Grain mill A (Yach 1985)	582	Asthma symptoms	Wheat	23–25%
Grain mill B (Bartie 2004)	84	Work-related asthma symptoms	Maize, storage pests (weevils)	7%
Soybean processing (Mansoor 2004)	115	Work-related soybean allergic asthma*	Soybean	IR: 2 per 1000 person months
Spice mill (Van der Walt 2013)	150	Work-related asthma symptoms Asthma**	Garlic, onion, chili pepper	17% 4–8%
Poultry processing (Ngajilo 2015)	230	Probable occupational asthma	Feed (sunflower seeds), storage mite, mould, poultry matter	4%
Poultry processing (Rees 1998)	134	Asthma symptoms	Feed, poultry matter (feathers, droppings, serum)	12%
Seafood processing (Jeebhay 2008)	594	Occupational asthma	Fish products, fish parasite (*Anisakis*)	2%
Table grape vineyards (Jeebhay 2007)	207	Work-related spider mite allergic asthma*	Spider mite	6%

CI: cumulative incidence, IR: incidence rate.

* Work-related asthma symptoms + antigen specific allergic sensitisation with or without spirometry changes.

** Spirometry changes (post bronchodilator increase) or high exhaled nitric oxide (>50 ppb)

(Adapted with permission from Jeebhay 2012 and references from van der Walt 2013 and Ngajilo 2015) lines, as well as shortcomings in occupational health and safety laws and interventions (Howse 2012).

Since most food allergens are high molecular weight proteins or glycoproteins capable of inducing an IgE-mediated response, atopy is an important host risk factor for the development of allergic sensitization and occupational asthma. Atopy is associated with an increased risk of sensitization in workers exposed to crabs, prawns, cuttlefish, pilchard, anchovy, green coffee beans, and bakery allergens including enzymes (Pacheco 2013, Nicholson 2005). An increased risk for occupational asthma among atopic workers has also been reported in workers exposed to flour (bakers), enzymes, and crabs, but this association has not been confirmed in other settings (e.g., exposure to salmon) (Nicholson 2005, Jeebhay and Cartier 2010). Data from a recently published study of supermarket bakery workers has demonstrated that atopy is more of an effect modifier in that non-atopic workers exposed to flour dust also demonstrated an increased risk for sensitisation to wheat (Figure 8.3) (Baatjies et al. 2015). The presence of rhinitis has also been associated with an increased risk of developing occupational asthma to a number of food proteins (Nicholson 2005). Finally, smoking has

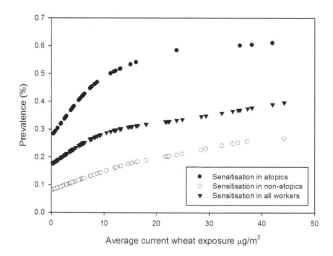

Figure 8.3 Relationship between wheat sensitisation and wheat allergen concentration among supermarket bakery workers, stratified by atopic status.
(Reproduced with permission from Baatjies et al. 2015)

been associated with an increased risk of sensitisation to various seafood including prawns, crab and fish (pilchard, anchovy and salmon) green coffee beans and flour (Nicholson 2005, Jeebhay and Cartier 2010).

8.4 CLINICAL FEATURES AND DIAGNOSTIC APPROACHES

Occupational allergy can arise as a result of *de novo* occupational inhalation of food products containing single (e.g., wheat flour) or multiple allergens (e.g., flour dust containing cereal flours, enzymes, mites); *cross-reactivity* between occupational allergens in already sensitised workers (e.g., wheat vs. rye, crab vs. lobster, pollen vs. spice); and re-exposure in a worker with known food allergy (e.g., seafood allergy).

Occupational allergic reactions as a result of inhalant exposures to food allergens in the workplace generally present with upper and/or lower airway symptoms. Rhinitis, conjunctivitis, and less frequently urticaria, are often associated and may precede the development of chest symptoms. Systemic anaphylactic reactions have also been reported but are rare, although there have been incidents of anaphylactic reactions in the domestic setting following work-related sensitisation to certain food allergens (Siracusa 2015). Most workers with occupational asthma to food can tolerate ingestion of the relevant food; however, some workers have subsequently developed clinical ingestion-allergic–related symptoms (Sikora 2008, Cartier 2010).

Diagnostic approaches for occupational allergy and asthma associated with food allergens are similar to the general investigative approaches used in the evaluation of the patient from other non-food causes (Jeebhay 2012, Sikora 2008, Cartier 2010, Gill 2002). That most food allergens are high molecular agents causing an IgE-mediated reaction lends itself to the use of traditional immunological techniques to identify the cause of the allergy. Specific allergic sensitization may be demonstrated by skin prick skin test or specific IgE to the offending allergens using either the natural raw extract or a standardised commercial extract of the food (van Kampen 2013).

However, the positive and negative predictive value of these tests in predicting occupational asthma vary depending on the allergen. For example, the negative predictive values of skin tests to flour and enzymes are very high, whereas the positive predictive value is lower, in that a sizeable proportion of individuals with positive skin tests have no evidence of clinical allergy (Cartier 2010). Studies among supermarket bakery workers show that although 25% of workers demonstrated specific IgE to wheat, only between 5 to 13% had allergic asthma or rhinitis (Baatjies et al. 2015). In crab processing workers, the positive predictive value of a positive skin prick test to crab extracts or positive specific IgE for occupational asthma confirmed by specific inhalation challenge (SIC) was 76% and 89%, respectively (Cartier 1986). A negative skin test therefore does not exclude the diagnosis of occupational asthma, whereas a positive test supports the diagnosis but is not definitive in and of itself. Other approaches such as component-resolved diagnostics using recombinant wheat flour proteins have recently been used to distinguish between wheat sensitization caused by inhalational flour exposure, cross-reactivity to grass pollen and ingestion related wheat allergy (Sander 2015). However, for the routine diagnosis of baker's allergy, allergen-specific IgE tests with whole wheat and rye flour extracts still remain the preferred method due to their superior diagnostic sensitivity. The work-relatedness of the asthma can be demonstrated using serial peak expiratory flow monitoring at and away from work or increased non-specific bronchial responsiveness (NSBH) on return to work after a period away from work. Specific inhalation challenges for high molecular weight proteins show that an early asthmatic reaction is the more commonly observed, although dual reactions are also possible. However, in crab processing workers, isolated late asthmatic reactions are more frequent (Cartier 1984). Finally, standardisation of exposure characterisation approaches for determination of environmental allergen presence and concentrations are important in making the link between allergen exposures and work-related allergic symptoms and adverse respiratory outcomes in relation to diagnosis as well as in evaluating the impact of interventions to reduce allergen exposures (Raulf 2014).

8.5 BIOLOGICAL AND BIOCHEMICAL CHARACTERISTICS OF KNOWN OCCUPATIONAL ALLERGENS

8.5.1 Seafood Allergens

Airway exposure to the main fish allergen *parvalbumin* has been documented in workplaces in the seafood industry (Table 8.1) (Lopata and Jeebhay 2013). Parvalbumin is a highly stable, low molecular weight protein (10–12 kDa), belonging to the EF-hand superfamily of proteins that contains a characteristic cation binding helix–loop–helix structural motif (Kawasaki 1998). Large amounts of this protein are expressed in fast skeletal muscles of lower vertebrates. Fish and frog parvalbumins, belonging to the beta-parvalbumins are confirmed allergens, whereas alpha parvalbumins, expressed in much lower amounts in skeletal muscles of higher vertebrates are apparently non-allergenic. Parvalbumins function in muscle relaxation by buffering and transporting calcium to the sarcoplasmatic reticulum. The allergenicity of different fish species corresponds with their parvalbumin content. This content varies considerably between species from <0.5 mg/g tissue in mackerel to >2 mg/g in cod, carp, redfish and herring (Kuehn 2010). In addition variations in amino acid sequence between species (55–95% identity), especially in the epitope regions affect allergenic potency (Kuehn 2014). Three epitope sets have been identified in parvalbumin. Interestingly, the specific epitope preferred by a given patient IgE, seem to correspond to symptom severity of the patient (Leung 2014). Several IgE-binding proteins other than parvalbumin have also been reported in the past decade, but the clinical relevance is uncertain for the majority (Kuehn 2014). Recently, the two muscle enzymes enolase (50 kDa) and aldolase (40 kDa) have been recognized as important allergens in fish species (Kuehn 2013). Although most patients displaying IgE reactivity to these proteins also show reactivity to parvalbumin, patients being monoallergic to aldolase and enolase have been identified. As opposed to parvalbumin, which is stable over a broad range of pH and temperatures, these enzyme allergens are heat sensitive. Thus handling of raw fish, as in occupational settings, may be of greater relative importance for sensitization to these allergens compared to parvalbumin, which show strong IgE-binding both in raw and processed forms (Saptarshi 2014). Cross sensitization between fish species is common but not absolute. Large variations between parvalbumin content and amino acid sequences between species may explain why patients may react to some fish species but tolerate others. Increased awareness of allergens other than parvalbumin and the relative importance of these in occupational settings are needed to understand exposure patterns, sensitization and tolerability in workplace environments.

Tropomyosin is the main crustacean allergen (Lopata 2010b) and exposure to tropomyosin has been documented in environmental air samples from different types of crab industries (Abdel Rahman 2010, Kamath 2014, Lopata and Jeebhay 2013). The 34–39 kDa protein belong to the highly conserved family of actin filament binding proteins, functioning in contraction of muscle cells. The secondary structure is a two stranded alpha-helical coiled coil and display up to eight conserved IgE-binding epitopes. Tropomyosin is highly resistant to heat, low pH and protease digestion. In addition to tropomyosin, arginine kinase (40 kDa enzyme), myosin light chain (20 kDa muscle protein) and sarcoplasmic calcium-binding protein (SCP, 20 kDa muscle protein) are reported allergens in crustacean species. Considerable IgE cross reactivity between crustaceans like crabs, shrimps and prawns has been documented, and is likely to be related to a highly conserved amino acid sequence with up to 98% homology between different crustacean species. Cross reactivity also extend to other arthropods (Ayuso 2002) such as insects, mites and notably the fish parasite *Anisakis*. Allergy and asthma among workers handling fish may also be related to *Anisakis*-allergens inhaled together with fish allergens present in workplace bioaerosols. In industries utilizing marine ingredients for taste or dietary supplements, IgE reactivity may extend to other crustacean species such as krill and calanus which are less likely to be used for human consumption.

Respiratory effects in workers exposed to mollusks such as octopus, squid, mussels and bivalves are well known. In mollusks, paramyosin, a 100 kDa myofibrillar protein is documented as an allergen in addition to tropomyosin. Several other proteins display IgE-reactivity but are not yet identified. The homology between crustacean and mollusk tropomyosin is less than 60%, indicating that cross-reactivity would be unlikely. However, a conserved epitope sequence shared by the two groups of marine organisms has been suggested to cause cross-reactivity in spite of the relatively low overall homology (Leung 2014).

8.5.2 Flour Allergens Including Enzyme Additions

Flour is the most important allergen source in the work environment of bakeries, associated with baker's asthma. A number of different allergens belonging to several protein classes are present in flour and there is a wide heterogeneity in sensitization patterns between individual patients with baker's asthma (Salcedo 2011). The alpha-amylase inhibitors are considered the major cereal allergens. These consist of 1 to 4 subunits (12–16 kDa) and are encoded by a multigene family expressed in wheat, barley and rye. Glycosylation seem to increase IgE-binding, at least for some subunits and species (Tatham and Shewry 2008). Other wheat proteins associated with IgE-reactivity and baker's asthma include peroxidase, lipid transfer proteins (LTP), thioredoxin, serine protease inhibitor, thaumatin-like protein, gliadins and glutenins. Homology in alpha-amylase inhibitor subunits are suspected to account, at least partially, for the observed cross-reactivity between wheat, rye and barley with amino acid sequence identities ranging from 30 to 95%. Cross reactivity between different grain flours and between grain flours and grass pollen is demonstrated (Sander 2015). There is, however, limited knowledge of specific common epitopes responsible for the cross-sensitization observed.

Enzymes, added to cereal flours to improve dough qualities, are also present in flour dust. IgE reactivity to fungal α-amylase (from *Aspergillus*) used to digest starch and provide sugar for the yeast, is well documented in patients with baker's asthma. Other enzymes used as flour additives, including xylanases and proteases added to digest cell walls and weaken the gluten network, respectively, are also shown to display IgE-reactivity in asthmatic patients (Tatham and Shewry 2008).

8.5.3 Spice Allergens

Production of spices from plants involve drying and crushing the raw materials, processes that give rise to dust with potential of being inhaled by workers. Originating from plants, many spices contain well known plant allergens, like profilins, lipid-transfer proteins or high molecular weight glycoproteins. Ubiquitous and cross-reactive plant allergens, as the birch pollen-associated Bet v1 and Bet v2 or mugwort Artv 4 profilins, may thus contribute to allergic reactions in workers handling spices. Specific spice allergens, best known to produce sensitization via the oral route, and are also likely to be airborne in work environments during spice production, packing, manufacturing and food preparation. Only few studies have presented molecular data on inhalable spice allergens causing sensitization in the occupational environment. In spice mills, two bands of 40 and 52 kDa in chili pepper have been identified as IgE-reactive proteins in immunoblots, using serum from airway sensitized spice mill workers (van der Walt 2010). Similarly, a 50 kDa IgE-reactive protein is identified in garlic and onion, sensitized asthmatic workers (Mansoor and Ramafi 2000). Sequencing is needed to identify the proteins involved. Being heat stable, enriched and more likely to be airborne after processing, handling dried garlic- or onion powder seem to cause stronger IgE-reactivity than working with the raw plant. It is also possible that the dry heating process itself may enhance the allergenicity due to structural rearrangements of the IgE-reactive molecules in so-called Maillard reactions (Toda 2014). This has been previously shown for other plant allergens, such as peanut.

8.6 PREVENTIVE APPROACHES

The only effective strategy to limit allergen related asthma in workplaces, is to control the environmental exposure to allergens. To achieve this various legislative, engineering, organizational and surveillance measures are required (Jeebhay 2002a, Sikora 2008, Cartier 2010).

Health and safety regulations to reduce allergen exposure of the total workforce or certain risk-associated worker-groups are needed. Presently, there are few occupational exposure limits for food allergens. In general, establishing threshold levels for allergen exposure is considered complicated due to large inter-individual variations in susceptibility to both sensitization and allergic response. Complex mixtures of allergens, as exemplified by the numerous allergens present in flour, further complicate the picture. There is also a need of better standardization of sampling methods and assays for the analyses of environmental allergens. As a result, the occupational exposure limit for flour is presented as a limit for exposure to inhalable flour dust only, without consideration of specific allergens (Cartier 2010, Baatjies and Jeebhay 2013).

Economic incentives may be used to reduce workplace exposures. Asthma, being a serious adverse health outcome, results not only in reduced quality of life for the individual but implies extensive use of the health care system by the patient, which may often be of life-long duration. The seriousness of this illness is only poorly reflected in taxes, economic sanctions and risk-based insurance premiums. At least for large companies, economic control measures could be considered to a greater extent to increase risk control in work environments with exposure to allergens.

Spreading of airborne allergens is best prevented at the source. Thus, the identification of main sources of allergen liberation to the air should be a primary focus of health and safety walk-troughs in workplaces. Departments, machineries and work tasks with high aerosol exposure should be prioritized for preventive measures. Substitution of food materials with less allergenic species may rarely be feasible, but changing the physical forms are sometimes possible, for instance the use of granulated or dissolved, instead of powdered ingredients. Change of processes, e.g., the use of water jets instead of air jets to remove shrimp shells; or modification of processes as reducing the pressure of water jet when cleaning, can reduce aerosol liberation (Cartier 2010, Pacheco 2013, Jeebhay and Cartier 2010).

Aerosol liberation should always be an issue when new machineries are evaluated, forcing supplier companies to minimize aerosol production from their products. Separation of workers from aerosol sources can be achieved by placing shields or isolating processes and machines in separate rooms. Improving local and general ventilation will remove airborne substances faster from the ventilated zone. Use of respirators in addition to other measures may be relevant in some situations. Air supplied respirators are the best option for safety, but are expensive and often inconvenient in the work situation. The effectiveness of respirators without air-supply greatly depends on the goodness of fit of the mask to the face, and should be tested to find the optimal mask for each worker.

Education and training of the workforce is important to make sure all employees understand the risks associated with allergen exposure, adopt good work practices aimed at minimizing the liberation of allergens to the environment. Ultimately, a multi-pronged strategy that combines engineering and improved work practices through training appear to be the most effective in reducing allergen exposures as has been recently demonstrated in supermarket bakeries (Baatjies 2014).

Finally, surveillance programs including risk assessment of environmental factors and medical surveillance (using questionnaires and skin prick tests/specific IgE) of the workforce should be performed regularly on high risk working populations. Studies have shown that early intervention, for instance relocation of sensitized workers, is crucial in preventing further development of allergic disease.

8.7 CONCLUSION

As new foods are developed, it is possible that new occupational reactions can occur during food processing activities. The constant need for increasing the global food production output has resulted in renewed approaches to encourage utilization of by-products, wastes and species not previously regarded as human food sources. More specific assays for airborne allergens are therefore needed to assess workplaces and tasks that may pose an increased risk with respect to allergic sensitization and development of respiratory disease.

Of special interest is the increasing use of biotechnology in food processing and the introduction of genetically modified crops that may contain novel proteins, not previously known, which may be capable of causing allergic reactions in the occupational setting well before these products are made available to the consumer market. It is therefore crucial that epidemiological surveillance programmes be initiated on sentinel groups such as workers in food processing plants to detect the emergence of new allergies and health risks at a very early stage. Food manufacturer responsibility for product stewardship should include, among others, product labelling and accurate information on allergenicity of these products in material safety data sheets provided to workers and consumers handling these foods, and in this way ensuring overall public health and safety.

Keywords: Occupational allergy; occupational asthma; inhaled food allergen; food allergy

REFERENCES

Abdel Rahman, A. M. et al. (2010). Analysis of the allergenic proteins in black tiger prawn (*Penaeus monodon*) and characterization of the major allergen tropomyosin using mass spectrometry. Rapid Commun Mass Spectrom. 24(16): 2462–70.

Ayuso, R. et al. (2002). Molecular basis of arthropod cross-reactivity: IgE-binding cross-reactive epitopes of shrimp, house dust mite and cockroach tropomyosins. Int Arch Allergy Immunol. 129(1): 38–48.

Baatjies, R., A. L. Lopata, I. Sander, M. Raulf-Heimsoth, E. Bateman, T. Meijster, D. Heederik, T. G. Robins and M. F. Jeebhay (2009). Determinants of asthma phenotypes among supermarket bakery workers. European Respiratory Journal. 34(4): 825–833.

Baatjies, R., T. Meijster, A. L. Lopata, I. Sander, M. Raulf-Heimsoth, D. Heederik and M. F. Jeebhay (2010). Exposure to flour dust in South African supermarket bakeries: modelling of baseline measurements of an intervention study. Annals of Occupational Hygiene. 54(3): 309–318.

Baatjies, R. and M. F. Jeebhay (2013). Baker's allergy and asthma—A review of the literature. Current Allergy and Clinical Immunology. 26(4): 232–243.

Baatjies, R., T. Meijster, D. Heederik, I. Sander and M. F. Jeebhay (2014). Effectiveness of interventions to reduce flour dust exposures in supermarket bakeries in South Africa. Occupational and Environmental Medicine. 71: 811–818.

Baatjies, R., T. Meijster, D. Heederik and M. F. Jeebhay (2015). Exposure- response relationships for inhalant wheat allergen exposure and asthma. Occupational and Environmental Medicine. 72: 200–207.

Bartie, C., A. E. Calverly and D. Rees (2004). Sensitisation to maize in the wet-milling industry. National Institute for Occupational Health Report 17/2004, Johannesburg.

Burches, E., A. Pelaez, C. Morales et al. (1996). Occupational allergy due to spider mites: *Tetranychus urticae* (Koch) and *Panonychus citri* (Koch). Clin Exp Allergy. 26(11): 1262–7.

Cartier, A., J. L. Malo, F. Forest et al. (1984). Occupational asthma in snow crab-processing workers. J Allergy Clin Immunol. 74: 261–9.

Cartier, A., J. L. Malo, H. Ghezzo, M. McCants and S. B. Lehrer (1986). IgE sensitization in snow crab-processing workers. J Allergy Clin Immunol. 78: 344–8.

Cartier, A. (2010). The role of inhalant food allergens in occupational asthma. Curr Allergy Asthma Rep. 10(5): 349–56.

Choi, G. S., J. H. Kim, H. N. Lee et al. (2009). Occupational asthma caused by inhalation of bovine serum albumin powder. Allergy Asthma Immunol Res. 1: 45–47.

de las Marinas, M. D., R. Felix, C. Martorell et al. (2013). Cit s 3 as an occupational aeroallergen in an orange farmer. J Investig Allergol Clin Immunol. 23: 510–512.

De Palma, G., P. Apostoli, G. Mistrello et al. (2014). Microbial transglutaminase: a new and emerging occupational allergen. Ann Allergy Asthma Immunol. 112: 553–554.

Esterhuizen, T. M. and D. Rees (2002). Occupational asthma in food handlers: Cases reported to SORDSA (1996–2002). Current Allergy and Clinical Immunology. 15(4): 156–158.

Felix, R., C. Martorell, A. Martorell et al. (2013). Induced bronchospasm after handling of orange flavedo (zest). J Allergy Clin Immunol. 131: 1423–1425.

Food and Agriculture Organisation: http://www.fao.org/docrep/015/i2490e/i2490e01c.pdf.

Gill, B. V., M. Aresery and S. B. Lehrer (2002). Occupational reactions to foods. Current Allergy and Clinical Immunology. 15(4): 148–154.

Hilger, C., K. Swiontek, F. Hentges, C. Donnay, F. de Blay and G. Pauli (2010). Occupational inhalant allergy to pork followed by food allergy to pork and chicken: Sensitization to hemoglobin and serum albumin. Int Arch Allergy Immunol. 151(2): 173–8.

Howse, D., M. F. Jeebhay and B. Neis (2012). The changing political economy of occupational health and safety in fisheries—lessons from Eastern Canada and South Africa. Journal of Agrarian Change. 12(2-3): 344–363.

James, J. M. and J. F. Crespo (2007). Allergic reactions to foods by inhalation. Curr Allergy Asthma Rep. 7(3): 167–74.

Jeebhay, M. F., J. Stark, A. Fourie, T. Robins and R. Ehrlich (2000). Grain dust allergy and asthma among grain mill workers in Cape Town. Current Allergy Clin Immunol. 13: 23–25.

Jeebhay, M. F., T. G. Robins, S. B. Lehrer and A. L. Lopata (2001). Occupational seafood allergy—a review. Occupational and Environmental Medicine. 58(9): 553–62.

Jeebhay, M. F. (2002a). Occupational allergy and asthma among food processing workers in South Africa. African Newsletter in Occupational Health and Safety. 12(3): 59–62.

Jeebhay, M. (2002b). Occupational food allergens. Current Allergy and Clinical Immunology. 15(4): 138–140.

Jeebhay, M. F., T. G. Robins, N. Seixas, R. Baatjies, D. A. George, E. Rusford, S. B. Lehrer and A. L. Lopata (2005a). Environmental exposure characterization of fish processing workers. Annals of Occupational Hygiene. 49(5): 423–37.

Jeebhay, M. F., R. Baatjies and A. Lopata (2005b). Work-related respiratory allergy associated with sensitisation to storage pests and mites among grain mill workers. Current Allergy and Clinical Immunology. 18(2): 72–76.

Jeebhay, M. F., R. Baatjies, Y. S. Chang, Y. K. Kim, Y. Y. Kim and A. L. Lopata (2007). Risk factors for allergy due to the two-spotted spider mite (*Tetranychus urticae*) among table grape farm workers. International Archives of Allergy and Immunology. 144: 143–149.

Jeebhay, M. F., T. G. Robins, M. E. Miller, E. Bateman, M. Smuts, R. Baatjies and A. L. Lopata (2008). Occupational allergy and asthma among salt water fish processing workers. Am J Ind Med. 51(12): 899–910.

Jeebhay, M. F. and A. Cartier (2010). Seafood workers and respiratory disease—an update. Current Opinions in Allergy and Clinical Immunology. 10: 104–113.

Jeebhay, M. F. (2012). An approach to work-related asthma in the South African setting. Current Allergy and Clinical Immunology. 25(3): 164–170.

Kamath, S. D. et al. (2014). Molecular and immunological approaches in quantifying the air-borne food allergen tropomyosin in crab processing facilities. Int J Hyg Environ Health. 217(7): 740–50.

Kawasaki, H., S. Nakayama and R. H. Kretsinger (1998). Classification and evolution of EF-hand proteins. Biometals. 11: 277–295.

Kim, Y. K., M. H. Lee, Y. K. Jee et al. (1999). Spider mite allergy in apple-cultivating farmers: European red mite (*Panonychus* ulmi) and two-spotted spider mite (*Tetranychus urticae*) may be important allergens in the development of work-related asthma and rhinitis symptoms. J Allergy Clin Immunol. 104: 1285–92.

Kuehn, A. et al. (2010). Important variations in parvalbumin content in common fish species: a factor possibly contributing to variable allergenicity. Int Arch Allergy Immunol. 153(4): 359–66.

Kuehn, A. et al. (2013). Identification of enolases and aldolases as important fish allergens in cod, salmon and tuna: component resolved diagnosis using parvalbumin and the new allergens. Clin Exp Allergy. 43(7): 811–22.

Kuehn, A. et al. (2014). Fish allergens at a glance: variable allergenicity of parvalbumins, the major fish allergens. Front Immunol. 5: 179.

Leung, N. Y. et al. (2014). Current immunological and molecular biological perspectives on seafood allergy: a comprehensive review. Clin Rev Allergy Immunol. 46(3): 180–97.

Lluch-Perez, M., R. M. Garcia-Rodriguez, A. Malet et al. (2009). Occupational allergy caused by marigold (Tagetes erecta) flour inhalation. Allergy. 64: 1100–1101.

Lopata, A. (2010a). Allergenicity of food and impact of processing. pp. 459–478. *In*: Ahmed, Jasim, Ramaswamy, Hosahalli S., Kasapis, Stefan and Boye, Joyce I (eds.). Novel Food Processing: Effects on Rheological and Functional Properties. Electro-Technologies for Food Processing Series. CRC Press, Boca Raton, FL, USA.

Lopata, A. L., R. E. O'Hehir and S. B. Lehrer (2010b). Shellfish allergy. Clin Exp Allergy. 40(6): 850–8.

Lopata, A. L. and M. F. Jeebhay (2013). Airborne seafood allergens as a cause of occupational allergy and asthma. Curr Allergy Asthma Rep. 13(3): 288–97.

Mansoor, N. and G. Ramafi (2000). Onion allergy—a case report. Curr Allergy Clin Immunol. 13: 14–15.

Mansoor, N., D. Rees, D. Kielkowski, A. Spies and A. Fourie (2004). Assessment of workers exposed to soybean at a soybean processing plant—a prospective cohort study. National Institute for Occupational Health Report 1/2004, Johannesburg.

Meredith, S. and H. Nordman (1996). Occupational asthma: measures of frequency from four countries. Thorax. 51(4): 435–40.

Miedinger, D., J. L. Malo, A. Cartier and M. Labrecque (2009). Malt can cause both occupational asthma and allergic alveolitis. Allergy. 64: 1228–1229.

Ngajilo, D., T. S. Singh, E. Ratshikhopha, R. Baatjies and M. F. Jeebhay (2015). Risk factors associated with asthma phenotypes in poultry farm workers (abstract 65). Public Health Association of South Africa Conference, Durban, Oct. 2015 (abstract book).

Nicholson, P. J., P. Cullinan, A. J. Taylor, P. S. Burge and C. Boyle (2005). Evidence based guidelines for the prevention, identification, and management of occupational asthma. Occup Environ Med. 62(5): 290–9.

Pacheco, K. A., M. F. Jeebhay, D. Gautrin and A. L. Lopata (2013). Asthma and allergy to animals. pp. 238–261. *In*: Jean-Luc Malo, Moira Chan Yeung and David I.

Bernstein (eds.). Asthma in the Workplace, 4th edition, CRC Press, Florida, 17. ISBN: 13: 978-1-84184-925-6 (eBook—PDF).

Raulf, M., J. Buters, M. Chapman, L. Cecchi, F. de Blay, G. Doekes, W. Eduard, D. Heederik, M. F. Jeebhay, S. Kespohl, E. Krop, G. Moscato, G. Pala, S. Quirce, I. Sander, V. Schlünssen, T. Sigsgaard, J. Walusiak- Skorupa, M. Wiszniewska, I. M. Wouters and I. Annesi-Maesano (2014). Monitoring of occupational and environmental aeroallergens—EAACI position paper. Allergy. 69(10): 1280–99.

Rees, D., G. Nelson, D. Kielkowski, C. Wasserfall and A. Da Costa (1998). Respiratory health and immunological profile of poultry workers. S Afr Med J. 88: 1110–7.

Rodriguez del Rio, P., J. I. Tudela Garcia, N. J. Narganes et al. (2012). Occupational asthma caused by the inhalation of *Tyrophagus putrescentiae* allergens in a dry-cured ham transporter allergic to shrimp. J Investig Allergol Clin Immunol. 22: 383 384.

Rosado, A., M. A. Tejedor, C. Benito et al. (2009). Occupational asthma caused by octopus particles. Allergy. 64: 1101–1102.

Salcedo, G., S. Quirce and A. Diaz-Perales (2011). Wheat allergens associated with Baker's asthma. J Investig Allergol Clin Immunol. 21(2): 81–92.

Sander, I., H. P. Rihs, G. Doekes, S. Quirce, E. Krop, P. Rozynek, V. van Kampen, R. Merget, U. Meurer, T. Brüning and M. Raulf (2015). Component-resolved diagnosis of baker's allergy based on specific IgE to recombinant wheat flour proteins. J Allergy Clin Immunol. 135(6): 1529–37.

Sander, I. et al. (2015). Component-resolved diagnosis of baker's allergy based on specific IgE to recombinant wheat flour proteins. J Allergy Clin Immunol. 135(6): 1529–37.

Saptarshi, S. R. et al. (2014). Antibody reactivity to the major fish allergen parvalbumin is determined by isoforms and impact of thermal processing. Food Chem. 148: 321–8.

Shiryaeva, O., L. Aasmoe, B. Straume, A. H. Olsen, A. Ovrum, E. Kramvik, M. Larsen, A. Renstrøm, A. S. Merritt, K. K. Heldal and B. E. Bang (2014). Respiratory effects of bioaerosols: Exposure-response study among salmon-processing workers. Am J Ind Med. 57: 276–285.

Sikora, M., A. Cartier, M. Aresery et al. (2008). Occupational reactions to food allergens. pp. 223–250. *In*: D. D. Metcalfe, H. A. Sampson, R. A. Simon and M. A. Malden (eds.). Food Allergy. Adverse Reactions to Foods and Food Additives, Edn 4. Blackwell Publishing.

Siracusa, A., I. Folletti, R. Gerth van Wijk, M. F. Jeebhay, G. Moscato, S. Quirce, M. Raulf, F. Ruëff, J. Walusiak-Skorupa, P. Whitaker and S. M. Tarlo (2015). Occupational anaphylaxis—An EAACI task force consensus statement. Allergy. 70: 141–152.

Talleu, C., J. Delourme, C. Dumas et al. (2009). Allergic asthma due to sausage mould. Rev Mal Respir. 26: 557–559.

Tatham, A. S. and P. R. Shewry (2008). Allergens to wheat and related cereals. Clin Exp Allergy. 38(11): 1712–26.

Toda, M. et al. (2014). The Maillard reaction and food allergies: is there a link? Clin Chem Lab Med. 52(1): 61–7.

van der Walt, A., A. L. Lopata, N. E. Nieuwenhuizen and M. F. Jeebhay (2010). Work-related allergy and asthma in spice mill workers—the impact of processing dried spices on IgE reactivity patterns. International Archives of Allergy and Immunology. 152(3): 271–278.

van der Walt, A., T. Singh, R. Baatjies, A. L. Lopata and M. F. Jeebhay (2013). Work-related allergic respiratory disease and asthma in spice mill workers is associated with inhalant chili pepper and garlic exposures. Occup Environ Med. 70(7): 446–52.

van Kampen, V., F. de Blay, I. Folletti, P. Kobierski, G. Moscato, M. Olivieri, S. Quirce, J. Sastre, J. Walusiak-Skorupa and M. Raulf-Heimsoth (2013). EAACI position paper: Skin prick testing in the diagnosis of occupational type I allergies. Allergy. 68(5): 580–4.

Wiszniewska, M., D. Tymoszuk, A. Pas-Wyroslak et al. (2013). Occupational allergy to squid (*Loligo vulgaris*). Occup Med (Lond). 63: 298–300.

Yach, D., J. Myers, D. Bradshaw and S. R. Benatar (1985). A respiratory epidemiologic survey of grain mill workers in Cape Town, South Africa. Am Rev Respir Dis. 131: 505–510.

The Influence of Dietary Protein Modification During Food Processing on Food Allergy

Anna Ondracek **and** *Eva Untersmayr**

CONTENTS

Department of Pathophysiology and Allergy Research, Center of Pathophysiology, Infectiology and Immunology, Medical University of Vienna, Austria.
* Corresponding author: eva.untersmayr@meduniwien.ac.at

9.1 INTRODUCTION

Although indications of adverse reactions of food can be found in antique literature, the first scientific evidence of food allergy was published in 1912 by the pediatrician Oscar M. Schloss. He introduced skin tests for the diagnosis of food allergies (Wüthrich 2014). In 1921 Heinz Prausnitz and Karl Küstner demonstrated that fish allergy could be passively transferred using serum of an allergic individual providing first evidence for the existence of an allergy mediating substance in serum, which was identified as Immunoglobulin (Ig) E in 1966 (Bergmann 2014).

Food allergies have been identified as a significant health problem during the past few decades (Prescott and Allen 2011). The evidence for increasing numbers of patients affected by food adverse reactions is growing. Approximately 5% of the adult population and 8% of children are affected currently (Sicherer and Sampson 2014). However, accurate assessment of food allergy prevalence is challenging since different diagnostic approaches and methodologies, geographic variation, differences in dietary habits, age, and other factors make it difficult to compare food allergy studies side by side (Sicherer and Sampson 2014). Recent literature reviews suggest that between 1% and 10% of the population suffer from food allergy (Chafen et al. 2010) reflecting the enormous variation between different studies.

The apparent increase in prevalence has been predominantly observed in the so-called western societies although certain food allergies seem to be specific for particular geographic regions. A variety of factors are discussed to influence sensitization and food allergic reactions including sex, genetics, increased hygiene, microbiota and diet but knowledge about risk factors for sensitization remains limited (Sicherer and Sampson 2014). The increased prevalence of food allergic patients in developed countries may be attributed to a different lifestyle with implications on genetically predisposed individuals (Sicherer 2011).

The impact of food and its nutritional composition is one focus of food allergy research. High fat diet shows an influence on the composition of the gut microbiota as well as on innate immune responses (Berin and Sampson 2013). Considering that the microbiome resident in the gastrointestinal tract is suspected to influence allergic sensitization (Noval Rivas et al. 2013, Stefka et al. 2014), different dietary habits are likely to contribute to development of food allergy. Obesity, which is mostly caused by an unhealthy lifestyle including high fat diet, has been linked to higher sensitization rates and was positively correlated to high total IgE levels (Visness et al. 2009). It is assumed that cow's milk, egg, wheat, soy, peanut, tree nuts, fish, and shellfish comprise the most important food allergens. However, the frequency of adverse reactions to certain allergens shows substantial variation between countries and continents (Nwaru et al. 2014). Moreover, it remains to be elucidated why certain food compounds seem to have enhanced allergenic properties compared to others. Diet is a complex mixture of diverse carbohydrates, fatty acids and proteins constituting the food matrix (McClain et al. 2014). The difficulty in investigating allergenicity of food proteins is, thus, caused by the complexity of food composition and its influence on antigen recognition. Food matrix and relative fat content is capable of influencing tolerated doses of allergens and severity of allergic reactions (Grimshaw et al. 2003). Food composition might even influence the binding state of pocket proteins with impact on the specific immune response (Roth-Walter et al. 2014). Allergens imbedded in complex matrices are likely to undergo a range of chemical modifications during food processing and storage influencing gastrointestinal digestion, uptake, and presentation to the immune system. Such modifications comprise loss of certain amino acids (AAs), nitration, oxidation, reduction, glycation, unfolding, aggregation, cross-linking, or degradation (Mills et al. 2009, Hilmenyuk et al. 2010, Untersmayr et al. 2010, Rocha et al. 2012, Toda et al. 2014).

Based on this knowledge, this chapter will provide an overview on currently available information regarding influence of food modification on protein allergenicity.

9.2 FOOD PROTEIN MODIFICATION: FROM PROCESSING TO DIGESTION

Food has been processed since mankind gained the ability to handle fire for the purpose of cooking, broadening the spectrum of hominid diet. Nowadays, food processing is established on an industrial level to ensure food quality and influences food safety as well as shelf life of food by destruction of food-borne pathogens, natural toxins or enzymes and allergenic structures (van Boekel et al. 2010, Bu et al. 2013).

Food processing includes a range of methods such as heat application, high pressure techniques, pulsed electric field, fermentation, membrane processing or dehydration processes with different effects on food quality (van Boekel et al. 2010).

However, food protein modification is not restricted to food preparation. Dietary compounds are processed as soon as they are ingested. Exposure to proteases, lipases and carbohydrate degrading enzymes of the gastrointestinal tract ensures digestion and enables absorption of nutrients in the intestine. Degradation of allergenic proteins is initiated by gastric pH promoting proteolytic activity of gastric enzymes and starts denaturation of dietary proteins by facilitating access to cleavage sites for efficient degradation. Upon passage into the small intestine remaining polypeptides are exposed to pancreatic and mucosal brush border peptidases. Resulting single AAs or short AA chains evade recognition by the immune system (Untersmayr and Jensen-Jarolim 2008).

As gastrointestinal digestion decreases the bioavailability of intact allergens at immune induction sites of the intestine, the potential of food allergens to elicit primary sensitization has been linked for a long time to stability to gastrointestinal enzymes (Astwood et al. 1996). However, there is growing evidence that major sensitizing food allergens are rapidly degraded simulating gastric digestion in *in vitro* experiments (Fu 2002, Fu et al. 2002, Untersmayr and Jensen-Jarolim 2006).

Notably, the physiological gastrointestinal digestion has an important gate keeping function with regards to allergenic food proteins. Elevation of gastric pH and subsequent impairment of peptic digestion was revealed to induce IgE formation against molecules, which would be degraded under normal conditions, and was additionally demonstrated to influence existing food allergies (Untersmayr et al. 2003, Untersmayr et al. 2005a, Untersmayr et al. 2005b, Untersmayr et al. 2007, Pali-Schöll et al. 2010, Pali-Schöll and Jensen-Jarolim 2011).

9.3 THERMAL FOOD PROCESSING

Modern thermal food processing includes industrial food preservation by pasteurization and sterilization but also domestic cooking methods such as boiling, steaming, baking, frying, stewing or roasting (Jay et al. 2005). Thermal treatment does not only influence protein digestibility but also improves texture and flavor (Davis and Williams 1998) and efficiently reduces allergenicity in a large number of food proteins. However, heating might also be associated with creation of so-called neo-antigens or protein aggregates harboring potentially increased immunogenicity (Davis et al. 2001).

Depending on the extent and duration of heat application, proteins undergo substantial physical and chemical changes (Davis et al. 2001, Mills et al. 2009), which might influence protein recognition by the immune system. Especially the Maillard reaction has been extensively studied. Here, AAs and reducing sugars form products with various types of glycation. In case of dietary proteins, the ε-amino group of lysines interacts with reducing carbohydrates (glucose, lactose, maltose, maltodextrin, etc.) leading to so-called Amadori products (aminoketoses), 1,2-dicarbonyls or advanced glycation endproducts (AGEs) in a series of sequential and parallel reactions during prolonged heating or storage (Henle 2005, Poulsen et al. 2013). Well-characterized AGEs formed during glycation reactions are, e.g., Nε-carboxyethyllysine, Nε-carboxymethyllysine (CML) or pyrraline (Poulsen et al. 2013). High temperature processing above 120°C such as cooking, frying, roasting and baking of carbohydrate-rich foods may also lead to formation of the carcinogen acrylamide as a byproduct of the Maillard reaction (Tareke et al. 2002, Xu et al. 2014).

Since the Maillard reaction is successfully exploited for generation of brown coloration as well as desired flavors by the food industry (Cerny 2008, van Boekel et al. 2010), its possible influence on the allergenicity and immunogenicity of food proteins has become an important research focus. Six receptors have been identified to bind and internalize molecules with AGE modifications, being expressed on mononuclear phagocytes, endothelial cells and other cell types (Hilmenyuk et al. 2010, Ilchmann et al. 2010): Receptor for AGE (RAGE) (Neeper et al. 1992, Schmidt et al. 1992), galectin-3 (Vlassara et al. 1995), macrophage scavenger receptor class A type I and II (SR-AI/II) (Suzuki et al. 1997), scavenger receptor class B (Ohgami et al. 2001), and CD36 (Ohgami et al. 2001). However, exact binding motifs and consequences of internalization remain to be elucidated.

While deficiency of SR-AI/II resulted in significantly reduced uptake of AGE modified model proteins and decreased T cell proliferation demonstrating enhanced T cell immunogenicity of Maillard reaction products via SR-AI/II (Ilchmann et al. 2010), Hilmenyuk et al. claimed that AGE-OVA is not able to induce enhanced T cell proliferation and uptake would also be mediated by RAGE inducing activation of transcription factor NF-κB (Hilmenyuk et al. 2010). However, uptake of AGE modified OVA by DCs was significantly higher than uptake of untreated OVA despite natural carbohydrate residues, and can induce a T helper (Th) 2 biased milieu (Hilmenyuk et al. 2010, Ilchmann et al. 2010) highlighting the enhanced immunogenicity of glycated proteins.

In contrast to unintentional modifications during food processing and storage, intended modifications and processing techniques could improve food quality. Food preparation can control solubility and digestibility of proteins. Generally, amino, carboxyl, disulfide, guanidine, imidazole, indole, phenolic, sulfhydryl or thioether side chains can be modified to induce changes of physicochemical characteristics to optimize texture or to change functional properties such as improved foaming or whipping capabilities, enhanced digestibility of legumes or increased stability of milk powder products (Feeney 1977).

9.4 SPECIFIC INFLUENCE OF FOOD PROCESSING METHODS ON ALLERGENIC FOOD COMPOUNDS

9.4.1 Peanut and Tree Nuts

The consumption of peanuts and tree nuts (almonds, walnuts, pecans, cashews, pistachios, hazelnuts, Brazil nuts, macadamia nuts, pine nuts, chestnuts, black walnuts, and coconuts) has increased over the last decades (Masthoff et al. 2013) and the frequency of allergic reactions is rising (Brough et al. 2015). Especially the effects of different processing methods have been studied, since peanuts and tree nuts are usually ingested after boiling/blanching, frying, and roasting or in baked products such as chocolate, cakes, cookies, peanut/almond butter and other processed foods (Teuber et al. 2003, Masthoff et al. 2013). Relative nutritional composition varies considerably between particular nut seeds (Table 9.1) (Venkatachalam and Sathe 2006) influencing food matrix composition and food processing (Nowak-Wegrzyn and Fiocchi 2009).

However, particularly roasting of peanuts is suspected to contribute to the increase of peanut allergy in western societies. In East Asia, where peanuts are mainly consumed after boiling or frying, the proportion of peanut allergic individuals is much lower suggesting an influence of peanut preparation on allergenicity (Beyer et al. 2001).

Several studies investigated the structural properties of roasted peanut extracts and purified *Arachis hypogaea* 1 (Ara h1) and Ara h2, two major peanut allergens. When heated with various sugars,

Table 9.1 Variation of nutritional components in different tree nuts.

Lipid	Protein	Moisture	Soluble sugars	Ash
42.88–62.71%	7.5–21.56%	1.47–9.51%	0.55–3.96%	1.16–3.28%

Table adapted from Venkatachalam and Sathe (Venkatachalam and Sathe 2006)

Table 9.2 Impact of food processing on different allergens.

Modification	Food	Allergen	Influence on allergenicity and other properties	References
Heating	apple	Mal d 1	activation of Bet v 1 specific T cells,	Bohle et al. 2006
	carrot	Dau c 1	temperature dependent reduction of mediator	
	celery	Api g 1	release	
	milk	BLG	aggregation during pasteurization increased uptake through Peyer's patches	Roth-Walter et al. 2008
Maillard reaction	milk	BLG	masking of IgE epitopes with arabinose or ribose	Taheri-Kafrani et al. 2009
	peanut	Ara h1	formation of trimers higher IgE binding capacity	Maleki et al. 2000 and Mondoulet et al. 2005
		Ara h2	increased trypsin inhibitor function higher IgE binding capacity	Maleki et al. 2000 and Mondoulet et al. 2005
Roasting	hazelnut	extract	reduced activation of basophils	Worm et al. 2009
			reduced response in skin prick tests	Worm et al. 2009
			reduced symptoms after oral provocation	Worm et al. 2009 and Hansen et al. 2003
	peanut	extract	increased binding capacity to IgE	Maleki et al. 2000 and Mondoulet et al. 2005
			highest cross-linking capacity in cell assays	Kroghsbo et al. 2014
			increases specific IgG1 and IgE	Moghaddam et al. 2014
Nitration	egg	OVA	increased mediator release in cell assays reduced stability to gastrointestinal enzymes	Gruijthuijsen et al. 2006
	milk	BLG	increased dimerization altered secondary structure enhanced anaphylactic potential	Diesner et al. 2015
Reduction	egg	OVM	exposes sequential epitopes increased reactivity in skin prick tests	Roth-Walter et al. 2013

Ara h1 was revealed to form higher order structures by covalent crosslinking with a molecular weight corresponding to a trimer. In contrast, this is not observed for Ara h2 where sugar crosslinks did not result in the formation of ordered structures (Maleki et al. 2000, Mondoulet et al. 2005) (Table 9.2). Extracts of roasted peanuts (PE) were reported to show a significantly higher binding of serum IgE from peanut allergic patients in competitive ELISA than raw PE (Maleki et al. 2000, Mondoulet et al. 2005). A correlation between the level of CML modifications and the increase in IgE binding was demonstrated (Maleki and Hurlburt 2004).

Preferential binding of IgE to glycated proteins might also be explained by dietary habits, as consumption of raw peanuts is rather unusual in western countries. However, experiments with RBL cells passively sensitized with serum IgE from differently immunized rats (roasted PE, blanched PE, peanut butter extract) revealed significantly higher mediator release after stimulation with roasted PE compared to blanched PE or peanut butter extract. Even though serum IgE produced in response to roasted products induced lower levels of mediator release roasted PE had highest cross-linking capacity in these cell assays (Kroghsbo et al. 2014) (Table 9.2).

In line, also dry-roasted (dR) PE was suggested to have enhanced allergenic properties compared to raw PE indicated by significantly higher titers of peanut specific IgG1 and IgE in mice after oral sensitization. Furthermore, subcutaneously primed mice were orally exposed to raw or dR peanut homogenates or raw peanut kernels. Feeding of dR homogenate to primed mice induced significantly elevated titers of specific IgG1 and IgE and robust proliferation of mesenteric lymph node cells (Moghaddam et al. 2014) (Table 9.2).

Interestingly, factors contributing to enhanced allergenicity might be the formation of neo-antigens due to glycation (Moghaddam et al. 2014) and enhanced stability to gastric enzymes (Maleki et al. 2003). Moreover, the trypsin inhibitor function of Ara h2, protecting Ara h1 from degradation in *in vitro* assays, was shown to be increased by roasting (Maleki et al. 2003).

In contrast to peanut, roasting seems to decrease allergenicity of hazelnut (HN). Basophil activation testing indicated a reduced activation capacity accompanied by a dramatic reduction of positive response in skin prick testing compared to native HN extract (Worm et al. 2009). Likewise, the frequency of symptoms to roasted HN extract was lower after oral provocation of HN allergic patients (Hansen et al. 2003, Worm et al. 2009) (Table 9.2).

9.4.2 Milk

Proteins of cow's milk are among the most important food allergens eliciting food adverse reactions in Europe, particularly in children.

As BLG is the most abundant protein in whey and has important nutritional and functional properties, the chemical basis for aggregation has been thoroughly investigated. Upon heating, dissociation of native dimers precedes the partial unfolding of BLG monomers. The resulting reactive monomers can form heat-induced dimers and small oligomers by thiol-disulphide exchange and to a lesser extent via thiol-thiol oxidation and non-covalent interactions characterizing the early stages of aggregation (Schokker et al. 1999). It is hypothesized that accumulation of an aggregation nucleus provides the basis for the formation of larger aggregates. However, the mechanisms of aggregation during heating of BLG are strongly dependent on temperature, salt content, protein concentration, pH and was shown to differ between two protein variants of BLG (Bauer et al. 2000).

Experimental protein models were also used to investigate the impact of glycation of milk whey proteins using different carbohydrates. The reaction of lactose with lysine residues results among other AGEs and Amadori products in the formation of lactulosyllysine, the predominant modification upon thermal treatment of milk proteins (Meltretter et al. 2013). Heating of lactose-free milk is likely to result in the formation of AGEs and Amadori products being different from those arising from the reaction with lactose (Chevalier et al. 2001), which may be of importance considering changing consumption habits regarding lactose-free milk.

Glycation upon heating has also been shown to influence aggregation of BLG by the formation of covalent sugar crosslinks in addition to the observed disulfide bonds and hydrophobic interactions (Chevalier et al. 2002). Taheri-Kafrani et al. investigated the binding capacity of BLG specific IgE from milk allergic patients to glycated BLG. Low or moderate glycation was comparable to the effect of heated BLG (72 h at 60°C) on IgE recognition and was associated with slightly decreased binding compared to native BLG. Arabinose and ribose could effectively mask IgE epitopes leading to significantly decreased binding in ELISA (Taheri-Kafrani et al. 2009) (Table 9.2).

Roth-Walter et al. reported that aggregated BLG had a decreased anaphylactic potential in a food allergy mouse model. Aggregation of BLG was shown to prevent transcytosis through the epithelial barrier reducing the risk of anaphylactic reactions in allergic mice compared to native soluble whey proteins that can be easily transported through the epithelium. In contrary, initial sensitization to milk whey proteins is promoted by aggregation as uptake of accumulated antigens takes place at Peyer's patches increasing the immunogenicity of pasteurized milk (Roth-Walter et al. 2008). Similar results concerning sensitization potential, uptake and anaphylactic reactions was obtained for enzymatically cross-linked BLG forming high molecular weight structures. In addition, cross-linked BLG was shown to increase gastric stability and to alter antigen uptake by DCs with differences in peptide profile upon endolysosomal degradation compared to untreated BLG (Stojadinovic et al. 2014).

9.4.3 Pollen Cross-reactive Food Allergens

Food proteins sharing high sequence and structural similarities with pollen allergens can elicit hypersensitivity reactions upon ingestion including the so-called pollen associated food allergy. About 60% of food allergies in adolescents and adults are linked to pollen allergies (Werfel et al. 2015). Symptoms usually appear in the oropharynx as local itching, swelling and tingling termed as oral allergy syndrome (Amlot et al. 1987). More severe or systemic reactions have been reported especially after consumption of celery or soybean (Ballmer-Weber et al. 2002, Kleine-Tebbe et al. 2002).

It is generally accepted that thermal treatment of many pollen cross-reactive food allergens induces irreversible denaturation of protein structure. The loss of conformation results in abrogated IgE binding and prevents immediate food adverse reactions (Bohle 2007). Accordingly, processing generally reduces the ability to trigger allergic reactions; however, thermostability is different comparing various Bet v 1 homologues (Mills et al. 2009). Reactions to hazelnut or celery have been observed in a considerable proportion of allergic individuals even after thermal processing (Ballmer-Weber et al. 2002, Hansen et al. 2003).

Although it has been found that heated allergens from apple and carrot were not able to induce immediate allergic reactions, they still activate birch-pollen specific T cells leading to T cell mediated symptoms (Bohle et al. 2006). In line, simulated gastric digestion leads to rapid degradation of hazelnut, celery and apple Bet v 1 homologues preventing specific IgE binding and mediator release in *in vitro* experiments. However, these food allergen digests were able to induce proliferation of PBMCs and were shown to activate Bet v 1 specific T cells of birch pollen allergic patients (Schimek et al. 2005).

9.5 CHEMICAL FOOD MODIFICATION: NITRATION OF DIETARY PROTEINS

Food production is a long procedure from harvesting, processing to preservation and packaging. Therefore, numerous factors substantially influence final food properties and quality. In the context of allergenicity, the influence of food processing on protein nitration has barely been investigated, although nitration of proteins has been revealed to influence sensitization and food allergic reactions (Gruijthuijsen et al. 2006, Untersmayr et al. 2010).

In comparison to other posttranslational protein modifications, nitration is not enzymatically mediated but a chemical reaction

(Bottari 2015). Protein nitration leads to 3-nitrotyrosine (3-NT) formation by addition of a nitro group (NO_2) to the aromatic ring of a tyrosine residue. Two predominant mechanisms have been described requiring the formation of an aromatic radical that can either rapidly combine with NO_2 or react with nitric oxide (NO) to 3-nitrosotyrosine, which is oxidized by two electrons to 3-NT (Ischiropoulos 2009) (Figure 9.1).

Several publications suggested traffic-related air pollution nitrating molecules of primary biological aerosol particles, e.g., pollens. The reaction is mediated by ozone and NO_2 (Franze et al. 2005, Shiraiwa et al. 2011) and, theoretically, could play a role for an even broader range of biological molecules in plants and their fruits. For this reason, food itself might already contain nitrated tyrosine residues. Furthermore, it is likely that meat products contain basal levels of 3-NT arising from (patho-) physiological nitrosative stress in animals. Apart from this, muscle proteins are sensitive to oxidative and nitrating events occurring after slaughtering, during food processing and storage (Lund et al. 2011). Indeed, some publications demonstrated presence of 3-NT in proteins of meat products (Stagsted et al. 2004, Villaverde et al. 2014, Villaverde et al. 2014, Vossen and De Smet 2015). Villaverde and coworkers examined the effect of curing agents added to meat products, such as nitrite, and their chemical impact on myofibrillar proteins and on fermented sausages. Nitrite increased the degree of nitration of isolated proteins

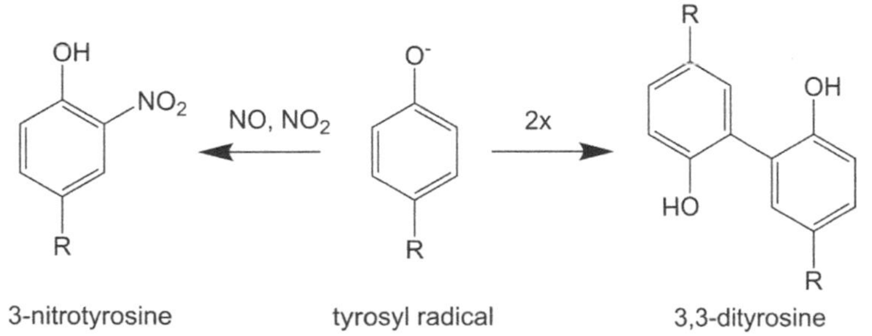

3-nitrotyrosine   tyrosyl radical   3,3-dityrosine

Figure 9.1 Formation of 3-NT. In the presence of radical species, tyrosine residues might get oxidized, nitrated or hydroxylated. Important nitrating agents are NO_2 or alternatively NO. Both can lead to the diffusion-controlled reaction to yield 3-NT. Two tyrosyl radicals may also combine to 3,3-dityrosine competing with the formation of 3-NT (Radi 2004). 3-NT, 3-nitrotyrosine; NO, nitric oxide; NO_2, nitrogen dioxide.

as wells as that of the processed meat product in a concentration dependent manner (Villaverde et al. 2014a, Villaverde et al. 2014b) highlighting the important influence of food additives.

Additional to exogenous nitration also the acidic environment of the stomach has been discussed to support several pathways leading to protein nitration. Recirculation of nitrate and nitrite between gut and saliva might play a major role in the process of nitration. Ingestion of nitrate contained in green leafy vegetables such as spinach or lettuce may even increase salivary levels of nitrite as a considerable proportion of nitrate can be reduced by facultative anaerobe bacteria in the oral cavity (Oldreive and Rice-Evans 2001, Lundberg et al. 2004). Protonation of nitrite in the stomach yields nitrous acid that can be decomposed to NO and NO_2 giving rise to other potent reactive nitrogen species (RNS). High concentrations of NO and O_2, a condition which might exclusively occur in the stomach, enable NO auto-oxidation to NO_2 (Rocha et al. 2012). Furthermore, dual oxidase 2 expressed in epithelial cells of the gastrointestinal tract (El Hassani et al. 2005) can provide O_2^- for the formation of peroxynitrite. Peroxynitrite exists in equilibrium with its protonated conjugate peroxynitrous acid (ONOOH, pKa = 6.8) at a pH dependent ratio (Radi et al. 2001). Homolytic cleavage of peroxynitrous acid results in formation of the strong nitrating species NO_2 and hydroxyl radicals (Yeo et al. 2008). Even in the gastric lumen, peroxynitrite could react with carbon dioxide that is present in the gastric headspace to form the intermediate nitroso-peroxocarbonate ($ONOOCO_2^-$). $ONOOCO_2^-$ is decomposed to NO_2 and a carbonate radical (CO_3^-) contributing to tyrosine nitration (Radi et al. 2001, Rocha et al. 2012). As a result, physiological conditions in the stomach allow the presence of both, oxidants and nitrating species, which might interact with proteins to form 3-NT.

9.6 NITRATION AS A CONCERN IN FOOD ALLERGY

Antigens containing modified tyrosine residues are known to change recognition by the immune system which is associated with an altered immune response (Birnboim et al. 2003). Depending on immunization routes, nitrated allergens showed an increased allergenic potential in murine models of allergy by influencing specific IgE, IgG1 and IgG2a levels and allergy-associated parameters (Untersmayr et al. 2010). Presence of 3-NT could also trigger enhanced mediator release in cell assays after passive sensitization with serum from allergic mice and was a more potent T cell stimulus. Overall, nitration is thought to enhance allergenicity (Gruijthuijsen et al. 2006, Untersmayr et al. 2010).

Recently, the influence of nitration on two important food proteins was investigated *in vivo* using an active anaphylaxis model. To determine the anaphylactic potential of nitrated tyrosine residues, BLG and ovomucoid (OVM) allergic mice were systemically challenged with untreated, sham-nitrated, and nitrated allergen, respectively. Nitrated BLG elicited significantly higher levels of the anaphylaxis marker mMCP-1 and induced a significant drop in core body temperature compared to untreated or sham-nitrated BLG. This effect was attributed to the structural characteristics of BLG favoring nitration of immunologically relevant tyrosine residues, which was not the case for OVM. Additionally, enhanced dimerization of nitrated BLG probably increased the crosslinking capacity (Diesner et al. 2015).

Posttranslational modifications of proteins may affect their steric properties influencing protein conformation as well as susceptibility to proteases with impact on processing by dendritic cells (DCs). Nitration exhibited a considerable enhancement of peptide presentation via HLA-DR. Membrane bound receptors recognizing 3-NT, resistance against endolysosomal proteases and altered secretion profiles of DCs after nitration have been suggested to influence allergenicity (Karle et al. 2012, Ackaert et al. 2014).

9.7 FURTHER CHEMICAL MODIFICATIONS: REDUCTION AND OXIDATION OF FOOD PROTEINS

The fact that proteins are targets of reactive oxygen species (ROS) is an important factor for food preservation (Lund et al. 2011). As a consequence, food additives such as the antioxidant glutathione (GSH) are used to counteract ROS induced damage. Milk, eggs, or fresh fruit and legumes naturally contain high levels of GSH being capable of reducing food proteins. Notably, reduction of OVM was shown to expose sequential epitopes being preferentially recognized by patients with persistent egg allergy (Roth-Walter et al. 2013). On the other hand, oxidative events can induce certain protein modifications such as cleavage of peptide bonds, modification of amino acid side chains and formation of covalent intermolecular cross links which were summarized elsewhere (Soladoye et al. 2015). Oxidation was especially addressed in muscle food and meat products as oxidation has been shown to influence water-holding capacity, tenderness, nutritional quality, digestibility and 3-NT content (Stagsted et al. 2004, Lund et al. 2011, Villaverde et al. 2014a, Villaverde et al. 2014b, Vossen and De Smet 2015). Another critical mechanism of oxidation in foods is believed to be photo-oxidation as many dairy products contain high amounts of photosensitizers such as riboflavin. Together with oxygen and unsaturated fatty acids, tryptophan and tyrosine residues are discussed to function as major targets of photo-oxidation by quenching triplet state riboflavin (Dalsgaard et al. 2011).

As outlined above, protein oxidation represents the first step of nitration by formation of tyrosyl radicals. However, these radicals can also combine to 3,3-dityrosine, one predominant modification upon protein oxidation, resulting in crosslinking of proteins (Radi 2004) (Figure 9.1). The relevance of photo-oxidation in terms of structural properties has been shown for different milk proteins after light exposure. Dityrosine content apparently varied depending on secondary structure revealing highest amounts in the random coil proteins α-casein and β-casein and lowest yields in the globular proteins BLG and lactoferrin. Other changes included partial loss of secondary structure, loss of tryptophan or altered tertiary structure (Dalsgaard et al. 2007).

Considering already discussed data showing important implications after dimerization/aggregation on uptake by epithelial cells (Roth-Walter et al. 2008), anaphylactic reactions (Diesner et al. 2015), and processing by DCs (Stojadinovic et al. 2014), exposure of dairy products to light could influence allergenicity. However, to the best of our knowledge there is currently no data available investigating the impact of oxidation on food allergy.

9.8 CONCLUSIONS

Based on the here reviewed studies, it is obvious that protein modification substantially influences food protein allergenicity. Most data were collected investigating particular proteins after experimental modification. It is well known that usage of model proteins represent an artificial approach with limited comparability if different reaction duration, protein concentration, temperature, or sample composition were used. In addition, effects of food matrices and dietary composition are often neglected in such models.

Modern food processing technologies ensure food safety on microbiological level and definitely improved food preservation. However, limited information regarding the influence on sensitization and allergic reactions is available highlighting the need for better understanding of how such modifications might change interaction with mucosal surfaces, receptors or immune cells. Desirable compound formation exploiting the Maillard reaction as well as unintentional modifications during processing, storage and digestion of food proteins imbedded in complex matrices influence the way food proteins are degraded and presented.

Additionally, the impact of chemical food protein modifications is hardly investigated and seems to be rather allergen specific and no general mechanism has been identified to date. Whereas roasting or glycation of peanut allergens result in enhanced IgE binding and increased allergenicity, the same procedures reduce immunological reactivity of hazelnut. Epitopes of the milk protein BLG may be masked by glycation, but increased aggregation facilitates allergic sensitization. In contrast, nitration leads to enhanced dimerization of BLG and might be associated more severe anaphylactic reactions in egg and milk allergy.

Therefore profound understanding of the impact of processing on a molecular level is essential and will improve food safety and adequate advice for food allergic individuals in the future.

ACKNOWLEDGMENTS

This work was supported by grants KLI284-B00 and WKP39 of the Austrian Science fund FWF.

Keywords: Allergen modification; food processing and allergy; nitration of allergens; Maillard reaction; food allergy

REFERENCES

Ackaert, C., S. Kofler, J. Horejs-Hoeck, N. Zulehner, C. Asam, S. von Grafenstein, J. E. Fuchs, P. Briza, K. R. Liedl, B. Bohle, F. Ferreira, H. Brandstetter, G. J. Oostingh and A. Duschl (2014). The impact of nitration on the structure and immunogenicity of the major birch pollen allergen Bet v 1.0101. PloS One. 9(8): e104520.

Amlot, P. L., D. M. Kemeny, C. Zachary, P. Parkes and M. H. Lessof (1987). Oral allergy syndrome (OAS): symptoms of IgE-mediated hypersensitivity to foods. Clinical Allergy. 17(1): 33–42.

Astwood, J. D., J. N. Leach and R. L. Fuchs (1996). Stability of food allergens to digestion *in vitro*. Nature Biotechnology. 14(10): 1269–1273.

Ballmer-Weber, B. K., A. Hoffmann, B. Wuthrich, D. Luttkopf, C. Pompei, A. Wangorsch, M. Kastner and S. Vieths (2002). Influence of food processing on the allergenicity of celery: DBPCFC with celery spice and cooked celery in patients with celery allergy. Allergy. 57(3): 228–235.

Bauer, R., R. Carrotta, C. Rischel and L. Øgendal (2000). Characterization and isolation of intermediates in β-Lactoglobulin heat aggregation at high pH. Biophysical Journal. 79(2): 1030–1038.

Bergmann, K. C. (2014). Milestones in the 20th century. History of Allergy. K. C. Bergmann and J. Ring. Basel, Karger. 100: 27–45.

Berin, M. C. and H. A. Sampson (2013). Mucosal immunology of food allergy. Current Biology. 23(9): R389–400.

Beyer, K., E. Morrow, X. M. Li, L. Bardina, G. A. Bannon, A. W. Burks and H. A. Sampson (2001). Effects of cooking methods on peanut allergenicity. Journal of Allergy and Clinical Immunology. 107(6): 1077–1081.

Birnboim, H. C., A. M. Lemay, D. K. Lam, R. Goldstein and J. R. Webb (2003). Cutting edge: MHC class II-restricted peptides containing the inflammation-associated marker 3-nitrotyrosine evade central tolerance and elicit a robust cell-mediated immune response. Journal of Immunology. 171(2): 528–532.

Bohle, B. (2007). The impact of pollen-related food allergens on pollen allergy. Allergy. 62(1): 3–10.

Bohle, B., B. Zwolfer, A. Heratizadeh, B. Jahn-Schmid, Y. D. Antonia, M. Alter, W. Keller, L. Zuidmeer, R. van Ree, T. Werfel and C. Ebner (2006). Cooking birch pollen-related food: divergent consequences for IgE- and T cell-mediated reactivity *in vitro* and *in vivo*. Journal of Allergy and Clinical Immunology. 118(1): 242–249.

Bottari, S. P. (2015). Protein tyrosine nitration: A signaling mechanism conserved from yeast to man. Proteomics. 15(2-3): 185–187.

Brough, H. A., P. J. Turner, T. Wright, A. T. Fox, S. L. Taylor, J. O. Warner and G. Lack (2015). Dietary management of peanut and tree nut allergy: what exactly should patients avoid? Clinical and Experimental Allergy. 45(5): 859–871.

Bu, G., Y. Luo, F. Chen, K. Liu and T. Zhu (2013). Milk processing as a tool to reduce cow's milk allergenicity: a mini-review. Dairy Sci Technol. 93(3): 211–223.

Cerny, C. (2008). The aroma side of the Maillard reaction. Annals of the New York Academy of Sciences. 1126: 66–71.

Chafen, J. J., S. J. Newberry, M. A. Riedl, D. M. Bravata, M. Maglione, M. J. Suttorp, V. Sundaram, N. M. Paige, A. Towfigh, B. J. Hulley and P. G. Shekelle (2010). Diagnosing and managing common food allergies: a systematic review. JAMA. 303(18): 1848–1856.

Chevalier, F., J.-M. Chobert, D. Molle, T. Haertlé. (2001). Maillard glycation of β-lactoglobulin with several sugars: comparative study of the properties of the obtained polymers and of the substituted sites. Le Lait, INRA Editions. 81(5): 651–666.

Chevalier, F., J. M. Chobert, M. Dalgalarrondo, Y. Choiset and T. Haertlé (2002). Maillard glycation of β-lactoglobulin induces conformation changes. Food/Nahrung. 46(2): 58–63.

Dalsgaard, T. K., J. H. Nielsen, B. E. Brown, N. Stadler and M. J. Davies (2011). Dityrosine, 3,4-dihydroxyphenylalanine (DOPA), and radical formation from tyrosine residues on milk proteins with globular and flexible structures as a result of riboflavin-mediated photo-oxidation. Journal of Agricultural and Food Chemistry. 59(14): 7939–7947.

Dalsgaard, T. K., D. Otzen, J. H. Nielsen and L. B. Larsen (2007). Changes in structures of milk proteins upon photo-oxidation. Journal of Agricultural and Food Chemistry. 55(26): 10968–10976.

Davis, P. J., C. M. Smales and D. C. James (2001). How can thermal processing modify the antigenicity of proteins? Allergy. 56 Suppl. 67: 56–60.

Davis, P. J. and S. C. Williams (1998). Protein modification by thermal processing. Allergy. 53(46 Suppl.): 102–105.

Diesner, S. C., C. Schultz, C. Ackaert, G. J. Oostingh, A. Ondracek, C. Stremnitzer, J. Singer, D. Heiden, F. Roth-Walter, J. Fazekas, V. E. Assmann, E. Jensen-Jarolim,

H. Stutz, A. Duschl and E. Untersmayr (2015). Nitration of beta-Lactoglobulin but not of ovomucoid enhances anaphylactic responses in food allergic mice. PloS One. 10(5): e0126279.

El Hassani, R. A., N. Benfares, B. Caillou, M. Talbot, J. C. Sabourin, V. Belotte, S. Morand, S. Gnidehou, D. Agnandji, R. Ohayon, J. Kaniewski, M. S. Noel-Hudson, J. M. Bidart, M. Schlumberger, A. Virion and C. Dupuy (2005). Dual oxidase2 is expressed all along the digestive tract. American Journal of Physiology: Gastrointestinal and Liver Physiology. 288(5): G933–942.

Feeney, R. E. (1977). Chemical modification of food proteins. Food Proteins, American Chemical Society. 160: 3–36.

Franze, T., M. G. Weller, R. Niessner and U. Pöschl (2005). Protein nitration by polluted air. Environmental Science & Technology. 39(6): 1673–1678.

Fu, T. -J. (2002). Digestion stability as a criterion for protein allergenicity assessment. Annals of the New York Academy of Sciences. 964(1): 99–110.

Fu, T. J., U. R. Abbott and C. Hatzos (2002). Digestibility of food allergens and nonallergenic proteins in simulated gastric fluid and simulated intestinal fluid-a comparative study. Journal of Agricultural and Food Chemistry. 50(24): 7154–7160.

Grimshaw, K. E., R. M. King, J. A. Nordlee, S. L. Hefle, J. O. Warner and J. O. Hourihane (2003). Presentation of allergen in different food preparations affects the nature of the allergic reaction—a case series. Clinical and Experimental Allergy. 33(11): 1581–1585.

Gruijthuijsen, Y. K., I. Grieshuber, A. Stöcklinger, U. Tischler, T. Fehrenbach, M. G. Weller, L. Vogel, S. Vieths, U. Pöschl and A. Duschl (2006). Nitration enhances the allergenic potential of proteins. International Archives of Allergy and Immunology. 141(3): 265–275.

Hansen, K. S., B. K. Ballmer-Weber, D. Luttkopf, P. S. Skov, B. Wuthrich, C. Bindslev-Jensen, S. Vieths and L. K. Poulsen (2003). Roasted hazelnuts—allergenic activity evaluated by double-blind, placebo-controlled food challenge. Allergy. 58(2): 132–138.

Henle, T. (2005). Protein-bound advanced glycation endproducts (AGEs) as bioactive amino acid derivatives in foods. Amino Acids. 29(4): 313–322.

Hilmenyuk, T., I. Bellinghausen, B. Heydenreich, A. Ilchmann, M. Toda, S. Grabbe and J. Saloga (2010). Effects of glycation of the model food allergen ovalbumin on antigen uptake and presentation by human dendritic cells. Immunology. 129(3): 437–445.

Ilchmann, A., S. Burgdorf, S. Scheurer, Z. Waibler, R. Nagai, A. Wellner, Y. Yamamoto, H. Yamamoto, T. Henle, C. Kurts, U. Kalinke, S. Vieths and M. Toda (2010). Glycation of a food allergen by the Maillard reaction enhances its T-cell immunogenicity: Role of macrophage scavenger receptor class A type I and II. Journal of Allergy and Clinical Immunology. 125(1): 175–183. e171-111.

Ischiropoulos, H. (2009). Protein tyrosine nitration—An update. Archives of Biochemistry and Biophysics. 484(2): 117–121.

Jay, J. M., M. J. Loessner and D. A. Golden (2005). Modern Food Microbiology. New York, Springer Science & Business Media, Inc. p.

Karle, A. C., G. J. Oostingh, S. Mutschlechner, F. Ferreira, P. Lackner, B. Bohle, G. F. Fischer, A. B. Vogt and A. Duschl (2012). Nitration of the pollen allergen Bet v 1.0101 enhances the presentation of Bet v 1-derived peptides by HLA-DR on human dendritic cells. PloS One. 7(2): e31483.

Kleine-Tebbe, J., L. Vogel, D. N. Crowell, U. F. Haustein and S. Vieths (2002). Severe oral allergy syndrome and anaphylactic reactions caused by a Bet v 1- related PR-10 protein in soybean, SAM22. Journal of Allergy and Clinical Immunology. 110(5): 797–804.

Kroghsbo, S., N. M. Rigby, P. E. Johnson, K. Adel-Patient, K. L. Bogh, L. J. Salt, E. N. Mills and C. B. Madsen (2014). Assessment of the sensitizing potential of processed peanut proteins in Brown Norway rats: roasting does not enhance allergenicity. PloS One. 9(5): e96475.

Lund, M. N., M. Heinonen, C. P. Baron and M. Estevez (2011). Protein oxidation in muscle foods: A review. Molecular Nutrition & Food Research. 55(1): 83–95.

Lundberg, J. O., E. Weitzberg, J. A. Cole and N. Benjamin (2004). Nitrate, bacteria and human health. Nature Reviews: Microbiology. 2(7): 593–602.

Maleki, S. J., S. -Y. Chung, E. T. Champagne and J. -P. Raufman (2000). The effects of roasting on the allergenic properties of peanut proteins. Journal of Allergy and Clinical Immunology. 106(4): 763–768.

Maleki, S. J. and B. K. Hurlburt (2004). Structural and functional alterations in major peanut allergens caused by thermal processing. Journal of AOAC International. 87(6): 1475–1479.

Maleki, S. J., O. Viquez, T. Jacks, H. Dodo, E. T. Champagne, S. -Y. Chung and S. J. Landry (2003). The major peanut allergen, Ara h 2, functions as a trypsin inhibitor, and roasting enhances this function. Journal of Allergy and Clinical Immunology. 112(1): 190–195.

Masthoff, L. J., R. Hoff, K. C. Verhoeckx, H. van Os-Medendorp, A. Michelsen-Huisman, J. L. Baumert, S. G. Pasmans, Y. Meijer and A. C. Knulst (2013). A systematic review of the effect of thermal processing on the allergenicity of tree nuts. Allergy. 68(8): 983–993.

McClain, S., C. Bowman, M. Fernandez-Rivas, G. S. Ladics and R. Ree (2014). Allergic sensitization: food- and protein-related factors. Clin Transl Allergy. 4(1): 11.

Meltretter, J., J. Wust and M. Pischetsrieder (2013). Comprehensive analysis of nonenzymatic post-translational beta-lactoglobulin modifications in processed milk by ultrahigh-performance liquid chromatography-tandem mass spectrometry. Journal of Agricultural and Food Chemistry. 61(28): 6971–6981.

Mills, E. N. C., A. I. Sancho, N. M. Rigby, J. A. Jenkins and A. R. Mackie (2009). Impact of food processing on the structural and allergenic properties of food allergens. Molecular Nutrition & Food Research. 53(8): 963–969.

Moghaddam, A. E., W. R. Hillson, M. Noti, K. H. Gartlan, S. Johnson, B. Thomas, D. Artis and Q. J. Sattentau (2014). Dry roasting enhances peanut-induced allergic sensitization across mucosal and cutaneous routes in mice. Journal of Allergy and Clinical Immunology. 134(6): 1453–1456.

Mondoulet, L., E. Paty, M. F. Drumare, S. Ah-Leung, P. Scheinmann, R. M. Willemot, J. M. Wal and H. Bernard (2005). Influence of thermal processing on the allergenicity of peanut proteins. Journal of Agricultural and Food Chemistry. 53(11): 4547–4553.

Neeper, M., A. M. Schmidt, J. Brett, S. D. Yan, F. Wang, Y. C. Pan, K. Elliston, D. Stern and A. Shaw (1992). Cloning and expression of a cell surface receptor for advanced glycosylation end products of proteins. Journal of Biological Chemistry. 267(21): 14998–15004.

Noval Rivas, M., O. T. Burton, P. Wise, Y. Q. Zhang, S. A. Hobson, M. Garcia Lloret, C. Chehoud, J. Kuczynski, T. DeSantis, J. Warrington, E. R. Hyde, J. F. Petrosino, G. K. Gerber, L. Bry, H. C. Oettgen, S. K. Mazmanian and T. A. Chatila (2013). A microbiota signature associated with experimental food allergy promotes allergic sensitization and anaphylaxis. Journal of Allergy and Clinical Immunology. 131(1): 201–212.

Nowak-Wegrzyn, A. and A. Fiocchi (2009). Rare, medium, or well done? The effect of heating and food matrix on food protein allergenicity. Current Opinion in Allergy and Clinical Immunology. 9(3): 234–237.

Nwaru, B. I., L. Hickstein, S. S. Panesar, G. Roberts, A. Muraro and A. Sheikh (2014). Prevalence of common food allergies in Europe: a systematic review and meta-analysis. Allergy. 69(8): 992–1007.

Ohgami, N., R. Nagai, M. Ikemoto, H. Arai, A. Kuniyasu, S. Horiuchi and H. Nakayama (2001). CD36, a member of class B scavenger receptor family, is a receptor for advanced glycation end products. Annals of the New York Academy of Sciences. 947: 350–355.

Ohgami, N., R. Nagai, A. Miyazaki, M. Ikemoto, H. Arai, S. Horiuchi and H. Nakayama (2001). Scavenger receptor class B type I-mediated reverse cholesterol transport is inhibited by advanced glycation end products. Journal of Biological Chemistry. 276(16): 13348–13355.

Oldreive, C. and C. Rice-Evans (2001). The mechanisms for nitration and nitrotyrosine formation *in vitro* and *in vivo*: Impact of diet. Free Radical Research. 35(3): 215–231.

Pali-Schöll, I., R. Herzog, J. Wallmann, K. Szalai, R. Brunner, A. Lukschal, P. Karagiannis, S. C. Diesner and E. Jensen-Jarolim (2010). Antacids and dietary supplements with an influence on the gastric pH increase the risk for food sensitization. Clinical and Experimental Allergy. 40(7): 1091–1098.

Pali-Schöll, I. and E. Jensen-Jarolim (2011). Anti-acid medication as a risk factor for food allergy. Allergy. 66(4): 469–477.

Poulsen, M. W., R. V. Hedegaard, J. M. Andersen, B. de Courten, S. Bugel, J. Nielsen, L. H. Skibsted and L. O. Dragsted (2013). Advanced glycation endproducts in food and their effects on health. Food and Chemical Toxicology. 60: 10–37.

Prescott, S. and K. J. Allen (2011). Food allergy: riding the second wave of the allergy epidemic. Pediatric Allergy and Immunology. 22(2): 155–160.

Radi, R. (2004). Nitric oxide, oxidants, and protein tyrosine nitration. Proceedings of the National Academy of Sciences. 101(12): 4003–4008.

Radi, R., G. Peluffo, M. N. Alvarez, M. Naviliat and A. Cayota (2001). Unraveling peroxynitrite formation in biological systems. Free Radical Biology and Medicine. 30(5): 463–488.

Rocha, B. S., B. Gago, R. M. Barbosa, J. O. Lundberg, R. Radi and J. Laranjinha (2012). Intragastric nitration by dietary nitrite: Implications for modulation of protein and lipid signaling. Free Radical Biology and Medicine. 52(3): 693–698.

Roth-Walter, F., M. C. Berin, P. Arnaboldi, C. R. Escalante, S. Dahan, J. Rauch, E. Jensen-Jarolim and L. Mayer (2008). Pasteurization of milk proteins promotes allergic sensitization by enhancing uptake through Peyer's patches. Allergy. 63(7): 882–890.

Roth-Walter, F., L. F. Pacios, C. Gomez-Casado, G. Hofstetter, G. A. Roth, J. Singer, A. Diaz-Perales and E. Jensen-Jarolim (2014). The major cow milk allergen Bos d 5 manipulates T-helper cells depending on its load with siderophore-bound iron. PloS One. 9(8): e104803.

Roth-Walter, F., P. Starkl, T. Zuberbier, K. Hummel, K. Nobauer, E. Razzazi-Fazeli, R. Brunner, I. Pali-Scholl, J. Kinkel, F. Felix, E. Jensen-Jarolim and T. Kinaciyan (2013). Glutathione exposes sequential IgE-epitopes in ovomucoid relevant in persistent egg allergy. Molecular Nutrition & Food Research. 57(3): 536–544.

Schimek, E. M., B. Zwolfer, P. Briza, B. Jahn-Schmid, L. Vogel, S. Vieths, C. Ebner and B. Bohle (2005). Gastrointestinal digestion of Bet v 1-homologous food allergens destroys their mediator-releasing, but not T cell-activating, capacity. Journal of Allergy and Clinical Immunology. 116(6): 1327–1333.

Schmidt, A. M., M. Vianna, M. Gerlach, J. Brett, J. Ryan, J. Kao, C. Esposito, H. Hegarty, W. Hurley, M. Clauss et al. (1992). Isolation and characterization of two binding proteins for advanced glycosylation end products from bovine lung which are present on the endothelial cell surface. Journal of Biological Chemistry. 267(21): 14987–14997.

Schokker, E. P., H. Singh, D. N. Pinder, G. E. Norris and L. K. Creamer (1999). Characterization of intermediates formed during heat-induced aggregation of β-lactoglobulin AB at neutral pH. International Dairy Journal. 9(11): 791–800.

Shiraiwa, M., Y. Sosedova, A. Rouviere, H. Yang, Y. Zhang, J. P. Abbatt, M. Ammann and U. Poschl (2011). The role of long-lived reactive oxygen intermediates in the reaction of ozone with aerosol particles. Nature Chemistry. 3(4): 291–295.

Sicherer, S. H. (2011). Epidemiology of food allergy. Journal of Allergy and Clinical Immunology. 127(3): 594–602.

Sicherer, S. H. and H. A. Sampson (2014). Food allergy: Epidemiology, pathogenesis, diagnosis, and treatment. Journal of Allergy and Clinical Immunology. 133(2): 291–307; quiz 308.

Soladoye, O. P., M. L. Juárez, J. L. Aalhus, P. Shand and M. Estévez (2015). Protein oxidation in processed meat: Mechanisms and potential implications on human health. Comprehensive Reviews in Food Science and Food Safety. 14(2): 106–122.

Stagsted, J., E. Bendixen and H. J. Andersen (2004). Identification of specific oxidatively modified proteins in chicken muscles using a combined immunologic and proteomic approach. Journal of Agricultural and Food Chemistry. 52(12): 3967–3974.

Stefka, A. T., T. Feehley, P. Tripathi, J. Qiu, K. McCoy, S. K. Mazmanian, M. Y. Tjota, G. Y. Seo, S. Cao, B. R. Theriault, D. A. Antonopoulos, L. Zhou, E. B. Chang, Y. X. Fu and C. R. Nagler (2014). Commensal bacteria protect against food allergen sensitization. Proceedings of the National Academy of Sciences of the United States of America. 111(36): 13145–13150.

Stojadinovic, M., R. Pieters, J. Smit and T. C. Velickovic (2014). Cross-linking of beta-lactoglobulin enhances allergic sensitization through changes in cellular uptake and processing. Toxicological Sciences. 140(1): 224–235.

Suzuki, H., Y. Kurihara, M. Takeya, N. Kamada, M. Kataoka, K. Jishage, O. Ueda, H. Sakaguchi, T. Higashi, T. Suzuki, Y. Takashima, Y. Kawabe, O. Cynshi, Y. Wada, M. Honda, H. Kurihara, H. Aburatani, T. Doi, A. Matsumoto, S. Azuma, T. Noda, Y. Toyoda, H. Itakura, Y. Yazaki, T. Kodama et al. (1997). A role for macrophage scavenger receptors in atherosclerosis and susceptibility to infection. Nature. 386(6622): 292–296.

Taheri-Kafrani, A., J. C. Gaudin, H. Rabesona, C. Nioi, D. Agarwal, M. Drouet, J. M. Chobert, A. K. Bordbar and T. Haertle (2009). Effects of heating and glycation of beta-lactoglobulin on its recognition by IgE of sera from cow milk allergy patients. Journal of Agricultural and Food Chemistry. 57(11): 4974–4982.

Tareke, E., P. Rydberg, P. Karlsson, S. Eriksson and M. Tornqvist (2002). Analysis of acrylamide, a carcinogen formed in heated foodstuffs. Journal of Agricultural and Food Chemistry. 50(17): 4998–5006.

Teuber, S. S., S. S. Comstock, S. K. Sathe and K. H. Roux (2003). Tree nut allergy. Current Allergy and Asthma Reports. 3(1): 54–61.

Toda, M., M. Heilmann, A. Ilchmann and S. Vieths (2014). The Maillard reaction and food allergies: is there a link? Clinical Chemistry and Laboratory Medicine. 52(1): 61–67.

Untersmayr, E., N. Bakos, I. Schöll, M. Kundi, F. Roth-Walter, K. Szalai, A. B. Riemer, H. J. Ankersmit, O. Scheiner, G. Boltz-Nitulescu and E. Jensen-Jarolim (2005a). Anti-ulcer drugs promote IgE formation toward dietary antigens in adult patients. FASEB Journal. 19(6): 656–658.

Untersmayr, E., S. C. Diesner, G. J. Oostingh, K. Selzle, T. Pfaller, C. Schultz, Y. Zhang, D. Krishnamurthy, P. Starkl, R. Knittelfelder, E. Förster-Waldl, A. Pollak, O. Scheiner, U. Pöschl, E. Jensen-Jarolim and A. Duschl (2010). Nitration of the egg-allergen ovalbumin enhances protein allergenicity but reduces the risk for oral sensitization in a murine model of food allergy. PloS One. 5(12): e14210.

Untersmayr, E. and E. Jensen-Jarolim (2006). Mechanisms of type I food allergy. Pharmacology and Therapeutics. 112(3): 787–798.

Untersmayr, E. and E. Jensen-Jarolim (2008). The role of protein digestibility and antacids on food allergy outcomes. Journal of Allergy and Clinical Immunology. 121(6): 1301–1308; quiz 1309–1310.

Untersmayr, E., L. K. Poulsen, M. H. Platzer, M. H. Pedersen, G. Boltz-Nitulescu, P. S. Skov and E. Jensen-Jarolim (2005b). The effects of gastric digestion on codfish allergenicity. Journal of Allergy and Clinical Immunology. 115(2): 377–382.

Untersmayr, E., I. Schöll, I. Swoboda, W. J. Beil, E. Förster-Waldl, F. Walter, A. Riemer, G. Kraml, T. Kinaciyan, S. Spitzauer, G. Boltz-Nitulescu, O. Scheiner and E. Jensen-Jarolim (2003). Antacid medication inhibits digestion of dietary proteins and causes food allergy: a fish allergy model in BALB/c mice. Journal of Allergy and Clinical Immunology. 112(3): 616–623.

Untersmayr, E., H. Vestergaard, H. J. Malling, L. B. Jensen, M. H. Platzer, G. Boltz-Nitulescu, O. Scheiner, P. S. Skov, E. Jensen-Jarolim and L. K. Poulsen (2007). Incomplete digestion of codfish represents a risk factor for anaphylaxis in patients with allergy. Journal of Allergy and Clinical Immunology. 119(3): 711–717.

van Boekel, M., V. Fogliano, N. Pellegrini, C. Stanton, G. Scholz, S. Lalljie, V. Somoza, D. Knorr, P. R. Jasti and G. Eisenbrand (2010). A review on the beneficial aspects of food processing. Molecular Nutrition & Food Research. 54(9): 1215–1247.

Venkatachalam, M. and S. K. Sathe (2006). Chemical composition of selected edible nut seeds. Journal of Agricultural and Food Chemistry. 54(13): 4705–4714.

Villaverde, A., D. Morcuende and M. Estevez (2014a). Effect of curing agents on the oxidative and nitrosative damage to meat proteins during processing of fermented sausages. Journal of Food Science. 79(7): C1331–1342.

Villaverde, A., V. Parra and M. Estevez (2014b). Oxidative and nitrosative stress induced in myofibrillar proteins by a hydroxyl-radical-generating system: impact of nitrite and ascorbate. Journal of Agricultural and Food Chemistry. 62(10): 2158–2164.

Visness, C. M., S. J. London, J. L. Daniels, J. S. Kaufman, K. B. Yeatts, A. M. Siega-Riz, A. H. Liu, A. Calatroni and D. C. Zeldin (2009). Association of obesity with IgE levels and allergy symptoms in children and adolescents: Results from the National Health and Nutrition Examination Survey 2005–2006. Journal of Allergy and Clinical Immunology. 123(5): 1163–1169, 1169. e1161-1164.

Vlassara, H., Y. M. Li, F. Imani, D. Wojciechowicz, Z. Yang, F. T. Liu and A. Cerami (1995). Identification of galectin-3 as a high-affinity binding protein for advanced glycation end products (AGE): a new member of the AGE-receptor complex. Molecular Medicine. 1(6): 634–646.

Vossen, E. and S. De Smet (2015). Protein oxidation and protein nitration influenced by sodium nitrite in two different meat model systems. Journal of Agricultural and Food Chemistry. 63(9): 2550–2556.

Werfel, T., R. Asero, B. K. Ballmer-Weber, K. Beyer, E. Enrique, A. C. Knulst, A. Mari, A. Muraro, M. Ollert, L. K. Poulsen, S. Vieths, M. Worm and K. Hoffmann-Sommergruber (2015). Position paper of the EAACI: Food allergy due to immunological cross-reactions with common inhalant allergens. Allergy.

Worm, M., S. Hompes, E. M. Fiedler, A. K. Illner, T. Zuberbier and S. Vieths (2009). Impact of native, heat-processed and encapsulated hazelnuts on the allergic response in hazelnut-allergic patients. Clinical and Experimental Allergy. 39(1): 159–166.

Wüthrich, B. (2014). History of food allergy. History of Allergy. K. C. Bergmann and J. Ring. Basel, Karger. 100: 109–119.

Xu, Y., B. Cui, R. Ran, Y. Liu, H. Chen, G. Kai and J. Shi (2014). Risk assessment, formation, and mitigation of dietary acrylamide: current status and future prospects. Food and Chemical Toxicology. 69: 1–12.

Yeo, W. S., S. J. Lee, J. R. Lee and K. P. Kim (2008). Nitrosative protein tyrosine modifications: biochemistry and functional significance. BMB Rep. 41(3): 194–203.

10

Detection of Food Allergen Residues by Immunoassays and Mass Spectrometry

Sridevi Muralidharan,[1] *Yiqing Zhao,*[1] *Steve L. Taylor*[2] *and Nanju A. Lee*[1,*]

CONTENTS

[1] ARC Training Centre for Advanced Technologies in Food Manufacture and School of Chemical Engineering, University of New South Wales, Sydney, Australia.

[2] Food Allergy Research and Resource Program, Department of Food Science and Technology, University of Nebraska-Lincoln, Nebraska, USA.

* Corresponding author: alice.lee@unsw.edu.au

229

10.1 INTRODUCTION

Food allergy is a type of non-toxic adverse food reaction, involving aberrant immune responses to food proteins that elicit various symptoms. Food allergy has developed into a significant health concern in many western countries. The World Allergy Organization (WAO) estimates that food allergy has touched the lives of approximately 220 to 250 million people, and primarily 5–8% of children of all ages are affected (WAO 2011). In 1995, immunological food allergy and non-immunological intolerance reactions were classified as two separate categories of non-toxic adverse food reactions, by the European Academy of Allergy and Clinical Immunology (EAACI) (Bruijnzeel-Koomen et al. 1995). Food allergies can be extremely severe and potentially fatal to a susceptible population. Traces of offending ingredients in food products can elicit immediate adverse reactions. In Australia, nine foods and derived food ingredients have been recognised to induce the majority of immunoglobulin E (IgE)-mediated food allergies. These are peanut, tree nuts, sesame, egg, cow's milk, soy, fish, shellfish and wheat (ATRS 2014). The true prevalence of food associated allergy in Australia remains unknown, but it is predicted to be 1–2% in the adult population and 4–6% in the paediatric population (FSANZ 2010). A 12-year (1995–2006) retrospective analysis reported that of 1,489 children between 0–5 years old, 47% had a food allergy, and most were sensitised by peanut, egg, cow's milk and cashew nut (Mullins 2007). A parental survey involving 4,173 children in South Australia indicated that 0.6% of children aged 3–17 years had histories of severe anaphylaxis while 7.3% had histories of food allergy (Boros et al. 2000). Food-induced anaphylaxis occurs more frequently in pre-school aged children compared to school-aged children. Food allergy prevalence in infants and pre-school aged children ranged from 1% (in Thailand) to 10% (in Australia), based on published data only available from 16 out of 89 countries across the world (Prescott et al. 2013). Furthermore food allergy prevalence in school aged children varied from 0.3% in Korea to 4.2% in Australia to 7.6% in Tanzania (Prescott et al. 2013).

The major food allergens causing approximately 90% of all food allergies found in a Longitudinal Study of Australian Children (LSAC) are from nine food groups, namely cow's milk, eggs, wheat, peanuts, tree nuts, soybean, sesame, fish and shellfish. Peanuts, tree nuts, wheat, eggs and cow's milk are the most common types of food to induce allergic reactions in earlier life (LSAC 2010), whilst seafood allergy is more frequent in teenage and adult life (ATRS 2014).

In this chapter, we review important factors affecting food allergen detection within a context of food allergen management. We review two important analytical methodologies for food allergen detection: (a) the ELISA immunoassay which has been primarily used as the gold standard technique (b) Mass Spectrometry which has a great potential to be developed into a confirmatory technique. The general principles, operating procedures for each of the two techniques are highlighted and recent applications and research developments relevant to detection of allergen residues in pre-packaged food are also discussed.

10.2 PRECAUTIONARY LABELLING OF FOOD ALLERGENS

Food allergy has increased in prevalence and the prevention of acute potentially fatal allergic reactions relies on the strict avoidance of offending foods. Food labelling, therefore, plays a significant role in improving the safety of food by providing credible and accurate information to food-allergic patients (van Hengel 2007). In 2003, food labelling legislation was implemented in Australia. Most of the common nine groups of allergens require mandatory labelling. Mandatory labelling of allergenic ingredients in pre-packaged foods was intended to help allergic consumers to manage their conditions. However, undeclared allergenic residues potentially presented in pre-packaged food products still remains a critical risk to allergic consumers (van Hengel 2007). In most food processing facilities, various manufactured products both containing and not containing allergenic ingredients are produced on the same production line using shared equipment (Clemente et al. 2004). Various factors including the inadequate cleaning of shared equipment, unsuitable handling of allergenic ingredients, and storage and transportation practices can lead to accidental contamination leaving a potential risk of hidden allergens to allergic consumers (Gaskin and Taylor 2011).

Uncertainty regarding the degree of risk prompts manufacturers to employ additional precautionary labelling (e.g., "may contain") for certain suspicious ingredients thereby communicating potential risk to consumers and enhance the safety and reliability of the food labelling system (Allen et al. 2014b, DunnGalvin et al. 2015). Uncertainty abounds in part because the public health authorities have not established safe limits for allergenic food residues. The overuse of precautionary labelling by food manufacturers without proper (evidenced-based) risk assessment has led clinicians and dieticians to give inconsistent recommendations to their patients (Brough et al. 2015). For the purpose of establishing a reliable labelling system with enhanced safety and credibility, and minimising overuse, a standardised risk management process with appropriate reference doses of allergens, has been developed as part of the Voluntary Incidental Trace Allergen Labelling (VITAL) program, of the Allergen Bureau of Australia & New Zealand (Taylor et al. 2014).

The VITAL procedure encourages manufacturers to intensively investigate the presence of possible allergenic food residues prior to their release to the market. It also factors in possible contamination of allergens among raw materials and equipments used in food processing plants. The VITAL program, in accordance with appropriate reference doses, provides a precautionary statement with rigour by evaluating the possible presence of allergenic residues using an interactive VITAL calculator; a precautionary declaration is recommended only if estimated cross-contact doses are above the reference doses (Taylor et al. 2014, Zurzolo et al. 2013). The reference doses have been developed from the statistical dose-distribution modelling of individual thresholds of patients with oral food challenges and are estimated as 0.2 mg of peanut protein, 0.1 mg of cow's milk, 0.03 mg of egg, 2.0 mg of cashew, 0.1 mg of hazelnut protein, 1.0 mg of wheat, 1.0 mg of soy flour, 0.2 mg of sesame seed, 0.05 mg of mustard, 4.0 mg of lupin and 10 mg of shrimp (Allen et al. 2014a, Taylor et al. 2014). Furthermore, the Allergen Bureau of Australia and New Zealand has set the action level at 0.1 mg protein for all tree nuts.

The identification of trace food allergens by reliable detection and quantification methods on processing equipment and finished food products is necessary to ensure quality management compliance with food labelling legislations as well as consumer safety (Taylor et al. 2007). To address this, immunoassays and mass spectrometry based methods have been developed for food allergen detection. The following sections in this chapter will focus on the working principles and applications of immunoassays and mass spectrometry in food allergen detection as well as address advantages and limitations of these two detection methods.

10.3 IMMUNOASSAYS

To detect and quantify undeclared food allergens, immunoassays offer adequate sensitivity and specificity, and have been commonly used in food industry (Wen et al. 2007). The fundamental basis of immunoassays is the inherent ability of an antibody to specifically bind to its antigen. Nearly all food allergens are proteins or glycoproteins and are sufficiently immunogenic to generate specific antibodies. Those highly specific antibodies that only bind to target molecules are ideal assay reagents for allergen detection. Both human serum and animal antisera make a contribution to the research and development of validated immunoassays for various intended purposes. Only animal antisera are used in immunoassays for allergen residue detection.

Sodium Dodecyl Sulfate Polyacrylamide Gel Electrophoresis (SDS-PAGE) and immunoblotting are also used for the identification and characterisation of individual allergens from foods, and have been used to confirm the presence of allergen residues in foods. Extracts of food proteins including allergens are separated by SDS-PAGE according to their molecular weight (one dimensional separation) or molecular weight and isoelectric points (two dimensional separation), followed by transfer to a polyvinylidene difluoride or nitrocellulose membrane. Enzyme-labelled antibodies are subsequently added for allergen identification. This technique was reported to successfully detect hazelnut and almond residues in complex food matrices such as chocolate (Kirsch et al. 2009).

10.3.1 Enzyme-linked Immunosorbent Assay (ELISA)

Enzyme-linked immunosorbent assay (ELISA) is the most commonly used method in routine food allergen detection. This method targets proteins from an allergenic source, which makes it ideal in validating the removal of allergenic proteins from shared equipment and confirm that pre-packed food products do not contain hidden allergen residue. An ELISA provides high sensitivity with a detection range at milligrams (mg) per kilogram (ppm: part per million). It is also suitable for rapid detection in a food processing facility (Baumert

2013). An ELISA involves the use of an antisera containing protein-specific IgG antibodies generated from animals to non-covalently bind to antigens extracted from foods or swabs of equipment surfaces. This antigen-antibody complex is later detected by an enzyme-conjugated secondary antibody. The enzyme component catalyses a reaction in the substrate for colour generation and the intensity of colour indicates the amount of allergen present in a food sample (Lee and Sun 2016). Common procedures in performing an ELISA are coating, washing, blocking, antigen-antibody interaction and colour development. Two ELISA formats known as non-competitive and competitive assay can be selected (Yeung 2006).

10.3.2 Non-competitive Assay for Food Analysis

A sandwich ELISA is the most popular and commonly used non-competitive format for food allergen detection and quantification (Figure 10.1). This type of assay involves a pair of antibodies, known as capture and detection antibodies, to bind two or more different epitopes on antigens to achieve sensitive and specific detection (Paulie et al. 2001). Basically, a capture antibody is primarily immobilised onto a hydrophobic surface such as polystyrene microwells, followed by a washing step to remove excess antibody. Blocking is the next step to block unoccupied binding sites on the surface and reduce non-specific binding. An antigen is then added, and if specific to the coated antibody, the antigen is captured by the immobilised capture antibody. After washing the unbound substances away, either the detection antibody linked with an enzyme or an anti-species antibody-enzyme conjugate, is applied to recognise the antigen-antibody complex. A substrate is then added to develop colour, which is proportional to the amount of antigen in the sample (Hornbeck 2001).

10.3.3 Competitive Inhibition ELISA

A competitive ELISA is based on the competition between a labelled and an unlabelled antigen for limited antibody binding sites (Figure 10.2). This assay is especially preferred for the detection of

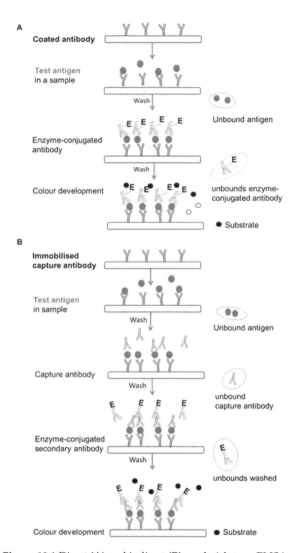

Figure 10.1 Direct **(A)** and indirect **(B)** sandwich type ELISAs.

relatively small proteins which contain only one antibody binding site (Immer and Lacorn 2015). For example, competitive ELISAs can be used to detect partially hydrolysed proteins. The microwell plate is coated with a known amount of antigen and incubated with an unlabelled antigen (i.e., allergen) as well as a fixed amount of

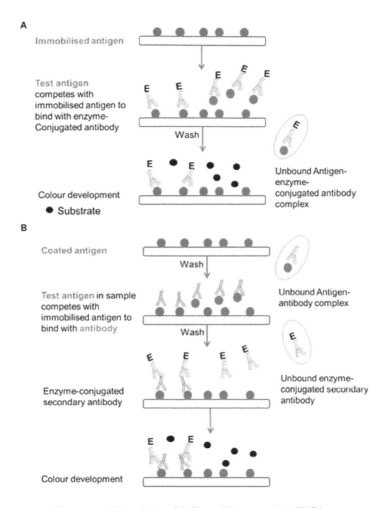

Figure 10.2 Direct **(A)** and indirect **(B)** competitive ELISAs.

the allergen-specific antibody. The (unlabelled) antigen in a sample competes with the immobilised antigen for the binding to antibodies and this interaction results in decreased colour development. The amount of antigen present in a sample is inversely proportional to the intensity of colour development (Paulie et al. 2001).

An ELISA is widely used as an analytical technique to detect traces of allergens in various food products. In the following section, we discuss ELISA relevant to tree nuts as an example. The sandwich ELISAs that have been developed for tree nuts, with the limit of quantification ranging from 0.3–2.5 mg protein kg^{-1}, is summarised in Table 10.1. To develop an ELISA for tree nut residue detection, besides selecting a suitable format, assay optimisation and validation are also crucial in evaluating assay efficiency and suitability for the intended purposes.

10.3.4 Lateral Flow Devices (LFDs)

Though commercial ELISA kits for tree nut allergen detection can deliver a fast outcome within 30 min, a strong demand exists for speedy, easy-to-use immunochemical tests for food manufacturers prompting the development of a new platform (Van Herwijnen and Baumgartner 2006). A lateral flow device, often used in concert with a swab, was introduced as a simple and rapid single-step test (Figure 10.3) that could offer quick indications of the effectiveness of allergen cleaning of shared equipment. It only takes a few minutes to perform the assay after sample extraction. The extract that may or may not be contaminated with nut proteins runs through a sample filter and then a conjugation pad that contain gold nanoparticles conjugated with antibodies specific to target proteins. The target proteins, if present in the sample, form a protein-antibody-nanoparticle complex and these then migrate further along a nylon or nitrocellulose membrane. A test line conjugated with a nut protein-specific antibody is used to capture the complex and generate a visible coloured line. No observed test line indicates the absence of nut proteins. A control line is included to ensure the test is visually validated to be functioning (Lipton et al. 2000, Van Herwijnen and Baumgartner 2006). To date, numerous LFDs have been developed and commercialised for food allergen detection; but as a case study we only present applications related to tree nuts where testing can be finished in approximately 10–15 mins with a detection limit of 1 to 10 mg protein kg^{-1} food sample (Table 10.2).

Table 10.1 Commercial sandwich ELISA kits for tree nut detection.

Tree nut	Limit of detection (mg protein kg^{-1})	Limit of quantification (mg protein kg^{-1})	Supplier
Pecan	< 1	1	BioFront Technologies
	-	0.7	Elution Technologies
	1.7	-	R-Biopharm
Almond	2.5	-	Neogen
	< 1	1	BioFront Technologies
	-	0.5	ELISA Systems
	-	1	Elution Technologies
	-	0.4	Romer Labs
	0.2	0.4	Diagnostic Automation/ Cortez Diagnostics
	1.5	2.5	R-Biopharm
Hazelnut	-	0.5	ELISA Systems
	< 1	1	BioFront Technologies
	0.3	1	Antibodies-online
	0.3	1	Diagnostic Automation/ Cortez Diagnostics
	-	1	Elution Technologies
	-	1	Romer Labs
	-	2.5	Neogen
	1.5	-	R-Biopharm
Macadamia	0.64	1	R-Biopharm
	-	0.3	Elution Technologies
Brazil nut	-	1	Elution Technologies
Cashew nut	< 1	1	BioFront Technologies
	-	0.9	Elution Technologies
	-	2	Romer Labs
Walnut	< 1	1	BioFront Technologies
	0.35	2	Diagnostic Automation/ Cortez Diagnostics
	-	2	Elution Technologies
	-	2	Romer Labs
Pistachio	< 1	1	BioFront Technologies
	-	1	Elution Technologies
	-	1	Romer Labs
Pine nut	-	1.5	Elution Technologies

Hyphen: data not shown in document of product. Data presented as mg whole nut protein kg^{-1}

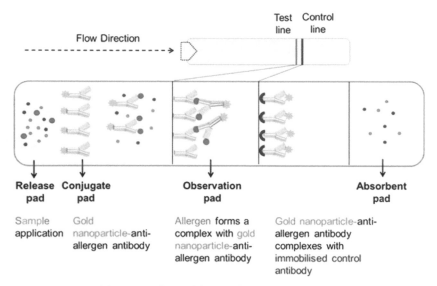

Figure 10.3 General design of a lateral flow test.

10.4 DEVELOPMENT OF AN ELISA

10.4.1 Immunogen Preparation—Tree Nut Protein Extraction and Purification

The molecules that can elicit immune responses after immunising laboratory animals are called immunogens, and this process, from where the antibodies derive, is referred to as immunisation. Both crude extracts from allergenic food ingredients and specific food allergens are excellent immunogens (Lee and Sun 2016). Immunogen may be in a form of undenatured (native) or denatured (processed) or mixed since allergic consumers may be exposed to raw and/or processed allergens (Sathe et al. 2009). The extraction of crude protein from tree nuts usually begins with mechanical homogenisation, for the purpose of disrupting cell walls and releasing water soluble protein. Defatting with *n*-hexane or acetone is helpful to remove unwanted interference such as lipids and free fatty acids without a loss of protein solubility (Bader et al. 2011, Neto et al. 2001). Aqueous solubility of tree nut proteins can be manipulated with extraction

Table 10.2 Commercial lateral flow tests for tree nut detection.

Tree nut	Detection limit (mg protein kg⁻¹)	Incubation time (min)	Supplier
Pecan	2	10	Elution Technologies
Almond	10	10	R-Biopharm
	5	10	Elution Technologies
	1	5	Neogen
	2	10	Romer Labs
Hazelnut	5	10	R-Biopharm
	5	10	Romer Labs
	5	5	Neogen
Macadamia	1	10	R-Biopharm
	2	10	Romer Labs
Brazil nut	1	10	R-Biopharm
	5	10	Romer Labs
Cashew nut	1	10	R-Biopharm
	2	10	Elution Technologies
	5	10	Romer Labs
Walnut	10	15	R-Biopharm
	10	10	Romer Labs
	2	10	Elution Technologies
Pistachio	1	10	R-Biopharm
	2	10	Elution Technologies
	5	10	Romer Labs
Multi-tree nut (almond, hazel nut, pecan, walnut, cashew & pistachio)	5–10	10	Neogen

buffers of different ionic strengths and pH values. Buffered sodium borate at a pH value above 8 along with a low concentration of sodium chloride gives the highest extraction yield for nine tree nuts (Sathe et al. 2009). The list of extraction buffers in Table 10.3 illustrates that low ionic strength (20–50 mM) of Tris-HCl and phosphate

Table 10.3 Protein extraction and purification conditions for five tree nuts.

Sample	Protein extraction buffer	Isolated protein	Purification methods	References[¥]
Brazil nut	Water	2S albumin (Ber e 1)	Gel filtration, gradient chromatofocusing on anion exchange column	1
Brazil nut	0.035 M phosphate buffer, 1 M NaCl, pH 7.5	2S albumin (Ber e 1)	Gel filtration and anion exchange chromatography	2
Cashew nut	0.02 M Tris pH 8.0, 200 mM NaCl, 1 mm EDTA	2S albumin (Ana o 3)	Ion exchange chromatography	3
Cashew nut	0.02 M Tris-HCl, 0.1 M NaCl, pH 8.1	11S globulin (Ana o 2)	Gel filtration	4
Hazelnut	0.05 M Tris-HCl pH 7.0, 500 mM NaCl	7S globulin (Cor a 11) 11S globulin (Cor a 9)	Affinity chromatography and gel filtration	5
Hazelnut	Water	2S albumin (Cor a 14)	Gel filtration and reverse phase chromatography	6
Almond	0.02 M Tris-HCl, pH 8.1	Amandin	Anion exchange and gel filtration	7
Walnut	0.02 M Tris-HCl, pH 8.4	2S albumin (Jug r 1)	Hydrophobic chromatography	8

[¥] References in Table 10.3 correspond to (1) (Moreno et al. 2004), (2) (Sharma et al. 2010), (3) (Mattison et al. 2014), (4) (Robotham et al. 2010), (5) (Rigby et al. 2008), (6) (Garino et al. 2010), (7) (Sathe et al. 2002), (8) (Doi et al. 2008)

buffered saline (0.01 M PBS) with a pH range of 8–9 are the most frequently used buffer solutions for tree nut protein extraction.

Those protein extracts are the source for subsequent purification of specific allergenic protein. Individual allergens are required to be highly pure since antibodies provide a response to any immunogenic epitopes in the immunised materials including allergens and unexpected impurities (He 2013). Therefore tree nut allergens are purified using strategies depending on their physicochemical properties such as molecular weight, polarity and ionic character (Ismail and Nielsen 2010). Some studies tend to concentrate proteins after initial extraction followed by various types of chromatographic separations. For instance, the 2S albumin (Ana o 3) in cashew nut, the 2S albumin (Jug r 1) in walnut and the 7S globulin (Cor a 11) in hazelnut protein were precipitated by the saturated ammonium sulphate method and then redissolved in a buffer solution for further purification (Doi et al. 2008, Iwan et al. 2011, Mattison et al. 2014). Alternatively, single-step sodium chloride precipitation was also demonstrated to separate cashew nut proteins into water-soluble albumins and salt-soluble globulins (Sathe 1994). Chromatography is the most efficient approach for protein separation and purification. The achievement of separation depends on protein physicochemical properties which include the isoelectric point (pI), water solubility, hydrophobicity and molecular weight. A number of tree nut allergens have been successfully purified by chromatographic techniques while conserving protein structure and allergenic activity. As can be seen in Table 10.3, most of the purification of allergens from tree nuts was fulfilled with ion exchange chromatography together with gel filtration. Circular dichroism (CD) spectroscopy analysis demonstrated that the purified allergens from hazelnut, cashew nut and Brazil nut are in a native folded state after purification (Mattison et al. 2014, Moreno et al. 2004, Rigby et al. 2008, Sharma et al. 2010). All purified allergens in each study retained IgE binding from allergic patients (Mattison et al. 2014, Rigby et al. 2008, Sharma et al. 2010).

10.4.2 Antibody Production

An antibody is the heart of an ELISA, since it dictates assay sensitivity and specificity. Both polyclonal and monoclonal antibodies produced by immunisation have been used for ELISA development. Polyclonal antibodies are often produced against a crude extract of the target allergenic food as an immunogen and are obtained from the serum of an immunised host animal. These polyclonal antibodies are able to recognise multiple epitopes on one or more proteins from the immunogen and their specificity relies on the purity of the immunogen. Purification of antiserum is necessary because other serum proteins may cause a high background via non-specific binding and lower signal to noise ratio. Monoclonal antibodies, differing from polyclonal antibodies, are produced as individual and unique antibodies where each antibody only recognises one epitope on the allergen (Karaszkiewicz 2005). This unique specificity can be challenged when the single antibody binding epitope is modified by processing resulting in protein denaturation, hydrolysis, unfolding and aggregation. Polyclonal antibodies are, therefore, preferred over monoclonal antibodies for detecting allergen residues in processed food since they are likely to be more tolerant to small structural changes occurring in the nature of the antigen (Koppelman and Hefle 2006). Rabbit polyclonal antibodies are the most frequent option and IgY antibodies produced in egg yolk are becoming more attractive. These antibodies share structural and functional similarities to mammalian IgGs with phylogenetic distance from a host to immunogens and with no invasive actions to the host animal (Spillner et al. 2012). A supplementary advantage is the higher production of polyclonal IgY, despite the difficulty of isolation and purification of IgY from egg yolk (De Meulenaer and Huyghebaert 2001).

Table 10.4 summarises the developed ELISAs for tree nut detection with antibodies raised in different host animals. For instance, a competitive ELISA for hazelnut detection by using anti-hazelnut-IgY was able to detect 30 mg (hazelnut protein) kg^{-1} (of cookies) (Cucu et al. 2012), which was not as sensitive as the sandwich ELISA developed with rabbit polyclonal antibody as a detection antibody. The ELISA with the rabbit polyclonal antibody has a quantification limit of 0.4 mg hazelnut protein kg^{-1} of cookie (Kiening et al. 2005).

The production of an antibody is typically carried out by subcutaneous and/or intramuscular injection of an antigen into a host animal. The application of adjuvant is necessary to enhance an immune response by improving the efficiency of antigen presentation and the amount of antibody-secreting B cells, as well as the affinity and avidity of antibodies. Freund's complete adjuvant is the most popular adjuvant used in the initial subcutaneous or intramuscular injection (Harboe and Ingild 1983). Specific antibody can be detected one week after the first immunisation. The initial type of antibody to be produced is IgM, followed by a switch to IgG which become

Table 10.4 A Summary of ELISAs for the detection of tree nut residues.

Sample	ELISA format	Antigen	Antibody matching pair	LOD, LOQ (mg protein kg⁻¹)	Cross-reactivity	References[v]
Brazil nut	Competitive	2S albumin	Detection: rabbit	LOD: 3.3	No reactivity	1
Hazelnut	Sandwich	Roasted	Capture: mouse (monoclonal) Detect: rabbit	LOD: 0.018 LOQ: 0.03–0.18	Walnut	2
Cashew nut	Sandwich	Raw and roasted	Capture: sheep Detect: rabbit	LOD: 0.025 LOQ: 0.097	Pistachio, hazelnut	3
Walnut	Sandwich	2S albumin	Capture/Detect: rabbit	LOD: 0.156 LOQ: 0.312	Pecan, hazelnut	4
Almond	Competitive	Amandin	Rabbit	LOD: 1.3 LOQ: 5–37	Cashew, rice, peanut, soybean	5
Hazelnut	Sandwich	Raw	Capture: rabbit Detect: mouse (monoclonal)	LOD: 1 LOQ: 50	Hazelnut, pecan, pistachio, cashew, Brazil nut, macadamia, almond	6
Walnut	Sandwich	Raw and roasted	Capture: sheep Detect: rabbit	LOQ: 1	Pecan, hazelnut, mace, poppy seed	7
Hazelnut	Competitive	Modified hazelnut protein	Chicken	LOD: 13.6 LOQ: 40.8	Walnut, pecan	8
Cashew	Sandwich	Cashew major protein	Capture: goat Detect: rabbit	LOD: 0.5 LOQ: 1	Sunflower seeds, pistachio, walnut, pecan	9
Brazil nut	Competitive	raw	rabbit	LOD: 0.1	Cinnamon	10

[v] References in Table 10.4 correspond to (1) (Clemente et al. 2004), (2) (Kiening et al. 2005), (3) (Gaskin and Taylor 2011), (4) (Doi et al. 2008), (5) (Roux et al. 2001), (6) (Costa et al. 2015), (7) (Niemann et al. 2009), (8) (Cucu et al. 2012), (9) (Wei et al. 2003), (10) (Sharma et al. 2009)

dominant after the extended immunisation (Lon 2013). The intervals between booster injections should be adequate so to allow affinity maturation. In general, the interval between the primary injection and a booster injection is three to four weeks and four to six weeks for subsequent injections. During immunisation, an antiserum is routinely collected and antibody titre needs to be monitored. Titre is an essential indicator of antibody quality that can be carried out by an indirect ELISA (Koppelman and Hefle 2006).

10.5 ELISA OPTIMISATION

10.5.1 Coating and Blocking

Developing an ELISA with high sensitivity requires selection of an appropriate capture-detection antibody pair with optimal concentrations and maximum signal-to-noise ratio. Assay optimisation is essential and imperative as it helps to attain the optimum performance of a developed ELISA (Karaszkiewicz 2005). Attachment of an antibody or an antigen to a solid phase is referred to as coating. This immobilisation is readily achieved by passive absorption between the hydrophobic regions of the protein (antigen and antibody) and the non-polar plastic surface. Polystyrene is the most widely used material for a microwell plate as it is very hydrophobic and provides a large capacity for protein binding (Gibbs and Kennebunk 2001a). The coating buffer should avoid detergents and other proteins, which may compete with an immobilising reagent for immobilisation. The pH of coating buffer should be at least one to two units higher than the isoelectric point of the coated protein. The most frequently used buffer for coating is 50 mM carbonate, pH 9.6. The coating temperature and time also affect the hydrophobic interaction rate in inverse proportions: the higher the temperature, the greater the interaction rate and the shorter incubation time. An in-house ELISA often requires incubation at 37°C for at least 2 h or 4°C overnight for coating (Crowther 2000).

The immobilisation of antibody or antigen onto the solid surface is obviously a crucial step. However, the possible excess spaces on the surface may be occupied by other reactants during subsequent steps.

Such non-specific binding can be detrimental to assay specificity and sensitivity (Gibbs and Kennebunk 2001b). An ideal blocking buffer is able to saturate unoccupied binding sites, to minimise background colour without altering antibody binding sites and to stabilise the immobilized protein by sterical support (Esser 1991). Proteins are effective blockers. Bovine serum albumin (BSA), non-fat dry milk, casein and fish gelatine are frequently used in ELISA. It is also a common practice that protein blockers are added to diluents of assay reactants to further reduce non-specific binding and stabilise assay reactants after surface binding (Huber et al. 2009). Apart from protein blockers, non-ionic detergents such as Tween 20 are able to hinder the irrelevant substances from absorbing onto the surface. If they are used as sole blocking reagents, they are should be added to all assay buffers (Gardas and Lewartowska 1988). Though detergents are inexpensive, easily stored and extremely stable, they may disrupt hydrophobic interactions and the residue after washing can interfere with enzymatic activity (Gibbs and Kennebunk 2001b).

10.5.2 Buffer System, Incubation Time and Colour Development

The fundamental buffer system in ELISA development is considered to be phosphate buffer or Tris buffer on account of the maintenance of protein (antigens and antibodies) stability (Koppelman and Hefle 2006). An antigen-antibody interaction requires sufficient time to allow their close proximity. A high affinity antibody can bind to an antigen in a short period of time, usually 10–30 min. This can be adjusted according to the temperature of incubation (room temperature and 37°C). For instance, a shorter incubation time can be utilised at high temperature (e.g., 37°C) but most developed commercial assays are optimised for 22–24°C (Crowther 2000). Detection system in ELISA requires colour generation. This reaction involves enzyme to be attached to an antibody and a suitable substrate system. Horseradish peroxidase or alkaline phosphatases are the most widely used enzymes in an antibody conjugation since it generates strong colour in a relative short time. Substrates such as 3,3′5,5′-teramethylbenzidine and o-phenylenediamine for peroxidase and *p*-nitro-phenylphosphate for alkaline phosphatase, are typically used (Koppelman and Hefle 2006).

10.5.3 Cross-reactivity

An ideal ELISA is one that detects the target analyte specifically with no potential interference during quantification. Cross-reactivity is defined as a false positive response to a sample in the absence of target analyte. Proteins from diverse sources sharing structural or amino acid sequence similarity are the basis of cross-reactivity in allergen detection (Ferreira et al. 2004). The cross-reactivity can occur either due to the presence of common epitopes, or the binding of structurally different determinants of the same antigen by the same antibody (van Hengel 2007). Walnut, pecan and hazelnut are strong cross-reactive tree nut groups, while hazelnut, cashew nut, Brazil nut, pistachio and almond constitute a moderately cross-reactive group. Notably, a pronounced cross-reactivity, based on botanical relationships, is presented between walnut and pecan in *Juglandaceae* along with cashew nut and pistachio from *Anacardiaceae* (Goetz et al. 2005). There are three major allergens in cashew nut, Ana o 1 (7S globulin, vicilin), Ana o 2 (11S globulin, legumin) and Ana o 3 (2S albumin) (Robotham et al. 2010, Wang et al. 2002, Wang et al. 2003). Vicilins were identified as allergens in walnut and hazelnut and some degree of cross-reactivity with cashew nut has been suggested (Willison et al. 2008). Pistachio vicilin, Pis v 3 presents a strong cross-reactivity with Ana o 1. Peanut vicilin, Ara h1 did not cross-react with Ana o 1 from cashew nut, though they share 27% similarity in amino acid sequences (Barre et al. 2008, Wang et al. 2002). Pistachio allergens Pis v 1 and Pis v 2 show 64% and 48% of amino acid sequence identity to cashew Ana o 3 and Ana o 2, respectively demonstrating potential cross-reactivity between each other (Ahn et al. 2009). Cross-reactivity also can be found between peanut and tree nuts such as almond, Brazil nut and hazelnut (de Leon et al. 2003). The generation of a positive response indicates that the antibody is also cross-reactive with other extracted protein to some extent (Taylor et al. 2009).

10.5.4 ELISA Validation

The development of ELISA allows determination of the possible presence of allergen residues in food products, provision of correct information on precautionary labelling, and prevention of occurrence of food allergy reactions (van Hengel 2007). The performance of a developed ELISA is assessed by assay validation. Assay validation is a process of demonstrating that the target analyte in a specific matrix will yield acceptably accurate, precise and reproducible results after a combined procedure of sample preparation and analysis (Lipton et al. 2000). In order to fully characterise assay performance, there are a number of parameters required to be assessed and these include accuracy, precision, sensitivity, specificity (cross reactivity) and matrix interference.

10.5.5 Accuracy and Precision

The accuracy of an assay expresses the closeness of the result of a test sample to the theoretical value (Hirst and Miguel 2015). It can be demonstrated by measuring the recovery of a target protein from spiked or incurred samples. An ideal percent recovery for an immunochemical method is between 80 and 120% (Lee and Kennedy 2007). With respect to an ELISA for food allergens, the acceptable recoveries are 50 to 150%, since difficult food matrices together with a wide range of food processing conditions may provide stronger interference to analysis (Abbott et al. 2010). The precision of an ELISA describes the closeness of analytical results when assays are repeated multiple times within an assay (intra-assay) and separate assays (inter-assay) performed on different days (Lee and Kennedy 2007). The variability between replicates is statistically expressed as the standard deviation or coefficient of variation (Lipton et al. 2000). According to the guidelines for assay validation from the U.S. Food and Drug Administration (USFDA), acceptable variation for precision should be less than 20% coefficient of variation (Selvarajah et al. 2014).

10.5.6 LOD, LOQ and Detection Range

The assay sensitivity of an ELISA is often regarded as the limit of detection (LOD). It is the lowest amount of an analyte that can be measured from a true blank matrix with an acceptable probability level. A limit of quantification (LOQ) is defined as the lowest amount of an analyte that can be quantified with an acceptable accuracy and precision. The dynamic working range of the assay for quantification gradually becomes imprecise with an increase in the allergens to be detected (Abbott et al. 2010, Koppelman and Hefle 2006). Table 10.4 summarises a few ELISAs that have been developed with an LOD range from 0.018–13.6 mg protein kg^{-1}. It is necessary to know the amount of allergenic food that can elicit an allergic reaction. Though clinical reactions and eliciting doses (EDs) are variable between individuals, threshold levels for specific allergens have been found to be less than 1 mg to more than 1 g according to double-blind placebo controlled food challenge (DBPCFC) studies (Taylor et al. 2014). This study seems to confirm the recommended detection limits of analytical methods for allergen detection should be at low milligram per kilogram levels (Hourihane 2001, Taylor et al. 2002). The detection of allergen residues can be challenged by the difficulty with adequate allergen extraction and co-extraction of interference from the food matrix (van Hengel 2007).

10.5.7 Food Matrix Interference

The overall performance of an ELISA is ultimately assessed by its capability to effectively detect food allergens in complex matrices (Abbott et al. 2010). Efficiency of extraction of a target protein from a test sample is a key determinant of immunoassay performance. In general, the non-specific responses generated from substances in the final extract are referred to as matrix effects (Lipton et al. 2000). Food matrices are complex mixtures in various forms including liquids, solids and powders. Extraction of allergen residues is affected by how complex they are presented in food matrices (van Hengel 2007). For instance, food allergens exist as non-defined protein mixtures with more or less denaturation, and/or proteins that are cross-linked or aggregated among themselves or with other matrix components by hydrogen bonds or disulfide bridges (Khuda et al. 2015, Taylor et al. 2009). Though it is possible to extract proteins from matrices under harsh conditions such as extracting with urea or the application of a reducing agent with high content of salts, protein structures can be altered and affect antigen-antibody reaction negatively if the detection antibody is specific to native form of protein. Simple phosphate buffered saline (PBS) or Tris buffer is not sufficient to extract protein from certain challenging matrices; additional additives are, therefore, necessary to improve the releasing of desired proteins (Immer and Lacorn 2015). Recently, use of sodium sulphite as a reducing agent in extraction buffer was noted to significantly improve extraction of allergenic protein(s) (Ito et al. 2016).

Table 10.5 presents a list of extraction protocols applied for representative matrices that are likely to contain tree nuts. Chocolate and caramel are analytically challenging matrices. The polyphenolic compounds and tannins make extraction more difficult due to non-specific interaction with proteins. The high fat content in chocolate influences crystallisation and potentially masks allergenic proteins (Khuda et al. 2015). Most of the extractions in Table 10.5 include the addition of skim milk in the extraction buffer, which is helpful in preventing the allergenic proteins from interacting with polyphenolic compounds (Khuda et al. 2015). Extraction at higher temperature (50–60°C) has proved to be more effective than lower temperature. For example, a higher yield of almond proteins was attained when treated at 60–70°C than at room temperature (Albillos et al. 2011).

10.5.8 Food Processing

Food processing can affect extraction and detection of allergen residues. Baking as an example of a thermal processing can physically and chemically modify the allergenic proteins, cause protein unfolding, aggregation, fragmentation, hydrolysis, or total degradation. Those modifications in food matrices result in a loss of protein solubility, which negatively affect extraction efficiency and consequently ELISA detectability (Khuda et al. 2015). For example,

Table 10.5 Extraction conditions of different food matrices for tree nut allergen detection.

Allergen	Test sample	Matrix extraction protocol	Recovery of protein from model foods*	References[¥]
Cashew (spiked)	Milk chocolate Chocolate-filled cookies Rice cereal	Borate saline buffer pH 8.2, Room temperature	77% 120% 110%	1
Cashew (spiked)	Milk chocolate Ice cream Cookies	10 mM 0.01 M PBS, pH 7.4, 1% nonfat dry milk Temperature: 60°C	74.7–98.7% 111–128% 100–110%	2
Peanut (spiked)	Milk chocolate Dark chocolate Cookie Cereals Ice cream	RIDASCREEN Allergen extraction buffer with skim milk powder Temperature: 60°C	113–123% 87–101% 91–107% 105–117% 94–110%	3
Hazelnut (spiked)	Milk chocolate Dark chocolate Cookie Cereals Ice cream	RIDASCREEN Allergen extraction buffer with skim milk powder Temperature: 60°C	95–115% 86–96% 95–127% 95–106% 93–111%	3
Hazelnut (spiked)	Cookies	2 mM urea in 0.01 M PBS, pH 7.4 Temperature: 50°C	73–107%	4
Hazelnut (spiked)	Cookie Cereal bar Milk chocolate	8 mM Tris, 25 mM Tricine, 2 mM calcium lactate, pH 8.6 Temperature: 37°C	108% 100% 103%	5
Hazelnut (spiked)	Chocolate	0.2 M 0.01 M PBS pH 7.4 10% milk powder Temperature: 60°C	77.3–111.2%	6

Table 10.5 contd. ...

...Table 10.5 contd.

Allergen	Test sample	Matrix extraction protocol	Recovery of protein from model foods*	References[¥]
Walnut (spiked)	Bread Biscuit	0.12 M Tris-HCl pH 7.4, 0.1% BSA, 0.05% Tween 20, 0.5% SDS, 2% 2-ME Room temperature	123% 83%	7
Walnut (both spiked and manufactured in to foods)	Milk chocolate	0.1 M 0.01 M PBS pH 7.4, 1% skim milk powder, 0.1% Tween 20 Temperature: 60°C	71.6–119%	8

* Model foods defined as the manufactured food products with incurred tree nuts
[¥] References in Table 10.5 correspond to (1) (Wei et al. 2003), (2) (Gaskin and Taylor 2011), (3) (Kiening et al. 2005), (4) (Cucu et al. 2012), (5) (Holzhauser and Vieths 1999), (6) (Costa et al. 2015), (7) (Doi et al. 2008), and (8) (Niemann et al. 2009)

hazelnut lost 30% of extractable proteins after baking (Cucu et al. 2012). Effect of thermal processing on peanut protein solubility, and separately on the performance of two commercial ELISA kits for peanut residue quantification was evaluated. Both moist heat (boiling, blanching and autoclaving) and dry roasting resulted in a decrease of protein solubility (Su et al. 2004). The two ELISA kits tended to reduce accuracy in quantifying the protein content in samples that were processed with autoclaving and dry roasting (Fu and Maks 2013). A similar result was observed when roasting at 120–190°C in which the yield of peanut protein was dramatically reduced by 75–80% (Poms et al. 2004b).

Food processing may reduce or increase the allergenic potential of an allergen either by altering the protein structure to destroy antibody binding sites, or by revealing the epitopes that were previously hidden within the three-dimensional structure, or by forming novel protein structures with allergenic potency (Maleki et al. 2000, van Hengel 2007). Maleki et al. (2000) found increased IgE binding of Ara h1 and Ara h2 in peanut after roasting, as opposed to the decrease in immunoreactivity that was observed in cashew nut allergens Ana o 1 and Ana o 3 observed in a different study (Venkatachalam et al. 2008). Ana o 2 remained stable during roasting despite a decline in stability under extreme conditions at 200°C. Other thermal processing techniques, including autoclaving, blanching and microwave heating, can affect allergenicity of certain food allergens. The extent of these effects is strongly depended on the stability of allergen structures. The negligible effects on cashew nut allergenicity indicate a superior structural stability of cashew nut allergens. Roasting, blanching, frying and autoclaving have no, or negligible effects on the immunoreactivity of Brazil nut protein while microwave heating increases the immunoreactivity by 32% (Sharma et al. 2009). The allergen Ber e 1 in Brazil nut is less immunogenic due to an irreversible denaturation at 80–110°C (van Boxtel et al. 2008). Almond and walnut proteins exhibit high antigenic stability under thermal processing and the protein profiles are barely affected (Su et al. 2004), suggesting that they may be excellent marker proteins for the detection of almond and walnut residues in processed foods by ELISA.

Additionally, the Maillard reaction can induce chemical modification of allergens. Maillard reaction is a well-defined heat-accelerated reaction between the free-amino group on protein and the carbonyl group on reducing sugars (Davis et al. 2001). Such glycation reactions subsequently undergo a series of chemical rearrangements involving oxidation, hydration and condensation, resulting in the formation of advanced glycation end products (AGEs) which contribute to the aroma, colour and flavour of processed foods (Iwan et al. 2011). The formation of AGEs is largely related to the cross-linking of food proteins and results in an alteration of allergenicity and immunoreactivity of allergenic proteins, and hence affects ELISA detectability (Mills et al. 2009). This has been observed in peanut allergens Ara h1 and Ara h2. The Maillard reaction prompted the cross-linking of the two allergens and formed a high molecular weight aggregate, leading to an enhanced allergenicity of peanut (Maleki et al. 2000). Glycation under 145°C, however, decreases the immunoreactivity of hazelnut allergen Cor a 11 (Iwan et al. 2011). The production of potential neoallergens in heated pecan as a result of Maillard-type degradation of ingredients during storage has been reported. Protein- and carbohydrate-rich foods have a greater tendency to undergo considerable degradation of proteins as a result of Maillard reaction during industrial or household handling and storage of such foods. The effects of Maillard reaction within the food preparation increase skin-active and allergenic potency of foods (Berrens 1996).

In the above section on immunoassays, more importantly we discussed different ELISA formats and applications for food allergen detection. We reviewed important factors that affect ELISA development and allergen detection in food products. In the following section, we present how mass spectrometry (MS) has gained importance in food allergen detection, and review some of the principles and factors that are crucial to using this method for food allergen detection. We illustrate applications in which this technique has shown some expedient outcomes. Nevertheless we also identify some potential areas where mass spectrometry would be a useful target for further analysis and improvement for food allergen detection.

10.6 MASS SPECTROMETRY FOR FOOD ALLERGEN DETECTION

Applying proteomics is valuable in understanding the structural and functional aspects of food allergen proteins including protein identification, their primary structure, allergenic functional groups and epitopes, post translational modifications, subcellular interactions, protein modifications arising from food processing and even the modelling of allergenic ingredients to make them safe in foods. The advent of MS instruments and increased availability of downstream analysis tools, have made MS analysis more cost effective and easier to access by a wider community. Over the last decade, MS applications for the characterisation and detection of allergens in food have continually risen and several review articles have summarised their applications (Cunsolo et al. 2014, Di Girolamo et al. 2015, Johnson et al. 2011).

10.6.1 Sample Complexity, Sample Preparation and Clean-up

Proteomics analysis facilitates high throughput analysis of several complex mixtures as well as targeted analysis of specific analytes such as intact proteins. Proteomics based mass spectrometry facilitates identification, detection and quantification by measuring several spectra corresponding to thousands of proteins and peptides simultaneously.

Food is a complex biological matrix and is one of the most difficult samples to investigate for allergens. This sample complexity occurs not only due to the intrinsic nutritional composition of the food but more importantly due to the presence of a complex mesh of thousands of proteins and peptides. Reducing this sample complexity is crucial for the detection of specific allergen proteins in food samples and warrants additional separation procedures such as gel electrophoresis and liquid chromatography during sample preparation, and needs to be undertaken prior to mass spectrometry analysis. Processing of stained protein bands or in-solution peptide digests requires optimal clean-up steps for the removal of stain, salts and detergents which result in ion suppression and loss of signal in addition to loss of spray. Clean-up steps may be achieved by solid phase extraction (Gobom et al. 1999), immunomagnetic bead pull down method (Careri et al. 2007) or adsorption chromatography prior to mass spectrometry (Faeste et al. 2011).

10.6.2 Allergen Detection—Intact Proteins and Complex Mixtures

Characterisation of the allergen proteins and their subsequent physicochemical properties may be achieved from the use of separation techniques such as 1D and/or 2D gel electrophoresis, ion-exchange or size exclusion chromatography, reverse-phase HPLC and mass spectrometry combined with biochemical methods. The biological activity of food allergens is predominantly investigated using immunoreactive assays and blots or using mediator release assays for, effector cells such as basophils. Although ELISAs allow quick routine analysis, ELISAs still face the issue of unspecific binding from cross-reactive species as well as effects of food processing.

Samples for subsequent MS analysis require upstream separation of proteins and peptides, and have been conventionally performed with gel electrophoresis or high performance liquid chromatography (HPLC). Gel electrophoresis has become a primary indispensable technique in most proteomics analysis. One-dimensional (1D) and two-dimensional (2D) gel electrophoresis are the most popular tools for immediate visual validation to ascertain the degree of separation and size of proteins in analysis.

However, one of their limitations includes detection of low abundant proteins. HPLC is the second most popular separation technique for protein purification and offers greater separation range, reproducibility and specificity. HPLC is a flexible tool with the ability to resolve large and small biomolecules with the selection of different stationary phases and offers greater sensitivity. LC coupled MS requires proper mobile phase selections devoid of salts or other ion supressing acids for successful ionisation of the analyte of interest without loss of signal and electrospray sensitivity. Protein analysis of food allergens using multidimensional protein identification technology (MudPIT) runs could provide greater separation of proteolytic peptide mixtures derived from a food sample. MudPIT is usually performed by combining two HPLC methods such as size-exclusion chromatography with reverse-phase liquid chromatography (RPLC), ion-exchange chromatography with RPLC, IEC with affinity chromatography, or affinity chromatography with RPLC. These combinations have been previously used in different food allergen studies (Fæste et al. 2011). For LC-MS/MS analysis, the amounts of starting materials, the proteolytic digestion of protein samples, and the reproducibility of identifications are critical and could affect quantification of proteins. Food allergen detection using proteomics is complicated and requires a thorough understanding and skill set to execute both instrumentation as well as downstream analysis.

10.6.3 Detection and Quantification of Allergen Peptides/ Proteins in Food Using Mass Spectrometry

10.6.3.1 Relative and absolute quantification of allergens

Proteomics techniques have become increasingly popular in allergen discovery and analysis (Akagawa et al. 2007, Picariello et al. 2011, Wagner et al. 2008), and facilitate comprehensive identification and quantification of food allergens at low parts per million (ppm) levels relatively comparable to ELISA. Both bottom-up and top-down mass spectrometry are powerful tools in proteomics for identification and characterisation of proteins and enables relative or absolute quantification respectively (Halim et al. 2015).

Mass spectrometry including tandem MS/MS offers highly sensitive and accurate approach for the identification, characterisation and quantification of allergenic proteins in food products (Figure 10.4). Mass spectrometry based detection of food allergen residues has become a superior confirmatory technique for both qualitative and quantitative analysis which complements immunoassays. Latest

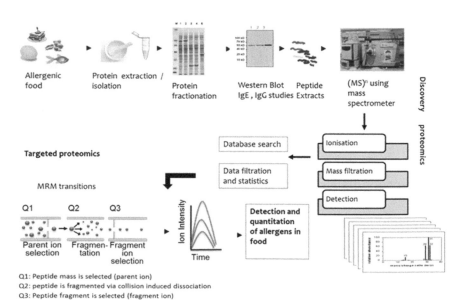

Q1: Peptide mass is selected (parent ion)
Q2: peptide is fragmented via collision induced dissociation
Q3: Peptide fragment is selected (fragment ion)

Figure 10.4 Mass spectrometry based detection and quantitation of food allergens.

hybrid mass spectrometers such as the Q exactive, triple quads and VelosPro instruments based on latest versions of orbitrap, time of flight (TOF) and linear-ion trap mass analysers show greater resolution, sensitivity and mass accuracy for improved quantification particularly useful for allergen screening (Neilson et al. 2011). LC-MS/MS applications in food allergen research using proteomics predominantly use the popular data-dependent MS/MS that scans for a series of the most intense ions following a full scan and is subjected to a dynamic exclusion window, thus generating thousands of fragmentation spectra over the chromatographic profile. In this mode, precursor mass information and spectra of fragmentation ions are used for peptide identification and quantification. This approach has been widely used to identify, analyse and confirm allergenic peptides and proteins of many priority allergenic foods (Pilolli et al. 2014).

The shotgun proteomics workflow is suited for discovery studies and typically adopts the bottom-up approach, where proteins are pre-fractionated, subjected to proteolytic digestion, followed by nanoflow-liquid chromatography- tandem mass spectrometry (nLC-MS/MS). This workflow is becoming ideal for identifying proteins and characterizing post translational modifications. Alternatively, top-down approach is more suited for characterizing intact proteins typically analysed by high resolution mass analysers, followed by MS/MS based on collision induced dissociation of ions. Gene ontology annotations and analysis have been possible from using databases such as UniProt, Interpro, KEGG, and new tools such as WEGO (Ye et al. 2006), MaxQuant (Cox and Mann 2008), STRAP (Bhatia et al. 2009), DAVID or MAPMAN. This approach is predominantly used for biomarker discovery across different areas of research but has recently been successfully demonstrated for discovery of potential allergens in uncharacterised novel allergenic foods (in house unpublished). As an alternative, data acquired using data-independent acquisition mode based on simultaneous fragmentation activation of all ions (all co-eluting peptides) could exponentially enhance the depth of data available for analysis, and has the potential for multiple retrospective analyses from a single experiment. This approach has been exploited in certain other areas of research, however it requires the generation of high quality spectral libraries for peptide identification and quantification and its application in food allergen research is still at infancy.

10.6.3.2 *Choosing suitable ionisation source and mass analyser*

Matrix assisted laser desorption ionisation (MALDI) (Tanaka et al. 1988) and electro spray ionisation (ESI) (Fenn et al. 1989) have long been the most widely used soft ionisation techniques for biomolecular analytes. Upon using MALDI, samples are co-crystallised with a matrix solution on a target plate and ionised with laser. Singly charged ions are typically detected by measuring their time of flight. MALDI MS/MS has been popular for analysis of simple mixtures of proteins, shorter analysis time, greater throughput, and has been commonly used for analysing purified allergens or those well separated by other techniques. On the other hand, ESI techniques such as LC-MS/MS are popular in analysing complex protein mixtures and generate multiply charged ions thereby increasing the range of detection. With ESI, analytes are ionised using high electric voltage and then subjected to collision induced fragmentation. This fragmentation and mass to charge ratios are captured in spectra over a certain run time.

All MS instruments measure ions by their mass-to-charge (m/z) ratios. Ion motion is regulated by electric and magnetic fields under vacuum. MALDI and ESI ionisation interfaces allow easy coupling with different mass analysers. Mass analysers types widely in use include quadrupoles (Q), linear ion-trap, quadrupole ion-trap, Orbitrap, time of flight TOF and Fourier-Transform Ion Cyclotron Resonance (FTICR) mass analysers. Multiple stage mass analysers such as triple quadrupole (QQQ), tandem TOF (TOF/TOF) or hybrid Q/TOF instruments are already popular for their increased accuracy and resolution. As extensively reviewed, different mass analysers show slightly varying sensitivity, resolution and mass accuracy in their MS/MS spectra and this is attributed to their technical improvements (Picariello et al. 2011, Rubert et al. 2015). Ion traps are sensitive and have high dynamic range but were limited to moderate mass accuracy and relative quantification but recent orbitrap and fourier transform (FT) instruments have shown excellent sensitivity and mass accuracy (Scigelova and Makarov 2006). Ion traps are capable of multi stage MS^n, useful for peptide characterisation, are sensitive and, whereas TOFs show high mass range but moderate resolution, and are suited for absolute quantification. Factors such as sensitivity and specificity (high resolution) are characteristic of the MS instrument being used and need to be chosen to fit for purpose. LC-ESI-MS/MS, MALDI-TOF TOF and QQQ instruments are predominant choices among the food allergen scientific community and have been used from identification to detection and quantification of food allergenic proteins and peptides from complex matrices (Johnson et al. 2011, Koeberl et al. 2014).

10.6.3.3 *Intensity and specificity of allergen signatures*

The main interest in the area of food safety is the measurement of the absolute quantity of an allergenic protein in a food matrix. Quantification of the allergens depends on the concentration of the analyte as well as the intensity of their signal detected. Owing to this issue, there are often other highly abundant proteins in the food matrix that supress the detection of the low abundant ones such as the target allergenic proteins/peptides. Peptides chosen as allergen protein signatures are often required to be global markers for the particular allergenic food, and this may warrant these signatures to be evaluated in different food matrices and under different processing conditions.

10.6.3.4 *Synthetic peptides and isotopic labelling*

For comparative relative quantification, measurements can be obtained by label-free methods or labeling methods including ICAT, ITRAQ or TMT. For absolute quantification of a particular allergen of interest, techniques involving isotope peptides spiking and selected reaction monitoring are available for use. Isotopic or isobaric labels incorporated into proteins offer a known shift in mass and differences between their intensities reflect differences in abundance across samples. When using synthetic peptides, samples are spiked with synthetic peptides of similar chemistry and absolute quantification is based on measuring and comparing the abundances of the spiked isotopic peptides to target analyte (Fæste et al. 2011). Validation may be achieved by using known quantities of synthetic peptides in the sample as a "pseudo" internal standard.

For allergen quantification, selected reaction monitoring including multiple reaction monitoring performed using a QQQ mass spectrometer is best suited and applications in food safety and diagnostics is gaining momentum.

10.6.4 Food Allergen Signatures for Mass Spectrometry Based Detection

Most major allergens in plant foods have been proteinases, transfer proteins, seed storage proteins, or enzymatic inhibitors, and have been grouped into just four major families, namely prolamins, cupins, profilins and latex proteins. Whereas in eggs, meat and seafood, allergens may be grouped in 3 major and 14 minor protein families (Jenkins et al. 2005, Radauer and Breiteneder 2007, Radauer et al. 2008) and include caseins, albumins, globulins, enzymes such as arginine kinase, glutathione synthase, sugar metabolic enzymes and muscle proteins among few others (Table 10.6). This small distribution of allergen protein families implies the possibility of other unknown factors that make up allergenic proteins, and novel allergens in addition to those presently suspected including homology and differences in 3D protein structures, distribution of conformational sites, protein folding and post translational modifications.

MS based research on food allergen characterisation and detection is still under development; however, for some priority food allergens such as chicken egg, cow's milk, peanut, fish, shrimp, soybean, and wheat gliadin, identification and quantitative methods for the detection of the allergen in typical food matrices have been developed (Bauermeister et al. 2011, Carrera et al. 2012, Cunsolo et al.

Table 10.6 Peptide MS signatures for notable plant and animal food allergens.

Allergen	Peptide biomarkers	Allergenic proteins	m/z (or) M+H	References ¥
Egg	FESNFNTQATNR NTDGSTDYGILQINSR SAVSASGTTETLK LTEWTSSNVMEER ISQAVHAAHAEINEAGR GGLEPINFQTAADQAR	Lysozyme Vitellogenin Ovalbumin	714.8^{2+} 877.42^{2+} 626.3^{2+} 791.3^{+2} 887.5^{2+} 844.42^{2+}	1, 2
Milk	DMPIQAFLLYQEPVL GPVR GPFPIIV LKPDPNTLCDEFK VPQVSTPTLVEVSR RHPEYAVSVLLR YLGYLEQLLR FFVAPFPEVFGK NAVPITPTLNR EGQCHV TPEVDDEALEK	β Casein Bovine Serum albumin α-casein β-lactoglobulin	1093.58^{2+} 742.45^{1+} $1,576.8^{+}$ $1,511.9^{+}$ $1,439.9^{+}$ 634.36^{+} 692.82^{+} 598.3^{2+} 671.3^{+} 623.32^{+}	2, 3, 4
Soy	TISSEDKPFNLR ESYFVDAQPK VFDGELQEGGVLIVP QNFAVAAK AIPSEVLAHSYNLR SQSDNFEYVSFK	β-conglycinin Glycinin G2 Glycinin G4 Glycinin	703.87^{2+} 592.29^{+2} 1201.14^{4+} 785.4^{+2} 725.7^{+2}	2
Wheat	SVYQELGVR EHGAQEGQAGTGAFPR YFIALPVPSQPVDPR	α-amylase inhibitor	525.78^{2+} 806.87^{2+} 849.96^{2+}	5
	NVANGASGGPYITR	Non-specific lipid transfer protein	688.849^{2+}	5
	VQVQIPFVHPSILQ QPFPQQPP VQIPFVHPSILQ	B3 hordein	802.99^{2+} 461.24^{2+} 689.43^{2+}	6
	PYVDPMAPLPRSGP PMAPLPRSGPE	β Amylase inhibitor	748.91^{2+} 576.313^{2+}	6
	KFPAAVFLK FPAAVFLK	thioredoxin	510.816^{2+} 446.76^{2+}	5
	QTQQPQQPFP LALQTLPAMC YIPPHCSTTI	γ-gliadin		
	AASFNIIPSSTGAAK	glyceraldehyde 3-phosphate dehydrogenase	717.88^{+}	5

Table 10.6 contd. ...

...Table 10.6 contd.

Allergen	Peptide biomarkers	Allergenic proteins	m/z (or) M+H	References ¥
Fish	SGFIEEDELK SGFIEEEELK LFLQNFSAGAR IGVEEFQALVK AFAIIDQDNSGFIEEEELK	parvalbumin	583.78[2+] 590.79[2+] 612.33[2+] 616.85[2+] 1084.52[2+]	7
Crustacea/moluscs	YTTAASKLEEASKA ADESER GANEDNIRSR RGLTXPIRM VSSTLSSLEGELK TFLVWVNEEDHLR	Tropomyosin Glutathione-S-transferase Arginine kinase	2156.5[+] 1130.5[+] 675.22[2+] 829.26[2+]	8
Peanuts	GTGNLELVAVR VLLEENAGGEQEER DLAFPGSGEQVEK WLGLSAEYGNLYR RPFYSNAPQEIFTQQGR SPDIYNPQAGSLK QIVQNLR AHVQVVDSNGDR	Ara h1 Ara h3/4	564.4[2+] 786.9[2+] 688.8[2+] 771.4[2+] 684.5[3+] 695.4[2+] 435.8[+] 432.5[3+]	9, 10, 11
Almond	GNLDFVQPPR GVLGAFSGCPETFEE SQQSSQQGR ALPDEVLANAYQISR NGLHLPSYSNAPQLI YIVQGR	Prunin	571.9 [2+] 896.1 [3+] 830.4[2+] 780.8[3+]	9
Lupin	IVEFQSKPNTLILPK ATITIVNPDRR VIIPPTMRPR GLEETLCTMK ALQQIYENQSEQCQGR ISGGVPSVDLIMDK VGFNTNSLK	β conglutin α conglutin δ-conglutin γ-conglutin	576.4[3+] 419.2[3+] 394.1[3+] 591.2[2+] 651.3[3+] 715.8[2+] 490.2[2+]	12
Buck wheat	EGVRDLKELPSK		1369.74	13, 14
Tomato	GQTWVINAPR	NP24	573.3[2+]	15
Hazelnut	ADIYTEQVGR INTVNSNTLPVLR QGQVLTIPQNFAVAK ALPDDVLANAFQISR	11S globulin	576.3[+2] 720.9[+2] 807.5[+2] 815.5 [+2]	10

Table 10.6 contd. ...

...Table 10.6 contd.

Allergen	Peptide biomarkers	Allergenic proteins	m/z (or) M+H	References [¥]
Sesame	WESRQCQMRHCMQ WMRSMRG QYEESFLRSAEANQ GQFEHFREC EANQGQFEHFRECC NELRDVK	beta-globulin		16
Northern Shrimp	SEEEVFGLQK QQLVDDHFLFVSGDR TIDVNGDGLVGVDEYR	Tropomyosin Arginine kinase Sarcoplasmic calcium binding protein	583[+2] 592[+3]	17
Snow crab	SQLVENELDHAQEQ LSAATHK LVSAVNEIEK	Tropomyosin Arginine kinase	588[4+] 551[2+]	18
Tiger prawn	ANIQLVEK	Tropomyosin	457.77[2+]	19

[¥] References in Table 10.6 correspond to (1) (Lee and Kim 2010), (2) (Monaci et al. 2013), (3) (Chen et al. 2015), (4) (Monaci et al. 2014), (5) (Rogniaux et al. 2015), (6) (Picariello et al. 2012), (7) (Carrera et al. 2012), (8) (Ortea et al. 2011), (9) (Heick et al. 2011a), (10) (Heick et al. 2011b), (11) (Pedreschi et al. 2012), (12) (Mattarozzi et al. 2012), (13) (Chen et al. 2011), (14) (Satoh et al. 2010), (15) (Ippoushi et al. 2015), (16) (Wolff et al. 2004), (17) (Abdel Rahman et al. 2013), (18) (Abdel Rahman et al. 2012), and (19) (Abdel Rahman et al. 2010)

2014, Fæste et al. 2011, Gomaa and Boye 2015, Heick et al. 2011b, Wolff et al. 2004). Marker peptides usable for detection of a particular food allergen differ considerably between species and type of allergens and rely on how well the allergen(s) has been characterised. There is considerable literature available on signature peptides for certain food allergens such as milk, eggs, peanut, and wheat among few others facilitating food allergen testing (Table 10.6) (Pedreschi et al. 2012). Efforts to characterise peptide sequences for some important seafood allergens have been difficult, but are increasing (Koeberl et al. 2014). However for other allergenic foods such as sesame, kiwifruit, certain nuts and seafoods, including novel allergenic foods, the allergenic proteins and their peptide signatures have not been sufficiently characterised. Further complicating this is the variation of allergenic proteins/peptides due to differences between cultivars, species, location and processing methods.

10.6.5 Effects of Food Processing on Food Allergen Detection

Processing techniques and conditions have the potential to reduce or enhance food allergenicity and clearly illustrate the complexity of elucidating the structural basis of allergenicity. For instance heat treatment or protein hydrolysis steps typically used in milk production have resulted in novel epitopes through the association of α-lactalbumin and κ-casein to β-lactoglobulin due to heat denaturation (Picariello et al. 2011). Such processing induced changes may affect the IgE or IgG antibody reactivity based allergen detection methods as they are dependent on the specificity of the protein/peptide sequence and functional groups. MS based detection methods such as the MRM approach can be adopted to suit these processing changes for one or more proteins/peptides and offers a more comprehensive and high throughput detection technique for simultaneous measurements of multiple allergens. Therefore, careful consideration must be given when choosing biomarker peptides which are susceptible to undergo modifications from food processing.

10.6.6 Protein Glycosylation in Food Allergens

Allergenic proteins with a glycan moiety have become suspected targets for recent investigations on what may cause allergenic reactions. Many food allergens are water-soluble glycoproteins of sizes 10–70 (kDa) in monomeric form observed on SDS-PAGE and for binding of patient IgE. Allergens bind to IgE receptors located on tissue mast cells initiating an efflux of mediators that elicit subsequent inflammatory response reactions (Altmann 2007, Shade et al. 2015). Studies revealed that the reactivity of most plant and insect glycoproteins with patient sera IgE involved N-glycan containing the α1, 3-fucose or the β1, 2-ylose moiety. However, the binding of IgE to such carbohydrate epitopes does not appear to be biologically relevant (Tretter et al. 1993).

Allergenic proteins often undergo Maillard reactions during the thermal processing of foods where the resulting glycation and glyco-oxidation modifications on certain amino acids of the proteins along with their advanced glycation end (AGE) products have often been suspected for the pronounced allergenicity of the processed foods. In heated milk, N-carboxymethyllysine is the principal glycosylation end product (Renzone et al. 2015), and is formed as a result of the Maillard reaction between lactose and the amino group of lysine residues of proteins producing lactulosyl-lysine, which upon further oxidative degradation forms N-carboxymethyllysine (Meltretter et al. 2007). The characterisation of glycosylation and lactosylation among other potential post translational modifications is essential for profiling food allergens, but needs to overcome limitations such as the glycopeptide ion suppression in complex peptide mixtures, and the complicated MS/MS fragmentation patterns arising from different glycation combinations.

Glycopeptide enrichment methods such as using the phenylboronate chromatography or by using lectin or hydrophilic resins prior to MS analysis help to overcome some of these limitations (Picariello et al. 2011). Shotgun proteomics based MS/MS of enriched glycopeptides have been successfully employed to study glycosylation sites, and identify intermediate and advanced glycation end products of proteins. Enriched glycopeptides and glycans upon deglycosylation can be profiled by MALDI or LC-MS/MS analysis to reveal linker amino acid(s), glycosylation sites, and identify glycosylation type into N-linked glycans (sugar at N-terminal of asparagine) or O-linked glycans (sugar at hydroxyl group of serine or threonine) or C-mannosylation (α-mannopyranosyl attached to tryptophan) or glycophosphatidylinositol (c-terminal anchored). Analysis of associated sugars, the intermediate and advanced glyco-oxidation end product adducts often involves enzymatic or chemical deglycosylation of the hydrolytic cleavage of glycopeptides to release glycans and this may be done with or without prior derivatisation (Halim et al. 2015). This approach had enabled characterisation of several glycosylation sites in glycoproteins of commercial milks, where hundreds of derivatives and adducts were assigned to 31 proteins (Renzone et al. 2015). Recently, the N-linked oligomannose glycan structure of domain 3 containing the N394 site in human IgE has been found to be a critical glycosylation site with allergenic potential. It has been recently suggested that alteration or removal of this oligomannose glycan gene would prevent mast cells from initiating the allergic cascade thereby preventing anaphylaxis (Shade et al. 2015). In milk as a result of Maillard reaction associated glycation, peptides 78–91 and 125–138 of β-lactoglobulin was identified as AGE-containing peptides and were found to undergo glycosylation modifications at the Lys83 site and Lys135 respectively. MS/MS analysis of the corresponding lactosylated peptides revealed rare b and y type ions from glycosidic bond cleavages fragmenting with unique mass shifts. Further novel glycosylation sites susceptible for modifications upon food processing were identified (Renzone et al. 2015).

Although glycoproteomics studies on milk and peanuts are gaining momentum, many of the known allergens lack such analysis which warrants glycoproteomic research towards further understanding. Together this has the potential to elucidate the properties of the allergenic moiety and could help in accurate food allergen management such as the development of more sensitive and specific allergen detection assays.

10.6.7 Multiplexed Allergen Detection

The analysis of food allergen residues could benefit from the development of multi-allergen detection methods. Many food-allergic individuals are reactive to multiple allergenic foods. When these consumers experience allergic reactions, the identity of the residues that triggered the reactions is often unclear. Thus, public health authorities in particular have been interested in the ability to detect and quantify the presence of multiple allergenic proteins and peptides at trace levels in foods. The detection of allergenic protein residues at trace levels is often impeded by complexities arising from the complex nature of the food matrix, difficulties in protein extraction, screening from other abundant proteins present, and modifications of any allergen proteins as a result of food processing. Further development of the capability for multi-allergen detection methods could improve effective allergen management and the safety of allergic consumers.

Typical ELISA methods are not ideally suited for simultaneous multi-allergen analysis, although more than one allergen assays can be performed simultaneously. Novel approaches using antisera bound to different beads may overcome this obstacle (Cho et al. 2015, Garber et al. 2016). Nevertheless, lateral flow tests with more than one allergen detection have started to appear in the market.

Mass spectrometry methods offer more promise for the development of multi-allergen detection approach (Monaci et al. 2014, Uvackova et al. 2013). Recent studies have demonstrated the ability of mass spectrometry in the simultaneous detection of many different allergens in raw as well as cooked food (Gomaa and Boye 2015, Heick et al. 2011a, Monaci et al. 2013). MS allows high throughput multiple allergen detection in a single experiment from one or more samples, being cost, time as well as labour efficient. With the application of proteomics and mass spectrometry, a combined approach using shotgun proteomics and multiple ion monitoring allows the detection of multiple allergenic proteins and peptides in complex food matrices.

Multi-allergen detection through MS/MS is best achieved using multiple peptide transitions, and tested based on transitions showing the highest allergen matches. For the detection and quantification of the allergenic peptides in food, the mass spectrometer can be operated in selective ion monitoring (SIM) mode. The SIM approach typically involves mass spectrometry based analysis of several selected ion transitions specific to each allergenic protein, where each selected ion refers to a different peptide, and several of these ion transitions can be chosen to represent a range of proteins that may be present in the allergenic foods. Specific mass/charge transitions of selected intense ions corresponding to specific peptides are subjected to ion isolation and subsequent MS/MS. The quantification of allergen proteins and peptides can be achieved by either spiking isotopically labelled synthetic peptides in samples resulting in absolute quantification or by employing methods such as spectral counting as in the case of relative quantification (Houston et al. 2011, Zybailov et al. 2006). This approach has been demonstrated successfully with suitable sensitivity in foods containing peanuts, walnuts, hazelnuts, almonds, eggs, milk, and strawberry, among other allergenic ingredients (Heick et al. 2011a, Heick et al. 2011b, Pedreschi et al. 2012). In this sense selected reaction monitoring analysis facilitated by using the triple quadrupole mass spectrometer, may be concluded as the most versatile and popular quantitative mass spectrometry methodology for food allergen detection using the measurement of specific proteotypic peptide(s) (peptide precursor(s) and respective fragments masses) corresponding to the specific protein(s) of interest. However, further development and validation of multi-analyte approaches by mass spectrometry is needed before it can be adopted for routine use in food allergen analysis.

10.7 CONCLUSIONS

Food safety requirements and clinical regulations warrant for more efficient food allergen management through product integrity. It is becoming increasingly necessary to reliably identify traces of food allergens not only in food products but also in food processing equipment for possible cross-contamination. In this chapter, emphasis was given to two most versatile methods of food allergen detection, firstly the immunoassays that are popular commercially for their sensitivity, reasonable specificity, ease of use and portability. ELISAs are currently considered as a gold standard method for food allergen detection. Secondly, we focused on mass spectrometry based methods, which are gaining considerable recognition for its unique multifaceted testing capability. Selected ion monitoring facilitated by the triple quadrupole mass spectrometer provides multi-analyte analysis and detects specific proteotypic peptide(s) (peptide precursor(s) and respective fragments masses) corresponding to the specific allergenic protein(s) of interest and is the latest quantitative mass spectrometry based method for food allergen detection. Upon adequate evaluation across several food matrices, mass spectrometry based multi-analyte approach may become adopted for routine use in food allergen testing. We also discussed about food processing induced changes to the allergenic proteins which in turn affect the IgG antibody reactivity based food allergen detection, and how mass spectrometry methods could complement ELISA for effectively addressing such gap and achieve more accurate multi-allergen detection. For consumers and food manufacturers concerned about the presence of one or two food allergens, ELISAs seem most appropriate method for testing individual food products, yet for manufacturers who require compliance for several allergenic foods, mass spectrometry may be an ideal way of testing. In parallel to the development of MS based multi-allergen analysis, development of a portable, easy to use assay platform with multi-allergen detection capability such as multi-allergen assay on a chip could become a breakthrough for more effective food allergen management.

Keywords: Detection of food allergen; quantification of allergens; mass spectrometry and allergen quantification; ELISA for allergen quantification; food allergy

REFERENCES

Abbott, M., S. Hayward, W. Ross, S. B. Godefroy, F. Ulberth, A. J. Van Hengel et al. (2010). Validation procedures for quantitative food allergen ELISA methods: community guidance and best practices. J AOAC Int. 93(2): 442–450.

Abdel Rahman, A. M., S. Gagne and R. J. Helleur (2012). Simultaneous determination of two major snow crab aeroallergens in processing plants by use of isotopic dilution tandem mass spectrometry. Anal Bioanal Chem. 403(3): 821–831.

Abdel Rahman, A. M., S. Kamath, A. L. Lopata and R. J. Helleur (2010). Analysis of the allergenic proteins in black tiger prawn (Penaeus monodon) and characterization of the major allergen tropomyosin using mass spectrometry. Rapid Commun Mass Spectrom. 24(16): 2462–2470.

Abdel Rahman, A. M., S. D. Kamath, S. Gagne, A. L. Lopata and R. Helleur (2013). Comprehensive proteomics approach in characterizing and quantifying allergenic proteins from northern shrimp: toward better occupational asthma prevention. J Proteome Res. 12(2): 647–656.

Ahn, K., L. Bardina, G. Grishina, K. Beyer and H. A. Sampson (2009). Identification of two pistachio allergens, Pis v 1 and Pis v 2, belonging to the 2S albumin and 11S globulin family. Clin Exp Allergy. 39(6): 926–934.

Akagawa, M., T. Handoyo, T. Ishii, S. Kumazawa, N. Morita and K. Suyama (2007). Proteomic analysis of wheat flour allergens. J Agric Food Chem. 55(17): 6863–6870.

Albillos, S. M., F. Al-Taher and N. Maks (2011). Increasing extractability of protein for allergen detection after food processing. Food Chemistry. 127(4): 1831–1834.

Allen, K. J., B. C. Remington, J. L. Baumert, R. W. Crevel, G. F. Houben, S. Brooke-Taylor et al. (2014a). Allergen reference doses for precautionary labeling (VITAL 2.0): clinical implications. J Allergy Clin Immunol. 133(1): 156–164.

Allen, K. J., P. J. Turner, R. Pawankar, S. Taylor, S. Sicherer, G. Lack et al. (2014b). Precautionary labelling of foods for allergen content: are we ready for a global framework? World Allergy Organ J. 7(1): 10.

Altmann, F. (2007). The role of protein glycosylation in allergy. International Archives of Allergy and Immunology. 142(2): 99–115.

ATRS. (2014). Allergic and Toxic Reactions to Seafood. The Australasian Society of Clinical Immunology and Allergy, from www.allergyfacts.org.au: 1–2.

Bader, S., J. P. Oviedo, C. Pickardt and P. Eisner (2011). Influence of different organic solvents on the functional and sensory properties of lupin (*Lupinus angustifolius* L.) proteins. LWT—Food Science and Technology. 44(6): 1396–1404.

Barre, A., C. Sordet, R. Culerrier, F. Rance, A. Didier and P. Rouge (2008). Vicilin allergens of peanut and tree nuts (walnut, hazelnut and cashew nut) share structurally related IgE-binding epitopes. Mol Immunol. 45(5): 1231–1240.

Bauermeister, K., A. Wangorsch, L. P. Garoffo, A. Reuter, A. Conti, S. L. Taylor et al. (2011). Generation of a comprehensive panel of crustacean allergens from the North Sea Shrimp Crangon crangon. Molecular Immunology. 48(15-16): 1983–1992.

Baumert, J. L. (2013). Detecting and measuring allergens in food. 215–226. Risk Management for Food Allergy.

Berrens, L. (1996). Neoallergens in heated pecan nut: products of Maillard-type degradation? Allergy. 51(4): 277–278.

Bhatia, V. N., D. H. Perlman, C. E. Costello and M. E. McComb (2009). Software tool for researching annotations of proteins: Open-source protein annotation software with data visualization. Anal Chem. 81(23): 9819–9823.

Boros, C. A., D. Kay and M. S. Gold (2000). Parent reported allergy and anaphylaxis in 4173 South Australian children. J Paediatr Child Health. 36(1): 36–40.

Brough, H. A., P. J. Turner, T. Wright, A. T. Fox, S. L. Taylor, J. O. Warner et al. (2015). Dietary management of peanut and tree nut allergy: What exactly should patients avoid? Clin Exp Allergy. 45(5): 859–871.

Bruijnzeel-Koomen, C., C. Ortolani, K. Aas, C. Bindslev-Jensen, B. Bjorksten, D. Moneret-Vautrin et al. 1995. Adverse reactions to food. European academy of allergology and clinical immunology subcommittee. Allergy. 50(8): 623–635.

Careri, M., A. Costa, L. Elviri, J. B. Lagos, A. Mangia, M. Terenghi et al. (2007). Use of specific peptide biomarkers for quantitative confirmation of hidden allergenic peanut proteins Ara h 2 and Ara h 3/4 for food control by liquid chromatography-tandem mass spectrometry. Anal Bioanal Chem. 389(6): 1901–1907.

Carrera, M., B. Canas and J. M. Gallardo (2012). Rapid direct detection of the major fish allergen, parvalbumin, by selected MS/MS ion monitoring mass spectrometry. J Proteomics. 75(11): 3211–3220.

Chen, P., Y. F. Guo, Q. Yan and Y. H. Li (2011). Molecular cloning and characterization of Fag t 2: a 16-kDa major allergen from Tartary buckwheat seeds. Allergy. 66(10): 1393–1395.

Chen, Q., J. Zhang, X. Ke, S. Lai, B. Tao, J. Yang et al. (2015). Quantification of bovine beta-casein allergen in baked foodstuffs based on ultra-performance liquid chromatography with tandem mass spectrometry. Food Addit Contam Part A Chem Anal Control Expo Risk Assess. 32(1): 25–34.

Cho, C. Y., W. Nowatzke, K. Oliver and E. A. Garber (2015). Multiplex detection of food allergens and gluten. Anal Bioanal Chem. 407(14): 4195–4206.

Clemente, A., S. J. Chambers, F. Lodi, C. Nicoletti and G. M. Brett (2004). Use of the indirect competitive ELISA for the detection of Brazil nut in food products. Food Control. 15(1): 65–69.

Costa, J., P. Ansari, I. Mafra, M. B. Oliveira and S. Baumgartner (2015). Development of a sandwich ELISA-type system for the detection and quantification of hazelnut in model chocolates. Food Chem. 173: 257–265.

Cox, J. and M. Mann (2008). MaxQuant enables high peptide identification rates, individualized p.p.b.-range mass accuracies and proteome-wide protein quantification. Nat Biotech. 26(12): 1367–1372.

Crowther, J. R. (2000). The ELISA Guidebook. Humana Press.

Cucu, T., B. Devreese, S. Trashin, B. Kerkaert, M. Rogge and B. De Meulenaer (2012). Detection of hazelnut in foods using ELISA: challenges related to the detectability in processed foodstuffs. J AOAC Int. 95(1): 149–156.

Cunsolo, V., V. Muccilli, R. Saletti and S. Foti (2014). Mass spectrometry in food proteomics: A tutorial. J Mass Spectrom. 49(9): 768–784.

Davis, P. J., C. M. Smales and D. C. James (2001). How can thermal processing modify the antigenicity of proteins? Allergy. 56 Suppl. 67: 56–60.

de Leon, M. P., I. N. Glaspole, A. C. Drew, J. M. Rolland, R. E. O'Hehir and C. Suphioglu (2003). Immunological analysis of allergenic cross-reactivity between peanut and tree nuts. Clin Exp Allergy. 33(9): 1273–1280.

De Meulenaer, B. and A. Huyghebaert (2001). Isolation and purification of chicken egg yolk immunoglobulins: A review. Food and Agricultural Immunology. 13(4): 275–288.

Di Girolamo, F., M. Muraca, O. Mazzina, I. Lante and L. Dahdah (2015). Proteomic applications in food allergy: food allergenomics. Curr Opin Allergy Clin Immunol. 15(3): 259–266.

Doi, H., Y. Touhata, H. Shibata, S. Sakai, A. Urisu, H. Akiyama et al. (2008). Reliable enzyme-linked immunosorbent assay for the determination of walnut proteins in processed foods. J Agric Food Chem. 56(17): 7625–7630.

DunnGalvin, A., C. H. Chan, R. Crevel, K. Grimshaw, R. Poms, S. Schnadt et al. (2015). Precautionary allergen labelling: perspectives from key stakeholder groups. Allergy. 70(9): 1039–1051.

Esser, P. (1991). Blocking agent and detergent in ELISA. Nunc Bull. 9: 1–4.

Fæste, C. K., H. T. Rønning, U. Christians and P. E. Granum (2011). Liquid chromatography and mass spectrometry in food allergen detection. Journal of Food Protection. 74(2): 316–345.

Fenn, J. B., M. Mann, C. K. Meng, S. F. Wong and C. M. Whitehouse (1989). Electrospray ionization for mass spectrometry of large biomolecules. Science. 246(4926): 64–71.

Ferreira, F., T. Hawranek, P. Gruber, N. Wopfner and A. Mari (2004). Allergic cross-reactivity: from gene to the clinic. Allergy. 59(3): 243–267.

FSANZ. (2010). Review of the regulatory management of food allergens. Food Standards Australia New Zealand, Food Standards Australia New Zealand.

Fu, T. J. and N. Maks (2013). Impact of thermal processing on ELISA detection of peanut allergens. J Agric Food Chem. 61(24): 5649–5658.

Garber, E. A., C. H. Parker, S. M. Handy, C. Y. Cho, R. Panda, M. Samadpour et al. (2016). Presence of undeclared food allergens in cumin: The need for multiplex methods. J Agric Food Chem. 64(5): 1202–1211.

Gardas, A. and A. Lewartowska (1988). Coating of proteins to polystyrene ELISA plates in the presence of detergents. J Immunol Methods. 106(2): 251–255.

Garino, C., L. Zuidmeer, J. Marsh, A. Lovegrove, M. Morati, S. Versteeg et al. (2010). Isolation, cloning, and characterization of the 2S albumin: a new allergen from hazelnut. Mol Nutr Food Res. 54(9): 1257–1265.

Gaskin, F. E. and S. L. Taylor (2011). Sandwich enzyme-linked immunosorbent assay (ELISA) for detection of cashew nut in foods. J Food Sci. 76(9): T218–226.

Gibbs, J. and M. Kennebunk (2001a). Immobilization Principles–Selecting the Surface. Sciences, C. I. L. New York.

Gibbs, J. and M. Kennebunk (2001b). Effective blocking procedures. ELISA Technical Bulletin No. Corp., C. C. 3.

Gobom, J., E. Nordhoff, E. Mirgorodskaya, R. Ekman and P. Roepstorff (1999). Sample purification and preparation technique based on nano-scale reversed-phase columns for the sensitive analysis of complex peptide mixtures by matrix-assisted laser desorption/ionization mass spectrometry. J Mass Spectrom. 34(2): 105–116.

Goetz, D. W., B. A. Whisman and A. D. Goetz (2005). Cross-reactivity among edible nuts: double immunodiffusion, crossed immunoelectrophoresis, and human specific igE serologic surveys. Ann Allergy Asthma Immunol. 95(1): 45–52.

Gomaa, A. and J. Boye (2015). Simultaneous detection of multi-allergens in an incurred food matrix using ELISA, multiplex flow cytometry and liquid chromatography mass spectrometry (LC-MS). Food Chem. 175: 585–592.

Halim, A., M. C. Carlsson, C. B. Madsen, S. Brand, S. R. Moller, C. E. Olsen et al. (2015). Glycoproteomic analysis of seven major allergenic proteins reveals novel post-translational modifications. Mol Cell Proteomics. 14(1): 191–204.

Harboe, N. M. G. and A. Ingild (1983). Immunization, isolation of immunoglobulins and antibody titre determination. Scand J Immunol. 17: 345–351.

He, J. (2013). Chapter 5.1. Practical Guide to ELISA Development A2 Wild, David. 381–393. The Immunoassay Handbook (Fourth Edition). Elsevier, Oxford.

Heick, J., M. Fischer, S. Kerbach, U. Tamm and B. Popping (2011a). Application of a liquid chromatography tandem mass spectrometry method for the simultaneous detection of seven allergenic foods in flour and bread and comparison of the method with commercially available ELISA test kits. J AOAC Int. 94(4): 1060–1068.

Heick, J., M. Fischer and B. Popping (2011b). First screening method for the simultaneous detection of seven allergens by liquid chromatography mass spectrometry. J Chromatogr A. 1218(7): 938–943.

Hirst, B. and S. J. Miguel (2015). 16—Validation, standardisation and harmonisation of analytical methods and test kits for detecting allergens in food A2—Flanagan, Simon. 295–309. Handbook of Food Allergen Detection and Control. Woodhead Publishing.

Holzhauser, T. and S. Vieths (1999). Quantitative sandwich ELISA for determination of traces of hazelnut (*Corylus avellana*) protein in complex food matrixes. J Agric Food Chem. 47(10): 4209–4218.

Hornbeck, P. (2001). Enzyme-linked immunosorbent assays. Curr Protoc Immunol. Chapter 2: Unit 2.1.

Hourihane, O. B. (2001). The threshold concept in food safety and its applicability to food allergy. Allergy. 56 Suppl. 67: 86–90.

Houston, N. L., D. G. Lee, S. E. Stevenson, G. S. Ladics, G. A. Bannon, S. McClain et al. (2011). Quantitation of soybean allergens using tandem mass spectrometry. J Proteome Res. 10(2): 763–773.

Huber, D., J. Rudolf, P. Ansari, B. Galler, M. Fuhrer, C. Hasenhindl et al. (2009). Effectiveness of natural and synthetic blocking reagents and their application for detecting food allergens in enzyme-linked immunosorbent assays. Anal Bioanal Chem. 394(2): 539–548.

Ito, K., T. Yamamoto, Y. Oyama, R. Tsuruma, E. Saito, Y. Saito, T. Ozu, T. Honjoh, R. Adachi, S. Sakai, H. Akiyama and M. Shoji (2016). Food allergen analysis for processed food using a novel extraction method to eliminate harmful reagents for both ELISA and lateral-flow tests. Anal Bioanal Chem. 408(22): 5973–5984.

Kaori Ito, Takayuki Yamamoto, Yuriko Oyama, Rieko Tsuruma.

Immer, U. and M. Lacorn (2015). 10 - Enzyme-linked immunosorbent assays (ELISAs) for detecting allergens in food A2 - Flanagan, Simon. 199–217. Handbook of Food Allergen Detection and Control. Woodhead Publishing.

Ippoushi, K., M. Sasanuma, H. Oike, M. Kobori and M. Maeda-Yamamoto (2015). Absolute quantification of protein NP24 in tomato fruit by liquid chromatography/tandem mass spectrometry using stable isotope-labelled tryptic peptide standard. Food Chem. 173: 238–242.

Ismail, B. and S. S. Nielsen (2010). Basic Principles of Chromatography. 473–498. Food Analysis. Springer US, Boston, MA.

Iwan, M., Y. M. Vissers, E. Fiedorowicz, H. Kostyra, E. Kostyra, H. F. Savelkoul et al. (2011). Impact of Maillard reaction on immunoreactivity and allergenicity of the hazelnut allergen Cor a 11. J Agric Food Chem. 59(13): 7163–7171.

Jenkins, J. A., S. Griffiths-Jones, P. R. Shewry, H. Breiteneder and E. N. C. Mills (2005). Structural relatedness of plant food allergens with specific reference to cross-reactive allergens: An in silico analysis. Journal of Allergy and Clinical Immunology. 115(1): 163–170.

Johnson, P. E., S. Baumgartner, T. Aldick, C. Bessant, V. Giosafatto, J. Heick et al. (2011). Current perspectives and recommendations for the development of mass spectrometry methods for the determination of allergens in foods. J AOAC Int. 94(4): 1026–1033.

Karaszkiewicz, J. W. (2005). Critical Factors in Immunoassay Optimization. Gaithersburg, MD: Kirkegaard & Perry Laboratories, Inc.

Khuda, S. E., L. S. Jackson, T. J. Fu and K. M. Williams (2015). Effects of processing on the recovery of food allergens from a model dark chocolate matrix. Food Chem. 168: 580–587.

Kiening, M., R. Niessner, E. Drs, S. Baumgartner, R. Krska, M. Bremer et al. (2005). Sandwich immunoassays for the determination of peanut and hazelnut traces in foods. J Agric Food Chem. 53(9): 3321–3327.

Kirsch, S., S. Fourdrilis, R. Dobson, M. L. Scippo, G. Maghuin-Rogister and E. De Pauw (2009). Quantitative methods for food allergens: a review. Anal Bioanal Chem. 395(1): 57–67.

Koeberl, M., D. Clarke and A. L. Lopata (2014). Next generation of food allergen quantification using mass spectrometric systems. J Proteome Res. 13(8): 3499–3509.

Koppelman, S. J. and S. L. Hefle (2006). Detecting Allergens in Food. Woodhead Pub.

Lee, J. Y. and C. J. Kim (2010). Determination of allergenic egg proteins in food by protein-, mass spectrometry-, and DNA-based methods. J AOAC Int. 93(2): 462–477.

Lee, N. A. and I. R. Kennedy (2007). Immunoassays. Chapter 5. *In*: Yolanda Picó (ed.). Food Toxicant Analysis, Techniques, Strategies and Developments. Elsevier Publication.

Lee, N. A. and X. Sun (2016). Immunoanalytical techniques for peanut allergen residue detection. Chapter 9. *In*: N. A. Lee, G. C. Wright and Rao Rachaputi (eds.). Peanuts: Bioactives and Allergens. Destech Publisher.

Lipton, C. R., J. X. Dautlick, C. D. Grothaus, P. L. Hunst, K. M. Magin, C. A. Mihaliak et al. (2000). Guidelines for the validation and use of immunoassays for determination of introduced proteins in biotechnology enhanced crops and derived food ingredients. Food and Agricultural Immunology. 12(2): 153–164.

Lon, V. K. (2013). Production of Polyclonal Antibodies. 33–65. Making and Using Antibodies. CRC Press.

Longitudinal Study of Australian Children Annual Statistical Report 2010. Annual Statistical Report. Maguire, B., Australian Institute of Family Studies.

Maleki, S. J., S. Y. Chung, E. T. Champagne and J. P. Raufman (2000). The effects of roasting on the allergenic properties of peanut proteins. J Allergy Clin Immunol. 106(4): 763–768.

Mattarozzi, M., C. Bignardi, L. Elviri and M. Careri (2012). Rapid shotgun proteomic liquid chromatography-electrospray ionization-tandem mass spectrometry-based method for the lupin (*Lupinus albus* L.) multi-allergen determination in foods. J Agric Food Chem. 60(23): 5841–5846.

Mattison, C. P., C. C. Grimm and R. L. Wasserman (2014). *In vitro* digestion of soluble cashew proteins and characterization of surviving IgE-reactive peptides. Mol Nutr Food Res. 58(4): 884–893.

Meltretter, J., S. Seeber, A. Humeny, C. -M. Becker and M. Pischetsrieder (2007). Site-specific formation of maillard, oxidation, and condensation products from whey proteins during reaction with lactose. Journal of Agricultural and Food Chemistry. 55(15): 6096–6103.

Mills, E. N., A. I. Sancho, N. M. Rigby, J. A. Jenkins and A. R. Mackie (2009). Impact of food processing on the structural and allergenic properties of food allergens. Mol Nutr Food Res. 53(8): 963–969.

Monaci, L., I. Losito, E. De Angelis, R. Pilolli and A. Visconti (2013). Multi-allergen quantification of fining-related egg and milk proteins in white wines by high-resolution mass spectrometry. Rapid Commun Mass Spectrom. 27(17): 2009–2018.

Monaci, L., R. Pilolli, E. De Angelis, M. Godula and A. Visconti (2014). Multi-allergen detection in food by micro high-performance liquid chromatography coupled to a dual cell linear ion trap mass spectrometry. Journal of Chromatography A. 1358: 136–144.

Moreno, F. J., J. A. Jenkins, F. A. Mellon, N. M. Rigby, J. A. Robertson, N. Wellner et al. (2004). Mass spectrometry and structural characterization of 2S albumin isoforms from Brazil nuts (Bertholletia excelsa). Biochim Biophys Acta. 1698(2): 175–186.

Mullins, R. J. (2007). Paediatric food allergy trends in a community-based specialist allergy practice, 1995–2006. Med J Aust. 186(12): 618–621.

Neilson, K. A., N. A. Ali, S. Muralidharan, M. Mirzaei, M. Mariani, G. Assadourian et al. (2011). Less label, more free: approaches in label-free quantitative mass spectrometry. Proteomics. 11(4): 535–553.

Neto, V. Q., N. Narain, J. B. Silva and P. S. Bora (2001). Functional properties of raw and heat processed cashew nut (Anacardium occidentale, L.) kernel protein isolates. Nahrung. 45(4): 258–262.

Niemann, L., S. L. Taylor and S. L. Hefle (2009). Detection of walnut residues in foods using an enzyme-linked immunosorbent assay. J Food Sci. 74(6): T51–57.

Ortea, I., B. Cañas and J. M. Gallardo (2011). Selected tandem mass spectrometry ion monitoring for the fast identification of seafood species. Journal of Chromatography A. 1218(28): 4445–4451.

Paulie, S., H. Perlmann and P. Perlmann (2001). Enzyme-linked Immunosorbent Assay. eLS. John Wiley & Sons, Ltd.

Pedreschi, R., J. Nørgaard and A. Maquet (2012). Current challenges in detecting food allergens by shotgun and targeted proteomic approaches: A case study on traces of peanut allergens in baked cookies. Nutrients. 4(2): 132.

Picariello, G., G. Mamone, F. Addeo and P. Ferranti (2011). The frontiers of mass spectrometry-based techniques in food allergenomics. Journal of Chromatography A. 1218(42): 7386–7398.

Picariello, G., G. Mamone, C. Nitride, F. Addeo, A. Camarca, I. Vocca et al. (2012). Shotgun proteome analysis of beer and the immunogenic potential of beer polypeptides. J Proteomics. 75(18): 5872–5882.

Pilolli, R., E. De Angelis, M. Godula, A. Visconti and L. Monaci (2014). Orbitrap monostage MS versus hybrid linear ion trap MS: application to multi-allergen screening in wine. J Mass Spectrom. 49(12): 1254–1263.

Poms, R. E., C. Capelletti and E. Anklam (2004b). Effect of roasting history and buffer composition on peanut protein extraction efficiency. Mol Nutr Food Res. 48(6): 459–464.

Poms, R. E., C. L. Klein and E. Anklam (2004a). Methods for allergen analysis in food: a review. Food Addit Contam. 21(1): 1–31.

Prescott, S. L., R. Pawankar, K. J. Allen, D. E. Campbell, J. K. Sinn, A. Fiocchi et al. (2013). A global survey of changing patterns of food allergy burden in children. World Allergy Organization Journal. 6(1): 1–12.

Radauer, C. and H. Breiteneder (2007). Evolutionary biology of plant food allergens. Journal of Allergy and Clinical Immunology. 120(3): 518–525.

Radauer, C., M. Bublin, S. Wagner, A. Mari and H. Breiteneder (2008). Allergens are distributed into few protein families and possess a restricted number of biochemical functions. Journal of Allergy and Clinical Immunology. 121(4): 847–852. e847.

Renzone, G., S. Arena and A. Scaloni (2015). Proteomic characterization of intermediate and advanced glycation end-products in commercial milk samples. Journal of Proteomics. 117: 12–23.

Rigby, N. M., J. Marsh, A. I. Sancho, K. Wellner, J. Akkerdaas, R. van Ree et al. (2008). The purification and characterisation of allergenic hazelnut seed proteins. Molecular Nutrition & Food Research 52(S2): S251–S261.

Robotham, J. M., L. Xia, L. N. Willison, S. S. Teuber, S. K. Sathe and K. H. Roux (2010). Characterization of a cashew allergen, 11S globulin (Ana o 2), conformational epitope. Mol Immunol. 47(9): 1830–1838.

Rogniaux, H., M. Pavlovic, R. Lupi, V. Lollier, M. Joint, H. Mameri et al. (2015). Allergen relative abundance in several wheat varieties as revealed via a targeted quantitative approach using MS. Proteomics. 15(10): 1736–1745.

Roux, K. H., S. S. Teuber, J. M. Robotham and S. K. Sathe (2001). Detection and stability of the major almond allergen in foods. J Agric Food Chem. 49(5): 2131–2136.

Rubert, J., M. Zachariasova and J. Hajslova (2015). Advances in high-resolution mass spectrometry based on metabolomics studies for food—a review. Food Addit Contam Part A Chem Anal Control Expo Risk Assess. 32(10): 1685–1708.

Sathe, S. K. (1994). Solubilization and electrophoretic characterization of cashew nut (Anacardium occidentale) proteins. Food Chemistry. 51(3): 319–324.

Sathe, S. K., M. Venkatachalam, G. M. Sharma, H. H. Kshirsagar, S. S. Teuber and K. H. Roux (2009). Solubilization and electrophoretic characterization of select edible nut seed proteins. J Agric Food Chem. 57(17): 7846–7856.

Sathe, S. K., W. J. Wolf, K. H. Roux, S. S. Teuber, M. Venkatachalam and K. W. Sze-Tao (2002). Biochemical characterization of amandin, the major storage protein in almond (Prunus dulcis L.). J Agric Food Chem. 50(15): 4333–4341.

Satoh, R., S. Koyano, K. Takagi, R. Nakamura and R. Teshima (2010). Identification of an IgE-binding epitope of a major buckwheat allergen, BWp16, by SPOTs assay and mimotope screening. Int Arch Allergy Immunol. 153(2): 133–140.

Scigelova, M. and A. Makarov (2006). Orbitrap mass analyzer—overview and applications in proteomics. Proteomics. 6 Suppl. 2: 16–21.

Selvarajah, S., O. H. Negm, M. R. Hamed, C. Tubby, I. Todd, P. J. Tighe et al. (2014). Development and validation of protein microarray technology for simultaneous inflammatory mediator detection in human sera. Mediators Inflamm. 2014: 820304.

Shade, K. T., B. Platzer, N. Washburn, V. Mani, Y. C. Bartsch, M. Conroy et al. (2015). A single glycan on IgE is indispensable for initiation of anaphylaxis. J Exp Med. 212(4): 457–467.

Sharma, G. M., C. Mundoma, M. Seavy, K. H. Roux and S. K. Sathe (2010). Purification and biochemical characterization of Brazil nut (Bertholletia excelsa L.) seed storage proteins. J Agric Food Chem. 58(9): 5714–5723.

Sharma, G. M., K. H. Roux and S. K. Sathe (2009). A sensitive and robust competitive enzyme-linked immunosorbent assay for Brazil nut (Bertholletia excelsa L.) detection. J Agric Food Chem. 57(2): 769–776.

Spillner, E., I. Braren, K. Greunke, H. Seismann, S. Blank and D. du Plessis (2012). Avian IgY antibodies and their recombinant equivalents in research, diagnostics and therapy. Biologicals. 40(5): 313–322.

Su, M., M. Venkatachalam, S. S. Teuber, K. H. Roux and S. K. Sathe (2004). Impact of γ-irradiation and thermal processing on the antigenicity of almond, cashew nut and walnut proteins. Journal of the Science of Food and Agriculture. 84(10): 1119–1125.

Tanaka, K., H. Waki, Y. Ido, S. Akita, Y. Yoshida, T. Yoshida et al. (1988). Protein and polymer analyses up to m/z 100000 by laser ionization time-of-flight mass spectrometry. Rapid Communications in Mass Spectrometry. 2(8): 151–153.

Taylor, S. L., J. L. Baumert, A. G. Kruizinga, B. C. Remington, R. W. Crevel, S. Brooke-Taylor et al. (2014). Establishment of Reference Doses for residues of allergenic foods: report of the VITAL Expert Panel. Food Chem Toxicol. 63: 9–17.

Taylor, S. L., S. L. Hefle, C. Bindslev-Jensen, S. A. Bock, A. W. Burks, Jr., L. Christie et al. (2002). Factors affecting the determination of threshold doses for allergenic foods: How much is too much? J Allergy Clin Immunol. 109(1): 24–30.

Taylor, S. L., S. L. Hefle, K. Farnum, S. W. Rizk, J. Yeung, M. E. Barnett et al. (2007). Survey and evaluation of Pre-FALCPA labeling practices used by food manufacturers to address allergen concerns. Comprehensive Reviews in Food Science and Food Safety. 6(2): 36–46.

Taylor, S. L., J. A. Nordlee, L. M. Niemann and D. M. Lambrecht (2009). Allergen immunoassays—considerations for use of naturally incurred standards. Anal Bioanal Chem. 395(1): 83–92.

Tretter, V., F. Altmann, V. Kubelka, L. Marz and W. M. Becker (1993). Fucose alpha 1,3-linked to the core region of glycoprotein N-glycans creates an important epitope for IgE from honeybee venom allergic individuals. Int Arch Allergy Immunol. 102(3): 259–266.

Uvackova, L., L. Skultety, S. Bekesova, S. McClain and M. Hajduch (2013). MS(E) based multiplex protein analysis quantified important allergenic proteins and detected relevant peptides carrying known epitopes in wheat grain extracts. J Proteome Res. 12(11): 4862–4869.

van Boxtel, E. L., S. J. Koppelman, L. A. van den Broek and H. Gruppen (2008). Heat denaturation of Brazil nut allergen Ber e 1 in relation to food processing. Food Chem. 110(4): 904–908.

van Hengel, A. J. (2007). Food allergen detection methods and the challenge to protect food-allergic consumers. Anal Bioanal Chem. 389(1): 111–118.

Van Herwijnen, R. and S. Baumgartner (2006). 10—The use of lateral flow devices to detect food allergens. 175–181. Detecting Allergens in Food. Woodhead Publishing.

Venkatachalam, M., E. K. Monaghan, H. H. Kshirsagar, J. M. Robotham, S. E. O'Donnell, M. S. Gerber et al. (2008). Effects of processing on immunoreactivity of cashew nut (Anacardium occidentale L.) seed flour proteins. J Agric Food Chem. 56(19): 8998–9005.

Wagner, K., K. Racaityte, K. K. Unger, T. Miliotis, L. E. Edholm, R. Bischoff et al. (2000). Protein mapping by two-dimensional high performance liquid chromatography. Journal of Chromatography A. 893(2): 293–305.

Wagner, S., C. Radauer, M. Bublin, K. Hoffmann-Sommergruber, T. Kopp, E. K. Greisenegger et al. (2008). Naturally occurring hypoallergenic Bet v 1 isoforms fail to induce IgE responses in individuals with birch pollen allergy. Journal of Allergy and Clinical Immunology. 121(1): 246–252.

Wang, F., J. M. Robotham, S. S. Teuber, S. K. Sathe and K. H. Roux (2003). Ana o 2, a major cashew (Anacardium occidentale L.) nut allergen of the legumin family. Int Arch Allergy Immunol. 132(1): 27–39.

Wang, F., J. M. Robotham, S. S. Teuber, P. Tawde, S. K. Sathe and K. H. Roux (2002). Ana o 1, a cashew (Anacardium occidental) allergen of the vicilin seed storage protein family. J Allergy Clin Immunol. 110(1): 160–166.

WAO, W. (2011). White Book on Allergy (WAO).

World Allergy Organization (WAO), Wisconsin, USA.

Wei, Y., S. K. Sathe, S. S. Teuber and K. H. Roux (2003). A sensitive sandwich ELISA for the detection of trace amounts of cashew (Anacardium occidentale L.) nut in foods. J Agric Food Chem. 51(11): 3215–3221.

Wen, H. -W., W. Borejsza-Wysocki, T. R. DeCory and R. A. Durst (2007). Peanut allergy, peanut allergens, and methods for the detection of peanut contamination in food products. Comprehensive Reviews in Food Science and Food Safety. 6(2): 47–58.

Willison, L. N., P. Tawde, J. M. Robotham, R. M. t. Penney, S. S. Teuber, S. K. Sathe et al. (2008). Pistachio vicilin, Pis v 3, is immunoglobulin E-reactive and cross-reacts with the homologous cashew allergen, Ana o 1. Clin Exp Allergy. 38(7): 1229–1238.

Wolff, N., S. Yannai, N. Karin, Y. Levy, R. Reifen, I. Dalal et al. (2004). Identification and characterization of linear B-cell epitopes of beta-globulin, a major allergen of sesame seeds. J Allergy Clin Immunol. 114(5): 1151–1158.

Ye, J., L. Fang, H. Zheng, Y. Zhang, J. Chen, Z. Zhang et al. (2006). WEGO: A web tool for plotting GO annotations. Nucleic Acids Res. 34(Web Server issue): W293–297.

Yeung, J. (2006). 6—Enzyme-linked immunosorbent assays (ELISAs) for detecting allergens in foods. 109–124. Detecting Allergens in Food. Woodhead Publishing.

Zurzolo, G. A., J. J. Koplin, M. L. Mathai, S. L. Taylor, D. Tey and K. J. Allen (2013). Foods with precautionary allergen labeling in Australia rarely contain detectable allergen. J Allergy Clin Immunol Pract. 1(4): 401–403.

Zurzolo, G. A., M. L. Mathai, J. J. Koplin and K. J. Allen (2012). Hidden allergens in foods and implications for labelling and clinical care of food allergic patients. Curr Allergy Asthma Rep. 12(4): 292–296.

Zybailov, B., A. L. Mosley, M. E. Sardiu, M. K. Coleman, L. Florens and M. P. Washburn (2006). Statistical analysis of membrane proteome expression changes in *Saccharomyces cerevisiae*. J Proteome Res. 5(9): 2339–2347.

Recombinant Food Allergens for Diagnosis and Therapy

*Heidi Hofer, Anargyros Roulias, Claudia Asam,
Stephanie Eichhorn, Fátima Ferreira, Gabriele
Gadermaier* and *Michael Wallner**

CONTENTS

Department of Molecular Biology, University of Salzburg, Salzburg, Austria.
* Corresponding author: michael-wallner@gmx.at

11.1 INTRODUCTION

IgE mediated food allergies affect approximately 3–8% of children and 1–3% of the adult population in industrialized countries. Allergic reactions to food can cause serious, sometimes life threatening reactions, thus food allergies require special medical attention, including accurate diagnostic as well as therapeutic strategies (Valenta et al. 2015). Depending on the sensitization process, food allergies can be classified into two categories which are (i) class I or classical food allergies and (ii) class II food allergies. In class I food allergies, the primary sensitizer is the food allergen that eventually also elicits the allergic reaction, whereas class II food allergies are characterized by a sensitization process, which is actually initiated by an inhalant allergen; the allergic reaction to the food allergen however, is triggered by cross-reactive IgE antibodies. Though epidemiologic data are rare, it is estimated that 60% of food allergies in older children or adults, respectively, are associated with inhalant allergies. This implies that an increase of pollen allergies will also negatively affect the number of food allergic individuals (Werfel et al. 2015). Since the cloning of the first allergens in 1988 (Cromwell et al. 2011), 287 food allergens from 98 different food sources have been identified and officially acknowledged by the WHO/IUIS allergen nomenclature sub-committee (www.allergen.org). Approximately 90% of class I food allergies in westernized countries are triggered by the so called "big 8" allergen sources, including milk, egg, fish, shellfish, tree nuts, peanuts, wheat, and soy (www.fda.gov). When looking at class II food allergies however, the picture is much different and rather heterogeneous. This high complexity of food allergies requires elaborated diagnostic as well as therapeutic methods to efficiently tackle the problem. Recombinant allergens have already established themselves as indispensable tools for allergy diagnosis accomplishing high sensitivity of allergy tests in parallel with superior specificity. Moreover, attempts have been made to substitute allergen extracts in therapeutic setups. Within this book chapter, we will provide a comprehensive overview on available recombinant food allergens, discuss production methods, as well as quality controls and will elaborate the use of recombinant allergens for food allergy diagnosis and therapy.

11.2 RECOMBINANT FOOD ALLERGENS

Of the 287 food allergens listed within the WHO/IUIS allergen nomenclature database, 157 have been produced as recombinant proteins, 151 in *E. coli*, 11 in the yeast *Pichia pastoris*, and two in other organisms. A summary of all food allergens entered into the WHO/IUIS allergen nomenclature database is given in Table 11.1. Unlike natural allergens or allergen extracts, the production of recombinant proteins is not dependent on biological source material (i.e., fruits, nuts, legumes, etc.) with variable allergen expression. Moreover, as shown for many food allergen families (i.e., PR-10 proteins, lipid transfer proteins, seed storage proteins, or parvalbumins, just to mention a few), natural allergen preparations are composed of complex mixtures of allergen isoforms, which sometimes show rather different immunologic properties (Son et al. 1999, Swoboda et al. 2002, Ramos et al. 2009, Ramazzina et al. 2012). Thus, one could imagine that the production processes of an allergen product based on natural allergens could bias the isoform composition, therefore influencing the properties of the preparation (Cromwell et al. 2011). A major advantage of recombinant proteins is that they can be fully characterized in terms of their physico-chemical as well as immunological properties. Both aspects will be discussed in detail within the next paragraphs of this book chapter. The tight characterization process enables excellent batch to batch reproducibility making the use of recombinant products attractive for pharmaceutical industry. Moreover, recombinant allergens can circumvent the problem of cross-contamination of natural allergen preparations with minute amounts of other allergens from the same source. The use of recombinant technology however, does not only allow tightly controlled production of wild-type allergens but further enables scientists to genetically engineer allergen derivatives meeting exactly the needs for medical applications. This implies proteins with reduced IgE binding to increase the safety profile of allergen products for therapeutic applications or allergen derivatives with increased immunogenicity. However, due to the often complex allergen composition, as well as individual patients' recognition patterns, questions concerning which allergens to use for therapeutic applications remain heavily debated. Moreover, many complex post-translational modifications (i.e., glycosylation) have been identified on food allergens. Therefore, it is extremely important to carefully select a suitable expression system for recombinant production and perform strict quality control on the product Figure 11.1.

11.3 PHYSICOCHEMICAL ANALYSIS OF RECOMBINANT FOOD ALLERGENS

Physicochemical data on recombinantly produced allergens is important to understand their properties and immunological behavior and to warrant their quality. Presently, there is a huge variety of well-established techniques available for such analyses. The following segment will describe and focus on a panel of techniques providing a solid and efficient physicochemical characterization of purified food allergens. **SDS-PAGE** is a good method to check protein samples for integrity and potential impurities and to estimate the size and quantity of a protein by its migration behavior in an electric field. Thereby, molecules can be either separated in their native folded structure or in a chemically linearized form where their mobility in the electric field depends only on length and mass-to-charge ratio (Laemmli 1970). **2D gel electrophoresis** even provides the possibility to resolve differently charged variants in different gel spots of the same protein, as these variants often show differences in their isoelectric point (Wallner et al. 2009). It is common practice to further analyze resolved spots by **mass spectrometry** (MS), the most accurate way to determine the identity of a protein. Thereby, analytes are ionized, accelerated in an electric field and sorted by their mass-to-charge ratio. Even post-translational modifications can be directly observed by mass shifts.

Table 11.1: List of food allergens approved by the IUIS which were purified recombinantly or from the native sorce. Recombinantly produced allergens were purified from E.coli unless otherwise stated ()

Allergen category	Allergen	Source	Species	Type/function	Recombinant	Native	MW [kDa]	PDB-code	Commercially available for diagnosis	References
Cow	Bos d 4	milk	Bos domesticus	α-lactalbumin	x		14.2		n (s,m)	(1)
	Bos d 5	milk		β-lactoglobulin	x		18.3		n (s,m)	(2)
	Bos d 6	milk		serum albumin		x	66.4		n (s,m)	(3)
	Bos d 7	beef, milk		immunoglobulin		x	160* /***		n (s,m)	(4)
	Bos d 8	milk		casein fraction		x	20-30*		n (s,m)	(5)
	Bos d 9	milk		αS1-casein	x		23.0			(6)
	Bos d 10	milk		αS2-casein		x	24.3			(5)
	Bos d 11	milk		β-casein	x (soybean)		24.6			(7)
	Bos d 12	milk		κ-casein		x	19.0			(5)
Hen	Gal d 1	egg (white)	Galus domesticus	ovomucoid	x		20.1	1HJA, 1CHO	n	(8)
	Gal d 2	egg (white)		ovalbumin	x		42.8	1OVA, 1UHG, 1JTI	n	(8)

Table 11.1 contd. ...

...*Table 11.1 contd.*

Allergen category	Allergen	Source	Species	Type/function	Recombinant	Native	MW [kDa]	PDB-code	Commercially available for diagnosis	References
	Gal d 3	egg (white)		ovotransferrin	x		75.8	2D3I, 1RYX, 1N04, 1IQ7, 1IEJ, 1NFT, 1TFA, 1AIV, 1OVT	n	(8)
	Gal d 4	egg (white)		lysozym C	x (loss of function)	x	14.3	2LYM, 4RLM		(8)
	Gal d 5	egg (yolk)		serum albumin		iden.	67.2		n (m)	(9)
	Gal d 6	egg (yolk)		YGP42		x	31.4			(10)
	Gal d 7	chicken (meat)		myosin light chain 1f	x		20.8			(4)
	Gal d 8	chicken (meat)		α-parvalbumin	x		11.8			(11)
	Gal d 9			endolase	x?		50*			(11)
	Gal d 10			aldolase	x?		40?			(11)
	Ran e 1	frog	Rana esculenta	α-parvalbumin	x		11.9			(12)
	Ran e 2	frog		β-parvalbumin	x		11.7			(12)

Wheat									
Fag e 2	common buckwheat	Fagopyrum esculentum	2S albumin	x		15.0			(13)
Fag e 3	common buckwheat		7S globulin	x		19*		n (m)	(14)
Fag t 2	tartarian buckwheat	Fagopyrum tataricum	2S albumin	x		15.3			(14)
Hor v 12	barley	Hordeum vulgare	profilin		iden.	14.2			(4)
Hor v 15	barley		α-amylase inhibitor BMAI-1 precursor		x	14.4**			(15, 16)
Hor v 16	barley		α-amylase	x (Pichia)	iden.	45.4	2XFF, 1RPK, 1P6W, 1HT6, 1BG9, 1AVA, 1AMY		(17, 18)
Hor v 17	barley		β-amylase		x	59.6	2XFR, 2XFY, 2XG9, 2XGB, 2XGI		(17, 19)
Hor v 20	barley		γ-hordein 3	x		34.0			(20)

Table 11.1 contd. ...

...Table 11.1 contd.

Allergen category	Allergen	Source	Species	Type/function	Recombinant	Native	MW [kDa]	PDB-code	Commercially available for diagnosis	References
	Sec c 20	rye	Secale cereal	γ-secalin		x	70*			(21)
	Tri a 12	wheat	Triticum aestivum	profilin	x		14.1			(22)
	Tri a 14	wheat		nsLTP	x		9.5**		r	(22)
	Tri a 18	wheat		agglutinin isolectin 1		x	17.1	1WGC, 2CWG, 7WGA	n (s)	(23, 24)
	Tri a 19	wheat		ω-5 gliadin	x		51.0		r	(22)
	Tri a 20	wheat		γ gliadin	x		32.0			(25)
	Tri a 25	wheat		thioredoxin	x		13.4			(22)
	Tri a 26	wheat		HMW Glutenin	x		88.1			(25)
	Tri a 36	wheat		LMW Glutenin GluB3-23	x		40.0			(25)
	Tri a 37	wheat		α-purothionin	x		11.8			(26)
	Tri a 41	wheat		mitochondrial ubiquitin ligase activator of NFKB 1		iden.	N.A			Unpublished (4)
	Tri a 42	wheat		hypothetical protein from cDNA		iden.	N.A			Unpublished (4)

	Allergen	Common name	Species	Protein			MW (kDa)	PDB	Form	Ref.
	Tri a 43	wheat		hypothetical protein from cDNA		iden.	N.A			Unpublished (4)
	Tri a 44	wheat		endosperm transfer cell specific PR60 precursor		iden.	N.A			Unpublished (4)
	Tri a 45	wheat		elongation factor 1 (EIF1)		iden.	N.A			Unpublished (4)
	Ory s 12	rice	Oryza sativa	profilin		x	14.1			(27)
	Zea m 14	maize	Zea mays	nsLTP		x	9.1	1AFH, 1FK0-7, 1MZL, 1MZM		(28)
Treenut	Ana o 1	cashew	Anacardium occidentale	7S globulin-like	x		59.1			(29)
	Ana o 2			11S globulin-like	x		50.5*		r (m)	(30)
	Ana o 3			2S albumin	x		14.2		r (s)	(31)
	Ber e 1	brazil nut	Bertholletia excelsa	2S albumin	x		12.2	1GYS	r	(32)
	Ber e 2			11S globulin		x	50.1			(33)
	Car i 1	pecan	Carya illiominensis	2S albumin	x	x	14.9			(34)

Table 11.1 contd. ...

...*Table 11.1 contd.*

Allergen category	Allergen	Source	Species	Type/function	Recombinant	Native	MW [kDa]	PDB-code	Commercially available for diagnosis	References
	Car i 2			7s globulin-like	x		92.4	5E1R		(35)
	Car i 4			11S globulin	x		55.4			(36)
	Cas s 5	chestnut	Castanea sativa	chitinase		x	32.0			(37)
	Cas s 8			nsLTP1		x	9.2			(38)
	Cas s 9			cytosolic class I small HSP	x		17.5			(39)
	Cor a 1.04	hazelnut	Corylus avellana	PR-10 protein	x		17.6		r	(40)
	Cor a 2			profilin	x		14.1			(41)
	Cor a 8			nsLTP1	x (E.coli, Pichia)		9.5	4XUW	r	(42, 43)
	Cor a 9			11S globulin	x		56.8		n	(44)
	Cor a 11			7S globulin	x		50.9			(45)
	Cor a 12			oleosin	x		16.7			(46)
	Cor a 13			oleosin			14.7			
	Cor a 14			2S albumin	x		14.9		r (s)	(47)
	Jug n 1	black walnut	Juglans nigra	2S albumin		iden.	18.9			(48)
	Jug n 2			7S globulin		iden.	55.7			(48)
	Jug n 4			11S globulin		x	34/22*			(49)

									r (s), n (m)	
	Jug r 1	English walnut	Juglans regia	2S albumin	x		15*		r (s), n (m)	(50)
	Jug r 2			7S globulin	x		70.0		n (m)	(51)
	Jug r 3			nsLTP1		x	9.2		r (s), n (m)	(52)
	Jug r 4			11S globulin	x		55.6			(53)
	Jug r 5			PR-10 protein	x		17.5			(54)
	Pis v 1	Pistachio	Pistacia vera	2S albumin	x		15.2			(55)
	Pis v 2			11S globulin subunit	x		54.0			(55)
	Pis v 3			7S globulin	x		59.7			(56)
	Pis v 4			manganese SOD	x		25.7			(57)
	Pis v 5			11S globulin subunit (acidic)		x	51.1			Unpublished (4)
	Pru du 3	Almond	Prunus dulcis	nsLTP1	x		10.3			(58)
	Pru du 4			profilin	x		13.9			(59)
	Pru du 5			60s acidic ribosomal prot. P2	x		11.4			(60)
	Pur du 6			11S globulin	x		60.9	3EHK		(61)
	Ric c 1	Castor bean	ricinus communis	2S albumin		x	23.7	1PSY		(62)
Legume	Ara h 1	peanut	Arachis hypogaea	7S globulin	x		68.8	3SMH, 3S7I, 3S7E	r	(63)
	Ara h 2			2S albumin	x		18.0		r	(64)

Table 11.1 contd. ....

...Table 11.1 contd.

Allergen category	Allergen	Source	Species	Type/function	Recombinant	Native	MW [kDa]	PDB-code	Commercially available for diagnosis	References
	Ara h 3			11S globulin	x		58.8	3C3V	r	(65)
	Ara h 5			profilin	x		13.9	4ESP		(66)
	Ara h 6			2S albumin	x		14.8	1W2Q	n (m)	(66)
	Ara h 7			2S albumin	x		16.3			(66)
	Ara h 8			PR-10 protein	x		17.0	4M9B, 4M9W, 4MAP, 4MA6	n (m)	(67)
	Ara h 9			nsLTP1	x (Pichia)		9.1		r	(68)
	Ara h 10			oleosin		x	17.8			(69)
	Ara h 11			oleosin		x	14.3			(69)
	Ara h 12			defensin		x	5.2*			(70)
	Ara h 13			defensin		x	5.5*			(70)
	Ara h 14			oleosin		x	18.4			(69)
	Ara h 15			oleosin		x	16.9			(69)
	Ara h 16			nsLTP2		x	7*			Unpublished (4)
	Ara h 17			nsLTP1		x	9.4*			Unpublished (4)
	Gly m 3			profilin	x		14.0			(71)

	Allergen	Common name	Species	Protein			MW	PDB		Ref
	Gly m 4			PR-10 protein	x		16.8	2K7H	r	(72)
	Gly m 5			7S globulin	x		63.2**	1IPJ, 1IPK	n	(73)
	Gly m 6			11S globulin	x (E.coli, Pichia)		52.6	1FXZ, 2D5F, 2D5H	n	(73, 74)
	Gly m 7			seed biotinylated protein		x	68.0			Unpublished (4)
	Gly m 8			2S albumin	x (Pichia)		14.0			(75)
	Len c 1	lentil	Lens culinaris	γ-7S globulin subunit		x	47.5**			(76)
	Len c 2			seed-specific biotinylated protein		x	66***			(77)
	Len c 3			nsLTP1	x		9.3	2MAL		(78)
	Lup an 1	blue lupin	Lupinus angustifolius	7S globulin		x	68.7			(79)
	Pha v 3	kidney bean	Phaseolus vulgaris	nsLTP1	x		9.6			(80)
Vegetable	Api g 1	celery	Apium graveolens	PR-10 protein	x		16.3	2BK0	r (s)	(81)
	Api g 2			nsLTP1	x		9.0			(82)
	Api g 3			chlorophyll a-b-binding protein		iden.	24.7			(39)
	Api g 4			profilin	x		14.1			(83)

Table 11.1 contd. ....

...Table 11.1 contd.

Allergen category	Allergen	Source	Species	Type/function	Recombinant	Native	MW [kDa]	PDB-code	Commercially available for diagnosis	References
	Api g 5			FAD-containing oxidase		x	53/57***			(84)
	Api g 6			nsLTP2		x	6.9			(85)
	Aspa o 1	asparagus	Asparagus officinalis	nsLTP1		x	9*			(86)
	Bra o 3	cabbage	Brassica oleracea	nsLTP1		x	9*			(87)
	Bra r 1	turnip	Brassica rapa	2S albumin		x	17.9			(88)
	Bra r 2			prohevein homologue		x	18.7			(89)
	Cap a 1	bell pepper	Capsicum annuum	osmotin-like protein (thaumatin-like)		iden.	23.9			(90)
	Cap a 2			profilin	x		14.1			(91)
	Dau c 1	carrot	Daucus carota	PR-10 protein	x		16.0	2WQL		(92)
	Dau c 4			profilin	x		14.3			(93)
	Dau c 5			isoflavone reductase like protein	x		33.3			(93)
	Lac s 1	cultivated lettuce	Lactuca sativa	nsLTP	x (loss of function / insoluble)	x	8.9			(94)

Allergen	Common name	Scientific name	Protein			MW (kDa)		Reference
Man e 5	cassava	manihot esculenta	glutamic-acid rich protein	x	x	18.8		(95)
Sola l 1	tomato	solanum lycopersicum / Lycopersicon esculentum	profilin	x	x	14.1		(91)
Sola l 2			β-fructofuranosidase	x	x	61.3		(96)
Sola l 3			nsLTP1	x	x	8.9		(97)
Sola l 4			PR-10 protein	x	x	17.9		(98)
Sola l 5			cyclophilin		iden	17.9		(99)
Sola l 6			nsLTP2		x	7.0		(100)
Sola l 7			nsLTP1		x	12.3*		Unpublished (4)
Sola t 1	potato	Solanum tuberosum	patatin		x	40.1		(101)
Sola t 2			cathepsin D inhibitor PDI		x	20.6		(102)
Sola t 3			cysteine protease inhibitor		x	20.1		(101)
Sola t 4			serine protease inhibitor 7		x	20.9		(101)
Fruits Act c 5	golden kiwi fruit	Actinidia chinensis	kiwellin		x	19.7		(103)
Act c 8			PR-10 protein	x		17.5		(104)

Table 11.1 contd. ....

...Table 11.1 contd.

Allergen category	Allergen	Source	Species	Type/function	Recombinant	Native	MW [kDa]	PDB-code	Commercially available for diagnosis	References
	Act c 10	kiwi fruit	Actinidia deliciosa	nsLTP1		x	10*			(105)
	Act d 1			cysteine protease		x	27.4	2ACT, 1AEC	n (m)	(106)
	Act d 2			thaumatin-like protein		x	21.6	4BCT	n (m)	(106)
	Act d 3			N.A.		x	40*			(106)
	Act d 4			phytocystatin		x	10.1			(106)
	Act d 5			kiwellin		x	16.0	4X9U	n (m)	(106) / (107)
	Act d 6			pectin methylesterase inhibitor		x	16.3			(108)
	Act d 7			pectin methylesterase		x	35.4			(108)
	Act d 8			PR-10 protein	x		16.9		r (s), n (m)	(106) / (104)
	Act d 9			profilin	x		11.7			(106)
	Act d 10			nsLTP1		x	9.5			(105)
	Act d 11			major latex protein/ ripening related protein (MLP/ RRP) Bet v 1 family member	x	x	17.4	4IGV, 4IGW, 4IGX, 4IGY, 4IH0, 4IH2, 4IHR		(109)

Act d 12			11S globulin		x	50.2		(110)
Act d 13			2S albumin		x	11.4		(110)
Ana c 1	pineapple	Ananas comosus	profilin	x		14.1		(111)
Ana c 2			bromelain		x	24.9		(112)
Cit l 3	lemon	Citrus limon	nsLTP1		x	9.6*		(113)
Cit r 3	tangerine	Citrus reticulata	nsLTP1		x	9*		(114)
Cit s 1	sweet orange	Citrus sinensis	germin-like protein		x	21.9		(115)
Cit s 2			profilin		x	13.9		(116)
Cit s 3			nsLTP1	x		9.46*		(113)
Cit s 7			gibberellin-regulated protein		x	8*		unpublished
Cuc m 1	muskmelon	Cucumin melo	alkaline serine protease	x		53.8	3VTA	(117)
Cuc m 2			profilin	x		13.8		(118)
Cuc m 3			PR-1 protein		x	16.1		(119)
Fra a 1	strawberry	Fragaria x ananassa	PR-10 protein	x		17.8	2LPX, 4C9C, 4C9I, 4C9J	(120)
Fra a 3			nsLTP1	x (Pichia)		9.0		(121)
Fra a 4			profilin		iden.	13.9		(4)

Table 11.1 contd. ...

...*Table 11.1 contd.*

Allergen category	Allergen	Source	Species	Type/function	Recombinant	Native	MW [kDa]	PDB-code	Commercially available for diagnosis	References
	Lit c 1	lychee nut	Litchi chinensis	profilin		iden.	13.9			(122)
	Mal d 1	apple	Malus x domestica	PR-10 protein	x		17.5		r (s)	(123)
	Mal d 2			thaumatin-like protein	x (Nicotiana benthamina)	x	23.2	3ZS3		(104, 124)
	Mal d 3			nsLTP1	x		9.1		r (s)	(125)
	Mal d 4			profilin	x		13.8			(104)
	Mor n 3	black mulberry	Morus nigra	nsLTP1		x	9.2			(126)
	Mus a 1	banana	Musa acuminate	profilin	x		13.9			(111)
	Mus a 2			class 1 chitinase		x	31.6			(127)
	Mus a 3			nsLTP1		iden.	9*			Unpublished (4)
	Mus a 4			thaumatin-like protein	x		21.2	1Z3Q		(128)
	Mus a 5			β-1,3-glucanase	x		37.1#	2CYG		(129)
	Mus a 6			ascorbate peroxidase	x	iden.	27*			Unpublished (4)
	Pers a 1	avocado	Persea americana	class 1 chitinase	x (Pichia)	x	32.0			(130)

					iden.				
Pru ar 1	apricot	Prunus armeniaca	PR-10 protein			17.3			(131)
Pur ar 3			nsLTP1		x	9.2			(132)
Pru av 1	cherry	Prunus avium	PR-10 protein	x		17.7	1H2O, 1E09		(133)
Pru av 2			thaumatin-like protein	x	x	23.3	2AHN		(134)
Pur av 3			nsLTP1	x		9.1			(135)
Pru av 4			profilin	x		13.8			(133)
Pru d 3	european plum	Prunus domestica	nsLTP1		x	9.2			(136)
Pru p 1	peach	Prunus persica	PR-10 protein	x		17.6	2ALG, 2B5S	r (s)	(137)
Pru p 2			thaumatin-like protein	X (Pichia)		23.3			(138)
Pru p 3			nsLTP1	x		9.2		r (s)	(139)
Pru p 4			profilin	x		14.0		r (s)	(140)
Pru p 7			gibberellin-regulated protein		x	6.9			(141)
Pun g 1	pomegranate	Punica granatum	nsLTP1		x	9.3			(142)
Pyr c 1	pear	Pyrus communis	PR-10 protein	x		17.6			(143)
Pyr c 3			nsLTP1	x		9.1			(144)

Table 11.1 contd. ...

...Table 11.1 contd.

Allergen category	Allergen	Source	Species	Type/function	Recombinant	Native	MW [kDa]	PDB-code	Commercially available for diagnosis	References
	Pyr c 4			profilin	x		13.9			(133)
	Pyr c 5			isoflavone reductase related protein	x		33.8			(145)
	Rub i 1	raspberry	Rubus idaeus	PR-10 protein		iden.	17*			(146)
	Rub i 3			nsLTP1		iden.	9.2			(146)
	Vit v 1	grape	Vitis vinifera	nsLTP1		x	9.1			(147)
	Ziz m 1	Chinese date	Ziziphus mauritiana	class 3 chitinase	x		34.0			(148)
Seeds	Bra j 1	oriental mustard	Brassica juncea	2S albumin		x	14.6			(149)
	Bra n 1	rapeseed	Brassica napus	2S albumin		x	14.0	1PNB		(88)
	Hel a 3	sunflower	Helianthus annuus	nsLTP1		x	9.3			(150)
	Ses i 1	sesame	Sesamum indicum	2S albumin	x		15.5		r (m)	(151)
	Ses i 2			2S albumin	x		12.5			(151)
	Ses i 3			7S globulin		iden.	49**			(152)
	Ses i 4			oleosin		x	17.3			(153)
	Ses i 5			oleosin		x	15.2			(153)

	Ses i 6			11S globulin	x		49.7	(154)e>
	Ses i 7			11S globulin	x		54.2	(154)e>
	Sin a 1	yellow mustard	Sinapis alba	2S albumin	x		14.5	(155)
	Sin a 2			11S globulin	x		54.2	(156)
	Sin a 3			nsLTP1	x (Pichia, E.coli)		9.5	(157)
	Sin a 4			profilin	x (Pichia, E.coli)		14.0	(157)
	Pin p 1	stone pine	Pinus pinea	2S albumin		x	19.0	(158)
Shrimp	Arc s 8	crustacean species	Archaeo-potamobius sibiriensis	triosephosphate isomerase		x	24.3	Unpublished (4)
	Art fr 5	brine shrimp	Artemia franciscana	myosin light chain 1		iden.	16.5	Unpublished (4)
	Cra c 1	North Sea shrimp	Crangon crangon	tropomyosin	x		32.7	(159)
	Cra c 2			arginine kinase	x		40.1	(159)
	Cra c 4			sarcoplasmic calcium-binding protein	x		22.1	(159)
	Cra c 5			myosin light chain 1	x		17.4	(159)

Table 11.1 contd. ....

...Table 11.1 contd.

Allergen category	Allergen	Source	Species	Type/function	Recombinant	Native	MW [kDa]	PDB-code	Commercially available for diagnosis	References
	Cra c 6			troponin C	x		16.8			(159)
	Cra c 8			triosephosphate isomerase	x		27.0			(159)
	Lit v 1	white shrimp	Litopenaeus vannamei	tropomyosin		x	32.8			(160)
	Lit v 2			arginine kinase		x	40.2	1BLH, 4AM1		(161)
	Lit v 3			myosin light chain 2	x		19.3			(162)
	Lit v 4			sarcoplasmic calcium-binding protein	x		22.1			(163)
	Mac r 1	giant freshwater prawn	Macrobrachium rosenbergii	tropomyosin		iden.	32.8			(164)
	Mel l 1	king prawn	Melicertus latisulcatus	tropomyosin	x		32.7			(165)
	Met e 1	sand shrimp	Metapenaeus ensis	tropomyosin	x		31.7			(166)
	Pan b 1	Northern shrimp	Pandalus borealis	tropomyosin	x		32.8			(167)
	Pen a 1	brown shrimp	Penaeus aztecus	tropomyosin	x		32.8		r (s)	(168)

	Allergen	Common name	Scientific name	Protein			MW (kDa)		Reference
	Pen i 1	Indian prawn	Penaeus indicus	tropomyosin		x	34.0		(169)Md.: 1950
	Pen m 1	black tiger shrimp	Penaeus monodon	tropomyosin	x		32.8	n (m)	(170)
	Pen m 2			arginine kinase		x	40.0	n (m)	(171)
	Pen m 3			myosin light chain 2	x*	iden.	19.2*		(172)
	Pen m 4			sarcoplasmic calcium-binding protein	x		22.1	n (m)	(173)
	Pen m 6			troponin C	x		16.8		(174)
Crab	Cha f 1	crab	Charybdis feriatus	tropomyosin	x		30.4		(175)
	Eri s 2	Chinese mitten crab	Eriocheir sinensis	ovary development-related protein		x	26.5		(160)
	Por p 1	blue swimmer crab	Porturus pelagicus	tropomyosin	x		32.8		(176)
Lobster	Hom a 1	American lobster	Homarus americanus	tropomyosin	x		32.9		(177)
	Hom a 3			myosin light chain 2		iden.	20.1		Unpublished (4)
	Hom a 6			troponin C		iden.	17.0		(178)
	Pan s 1	spiny lobster	Panulirus stimpsoni	tropomyosin	x		31.7		(177)

Table 11.1 contd. …

...Table 11.1 contd.

Allergen category	Allergen	Source	Species	Type/function	Recombinant	Native	MW [kDa]	PDB-code	Commercially available for diagnosis	References
Crayfish	Pon l 4	Danube crayfish	Pontastacus leptodactylus	sarcoplasmic calcium-binding protein		x	21.6			(179)
	Pon l 7			troponin I		iden.	23.5			Unpublished (4)
Mollusks	Hal m 1	abalone	Haliotis midae	N.A.		iden.	49*			(180)
	Hel as 1	brown garden snail	Helix aspersa	tropomyosin	x		32.6			(181)
	Tod p 1	Japanese common squid	Todarodes pacificus	tropomyosin		x	38*/***			(182)
Fish	Clu h 1	Atlantic herring	Clupea harengus	β-parvalbumin	x		11.7			(183)
	Cyp c 1	carb	Cyprinus cario	β-parvalbumin	x		11.5		r (s)	(183)
	Gad c 1	Baltic cod	Gadus callarias	β-parvalbumin	x		12.1		r	(183)
	Gad m 1	Atlantic cod	Gadus morhua	β-parvalbumin	x		11.5	2MBX		(184)
	Gad m 2			β-endolase		x	47.3*			(185)
	Gad m 3			aldolase A		x	40*			(185)

Code	Common name	Species	Allergen				Ref
Lat c 1	Asian sea bass / barramundi	Lates calcarifer	β-parvalbumin	x		11.6	(186)
Lep w 1	flat fish	Lepidorhombus whiffiagonis	β-parvalbumin		x	11.7	(187)
Onc k 5	chum salmon	Oncorhynchus keta	β-prime-component of vitellogenin	x		19.3	(188)
Onc m 1	rainbow trout	Oncorhynchus mykiss	β-parvalbumin		iden.	11.8	(39)
Ore m 4	Mozambique tilapia	Oreochromis mossambicus	tropomyosin		x	32.7	(189)
Sal s 1	Atlantic salmon	Salmo salar	β-parvalbumin 1	x		11.9	(183)
Sal s 2			β-endolase		x	47.3	(185)
Sal s 3			aldolase A		x	39.4	(185)
Sar sa 1	south American pilchard	Sardinops sagax	β-parvalbumin		x	11.9	(190)
Seb m 1	ocean perch (red fish, snapper)	Sebastes marinus	β-parvalbumin		x	11.4	(191)
Thu a 1	yellowfin tuna	Thunnus albacares	β-parvalbumin	x		11.5	(183)
Thu a 2			β-endolase		x	50*	(185)
Thu a 3			aldolase A		x	40*	(185)
Xip g 1	swordfish	Xiphias gladius	β-parvalbumin		x	11.5	(192)

Table 11.1 contd. ...

...Table 11.1 contd.

Allergen category	Allergen	Source	Species	Type/function	Recombinant	Native	MW [kDa]	PDB-code	Commercially available for diagnosis	References
Nematode	Ani s 1	Pacific fish worm	Anisakis simplex	unknown function, similar to Kunitz serine protease inhibitor	x		18.9			(193, 194)
	Ani s 2			paramyosin	x		100.5			(195)
	Ani s 3			tropomyosin	x		33.2			(196)
	Ani s 4			cysteine protease inhibitor	x		10.3			(197)
	Ani s 5			SXP/RAL-2 family protein	x (E.coli, Pichia)		14.7	2MAR		(196, 198, 199)
	Ani s 6			serine protease inhibitor	x		7.2			(198)
	Ani s 7			UA3-recognized allergen	x		119.4			(194)
	Ani s 8			SXP/RAL-2 family protein		x	15*/***			(200)
	Ani s 9			SXP/RAL-2 family protein	x		13.8			(201)
	Ani s 10			unknown function	x		21.1			(196)
	Ani s 11			unknown function	x		27.8			(202)

Ani s 12			unknown function	x		30.8		(202)
Ani s 13			hemoglobin		iden.	36.7		(203)(199)
Ani s 14			N.A.	x		24*		Unpublished (4)

MW = molecular weight, ref = reference, iden. = identified as allergen from the natural source, which was not purified. Commercially available proteins for diagnostics: n = native source; r = recombinantly produced; m = available on multiplex format; s = available in singleplex format. PDB-code: structure of protein available on www.rcsb.org with the regarding 4-letter code * data from IUIS (allergen.org); ** fragment of protein; *** MW identified by SDS-PAGE; # Accession number: AGT8193

References to table 11.1

1. Ishikawa, N., T. Chiba, L. T. Chen, A. Shimizu, M. Ikeguchi and S. Sugai (1998). Remarkable destabilization of recombinant alpha-lactalbumin by an extraneous N-terminal methionyl residue. Protein Engineering. 11(5): 333–5.

2. Chatel, J.M., H. Bernard, G. Clement, Y. Frobert, C. A. Batt, J. Gavalchin et al. (1996). Expression, purification and immunochemical characterization of recombinant bovine beta-lactoglobulin, a major cow milk allergen. Molecular immunology. 33(14): 1113–8.

3. Gjesing, B., O. Osterballe, B. Schwartz, U. Wahn and H. Lowenstein (1986). Allergen-specifc IgE antibodies against antigenic components in cow milk and milk substitutes. Allergy. 41(1): 51–6.

4. list IUofSANlo. [2015-10-28]. Available from: http://allergen.org/.

5. Bernard, H., C. Creminon, M. Yvon and J. M. Wal (1998). Specificity of the human IgE response to the different purified caseins in allergy to cow's milk proteins. International Archives of Allergy and Immunology. 115(3): 235–44.

6. Schulmeister, U., H. Hochwallner, I. Swoboda, M. Focke-Tejkl, B. Geller, M. Nystrand et al. (2009). Cloning, expression, and mapping of allergenic determinants of alphaS1-casein, a major cow's milk allergen. Journal of Immunology (Baltimore, Md : 1950). 182(11): 7019–29.

7. Philip, R., D. W. Darnowski, P. J. Maughan and L. O. Vodkin (2001). Processing and localization of bovine beta-casein expressed in transgenic soybean seeds under control of a soybean lectin expression cassette. Plant Science : An International Journal of Experimental Plant Biology. 161(2): 323–35.

8. Dhanapala, P., T. Doran, M. L. Tang and C. Suphioglu (2015). Production and immunological analysis of IgE reactive recombinant egg white allergens expressed in Escherichia coli. Molecular Immunology. 65(1): 104–12.

9. Egger, M., C., M. Wallner, P. Briza, D. Zennaro, A. Mari et al. (2011). Is aboriginal food less allergenic? Comparing IgE-reactivity of eggs from modern and ancient chicken breeds in a cohort of allergic children. PloS one. 6(4): e19062.

10. Amo, A., R. Rodriguez-Perez, J. Blanco, J. Villota, S. Juste, I. Moneo et al. (2010). Gal d 6 is the second allergen characterized from egg yolk. Journal of Agricultural and Food Chemistry. 58(12): 7453–7.

11. Kuehn, A., C. Lehners, C. Hilger and F. Hentges (2009). Food allergy to chicken meat with IgE reactivity to muscle alpha-parvalbumin. Allergy. 64(10): 1557–8.

Table 11.1 contd. ...

...Table 11.1 contd.

12. Hilger, C., F. Grigioni, L. Thill, L. Mertens and F. Hentges (2002). Severe IgE-mediated anaphylaxis following consumption of fried frog legs: definition of alpha-parvalbumin as the allergen in cause. Allergy. 57(11): 1053–8.

13. Satoh, R., S. Koyano, K. Takagi, R. Nakamura, R. Teshima and J. Sawada (2008). Immunological characterization and mutational analysis of the recombinant protein BWp16, a major allergen in buckwheat. Biological & Pharmaceutical Bulletin. 31(6): 1079–85.

14. Chen, P., Y. F. Guo, Q. Yan and Y. H. Li (2011). Molecular cloning and characterization of Fag t 2: a 16-kDa major allergen from Tartary buckwheat seeds. Allergy. 66(10): 1393–5.

15. Barber, D., R. Sanchez-Monge, L. Gomez, J. Carpizo, A. Armentia, C. Lopez-Otin et al. (1989). A barley flour inhibitor of insect alpha-amylase is a major allergen associated with baker's asthma disease. FEBS letters. 248(1-2): 119–22.

16. Lopez-Rico, R., I. Moneo, A. Rico, G. Curiel, R. Sanchez-Monge and G. Salcedo (1998). Cereal alpha-amylase inhibitors cause occupational sensitization in the wood industry. Clinical and Experimental Allergy: Journal of the British Society for Allergy and Clinical Immunology. 28(10): 1286–91.

17. Sandiford, C. P., R. D. Tee and A. J. Taylor (1994). The role of cereal and fungal amylases in cereal flour hypersensitivity. Clinical and Experimental Allergy: Journal of the British Society for Allergy and Clinical Immunology. 24(6): 549–57.

18. Robert, X., R. Haser, H. Mori, B. Svensson and N. Aghajari (2005). Oligosaccharide binding to barley alpha-amylase 1. The Journal of Biological Chemistry. 280(38): 32968–78.

19. Rejzek, M., C. E. Stevenson, A. M. Southard, D. Stanley, K. Denyer, A. M. Smith et al. (2011). Chemical genetics and cereal starch metabolism: structural basis of the non-covalent and covalent inhibition of barley beta-amylase. Molecular bioSystems. 7(3): 718–30.

20. Snegaroff, J., I. Bouchez, A. Smaali Mel, C. Pecquet, N. Raison-Peyron, P. Jolivet et al. (2013). Barley gamma3-hordein: glycosylation at an atypical site, disulfide bridge analysis, and reactivity with IgE from patients allergic to wheat. Biochimica et Biophysica Acta. 1834(1): 395–403.

21. Palosuo, K., H. Alenius, E. Varjonen, N. Kalkkinen and T. Reunala (2001). Rye gamma-70 and gamma-35 secalins and barley gamma-3 hordein cross-react with omega-5 gliadin, a major allergen in wheat-dependent, exercise-induced anaphylaxis. Clinical and Experimental Allergy: Journal of the British Society for Allergy and Clinical Immunology. 31(3): 466–73.

22. Sander, L. P. Rozynek, H. P. Rihs, V. van Kampen, F. T. Chew, W. S. Lee et al. (2011). Multiple wheat flour allergens and cross-reactive carbohydrate determinants bind IgE in baker's asthma. Allergy. 66(9): 1208–15.

23. Sutton, R., J. H. Skerritt, B. A. Baldo and C. W. Wrigley (1984). The diversity of allergens involved in bakers' asthma. Clinical Allergy. 14(1): 93–107.

24. Wright, C. S. (1990). 2.2 A resolution structure analysis of two refined N-acetylneuraminyl-lactose—wheat germ agglutinin isolectin complexes. Journal of Molecular Biology. 215(4): 635–51.

25. Maruyama, N., K. Ichise, T. Katsube, T. Kishimoto, S. Kawase, Y. Matsumura et al. (1998). Identification of major wheat allergens by means of the *Escherichia coli* expression system. European Journal of Biochemistry / FEBS. 255(3): 739–45.

26. Pahr, S., R. Selb, M. Weber, M. Focke-Tejkl, G. Hofer, A. Dordic et al. (2014). Biochemical, biophysical and IgE-epitope characterization of the wheat food allergen, Tri a 37. PloS one. 9(11): e111483.

27. Hirano, K., S. Hino, K. Oshima, T. Okajima, D. Nadano, A. Urisu et al. (2013). Allergenic potential of rice-pollen proteins: expression, immuno-cross reactivity and IgE-binding. Journal of Biochemistry. 154(2): 195–205.

28. Pastorello, E. A., L. Farioli, V. Pravettoni, M. Ispano, F. Scibola, C. Trambaioli et al. (2000). The maize major allergen, which is responsible for food-induced allergic reactions, is a lipid transfer protein. The Journal of Allergy and Clinical Immunology. 106(4): 744–51.

29. Wang, F., J. M. Robotham, S. S. Teuber, P. Tawde, S. K. Sathe, K. H. Roux (2002). Ana o 1, a cashew (*Anacardium occidental*) allergen of the vicilin seed storage protein family. The Journal of Allergy and Clinical Immunology. 110(1): 160–6.

30. Wang, F., J. M. Robotham, S. S. Teuber, S. K. Sathe and K. H. Roux (2003). Ana o 2, a major cashew (*Anacardium occidentale* L.) nut allergen of the legumin family. International Archives of Allergy and Immunology. 132(1): 27–39.

31. Robotham, J. M., F. Wang, V. Seamon, S. S. Teuber, S. K. Sathe, H. A. Sampson et al. (2005). Ana o 3, an important cashew nut (*Anacardium occidentale* L.) allergen of the 2S albumin family. The Journal of Allergy and Clinical Immunology. 115(6): 1284–90.

32. Gander, E. S., K. O. Holmstroem, G. R. De Paiva, L. A. De Castro, M. Carneiro and M. F. Grossi de Sa (1991). Isolation, characterization and expression of a gene coding for a 2S albumin from Bertholletia excelsa (Brazil nut). Plant Molecular Biology. 16(3): 437–48.

33. Identification of an 11S globulin as a Brazil nut food allergen [Internet]. 2003.

34. Sharma, G. M., A. Irsigler, P. Dhanarajan, R. Ayuso, L. Bardina, H. A. Sampson et al. (2011). Cloning and characterization of 2S albumin, Car i 1, a major allergen in pecan. Journal of Agricultural and Food Chemistry. 59(8): 4130–9.

35. Zhang, Y., B. Lee, W. X. Du, S. C. Lyu, K. C., L. J. Grauke et al. (2016). Identification and Characterization of a New Pecan [Carya illinoinensis (Wangenh.) K. Koch] Allergen, Car i 2. Journal of Agricultural and Food Chemistry. 64(20): 4146–51.

36. Sharma, G. M., A. Irsigler, P. Dhanarajan, R. Ayuso, L. Bardina, H. A. Sampson et al. (2011). Cloning and characterization of an 11S legumin, Car i 4, a major allergen in pecan. Journal of Agricultural and Food Chemistry. 59(17): 9542–52.

37. Diaz-Perales, A., C. Collada, C. Blanco, R. Sanchez-Monge, T. Carrillo, C. Aragoncillo et al. (1998). Class I chitinases with hevein-like domain, but not class II enzymes, are relevant chestnut and avocado allergens. The Journal of Allergy and Clinical Immunology. 102(1): 127–33.

38. Diaz-Perales, A., M. Lombardero, R. Sanchez-Monge, F. J. Garcia-Selles, M. Pernas, M. Fernandez-Rivas et al. (2000). Lipid-transfer proteins as potential plant panallergens: cross-reactivity among proteins of Artemisia pollen, Castanea nut and Rosaceae fruits, with different IgE-binding capacities. Clinical and Experimental Allergy: Journal of the British Society for Allergy and Clinical Immunology. 30(10): 1403–10.

39. allergome.org. 2015 [cited 2015 2015-10-28]. Available from: http://www.allergome.org/.

40. Weiss, C., B. Kramer, C. Ebner, M. Susani, P. Briza, K. Hoffmann-Sommergruber et al. (1996). High-level expression of tree pollen isoallergens in *Escherichia coli*. International Archives of Allergy and Immunology. 110(3): 282–7.

41. Hirschwehr, R., R. Valenta, C. Ebner, F. Ferreira, W. R. Sperr, P. Valent et al. (1992). Identification of common allergenic structures in hazel pollen and hazelnuts: a possible explanation for sensitivity to hazelnuts in patients allergic to tree pollen. The Journal of Allergy and Clinical Immunology. 90(6 Pt 1): 927–36.

42. Pokoj, S., I. Lauer, K. Fotisch, M. Himly, A. Mari, E. Enrique et al. (2010). Pichia pastoris is superior to *E. coli* for the production of recombinant allergenic non-specific lipid-transfer proteins. Protein Expression and Purification. 69(1): 68–75.

43. Schocker, F., D. Luttkopf, S. Scheurer, A. Petersen, A. Cistero-Bahima, E. Enrique et al. (2004). Recombinant lipid transfer protein Cor a 8 from hazelnut: a new tool for *in vitro* diagnosis of potentially severe hazelnut allergy. The Journal of Allergy and Clinical Immunology. 113(1): 141–7.

Table 11.1 contd. ...

...*Table 11.1 contd.*

44. Beyer, K., G. Grishina, L. Bardina, A. Grishin and H. A. Sampson (2002). Identification of an 11S globulin as a major hazelnut food allergen in hazelnut-induced systemic reactions. The Journal of Allergy and Clinical Immunology. 110(3): 517–23.

45. Lauer, I., K. Foetisch, D. Kolarich, B. K. Ballmer-Weber, A. Conti, F. Altmann et al. (2004). Hazelnut (Corylus avellana) vicilin Cor a 11: molecular characterization of a glycoprotein and its allergenic activity. The Biochemical Journal. 383(Pt 2): 327–34.

46. Akkerdaas, J. H., F. Schocker, S. Vieths, S. Versteeg, L. Zuidmeer, S. L. Hefle et al. (2006). Cloning of oleosin, a putative new hazelnut allergen, using a hazelnut cDNA library. Molecular Nutrition & Food Research. 50(1): 18–23.

47. Garino, C., L. Zuidmeer, J. Marsh, A. Lovegrove, M. Morati, S. Versteeg et al. (2010). Isolation, cloning, and characterization of the 2S albumin: a new allergen from hazelnut. Molecular Nutrition & Food Research. 54(9): 1257–65.

48. Bannon, G., M. Ling, G. Cockrell and H. Sampson. (2001). Cloning, expression and characterization of two major allergens, Jug n 1 and Jug n 2, from the black walnut, Juglans niger. Allergy Clinical Immunology: 107(140).

49. Zhang, Y. Z., W. X. Du, Y. Fan, J. Yi, S. C. Lyu, K. C. Nadeau et al. (2017). Purification and Characterization of a Black Walnut (Juglans nigra) Allergen, Jug n 4. Journal of Agricultural and Food Chemistry. 65(2): 454–62.

50. Teuber, S. S., A. M. Dandekar, W. R. Peterson and C. L. Sellers (1998). Cloning and sequencing of a gene encoding a 2S albumin seed storage protein precursor from English walnut (Juglans regia), a major food allergen. The Journal of Allergy and Clinical Immunology. 101(6 Pt 1): 807–14.

51. Teuber, S. S., K. C. Jarvis, A. M. Dandekar, W. R. Peterson and A. A. Ansari (1999). Identification and cloning of a complementary DNA encoding a vicilin-like proprotein, jug r 2, from english walnut kernel (Juglans regia), a major food allergen. The Journal of Allergy and Clinical Immunology. 104(6): 1311–20.

52. Pastorello, E. A., L. Farioli, V. Pravettoni, A. M. Robino, J. Scibilia, D. Fortunato et al. (2004). Lipid transfer protein and vicilin are important walnut allergens in patients not allergic to pollen. The Journal of Allergy and Clinical Immunology. 114(4): 908–14.

53. Wallowitz, M., W. R. Peterson, S. Uratsu, S. S. Comstock, A. M. Dandekar and S. S. Teuber (2006). Jug r 4, a legumin group food allergen from walnut (Juglans regia Cv. Chandler). Journal of Agricultural and Food Chemistry. 54(21): 8369–75.

54. Wangorsch, A., A. Jamin, J. Lidholm, N. Grani, C. Lang, B. Ballmer-Weber et al. (2017). Identification and implication of an allergenic PR-10 protein from walnut in birch pollen associated walnut allergy. Molecular Nutrition & Food Research.

55. Ahn, K., L. Bardina, G. Grishina, K. Beyer and H. A. Sampson (2009). Identification of two pistachio allergens, Pis v 1 and Pis v 2, belonging to the 2S albumin and 11S globulin family. Clinical and Experimental Allergy: Journal of the British Society for Allergy and Clinical Immunology. 39(6): 926–34.

56. Willison, L. N., P. Tawde, J. M. Robotham, R. Mt. Penney, S. S. Teuber, S. K. Sathe et al. (2008). Pistachio vicilin, Pis v 3, is immunoglobulin E-reactive and cross-reacts with the homologous cashew allergen, Ana o 1. Clinical and Experimental Allergy: Journal of the British Society for Allergy and Clinical Immunology. 38(7): 1229–38.

57. Noorbakhsh, R., S. A. Mortazavi, M. Sankian, F. Shahidi, M. A. Assarehzadegan and A. Varasteh (2010). Cloning, expression, characterization, and computational approach for cross-reactivity prediction of manganese superoxide dismutase allergen from pistachio nut. Allergology International: Official Journal of the Japanese Society of Allergology. 59(3): 295–304.

58. Identification of almond lipid transfer protein (LTP) isoform [Internet]. 2009.

59. Tawde, P., Y. P. Venkatesh, F. Wang, S. S. Teuber, S. K. Sathe and K. H. Roux (2006). Cloning and characterization of profilin (Pru du 4), a cross-reactive almond (Prunus dulcis) allergen. The Journal of Allergy and Clinical Immunology. 118(4): 915–22.

60. Abolhassani, M. and K. H. Roux (2009). cDNA Cloning, expression and characterization of an allergenic 60s ribosomal protein of almond (prunus dulcis). Iranian Journal of Allergy, Asthma, and Immunology. 8(2): 77–84.

61. Willison, L. N., P. Tripathi, G. Sharma, S. S. Teuber, S. K. Sathe and K. H. Roux (2011). Cloning, expression and patient IgE reactivity of recombinant Pru du 6, an 11S globulin from almond. International Archives of Allergy and Immunology. 156(3): 267–81.

62. Felix, S. P., R. O. Mayerhoffer, R. A. Damatta, M. A. Vericimo, V. V. Nascimento, and O. L. Machado (2008). Mapping IgE-binding epitopes of Ric c 1 and Ric c 3, allergens from Ricinus communis, by mast cell degranulation assay. Peptides. 29(4): 497–504.

63. Burks, A. W., G. Cockrell, J. S. Stanley, R. M. Helm and G. A. Bannon (1995). Recombinant peanut allergen Ara h I expression and IgE binding in patients with peanut hypersensitivity. The Journal of Clinical Investigation. 96(4): 1715–21.

64. Stanley, J. S., N. King, A. W. Burks, S. K. Huang, H. Sampson, G. Cockrell et al. (1997). Identification and mutational analysis of the immunodominant IgE binding epitopes of the major peanut allergen Ara h 2. Archives of Biochemistry and Biophysics. 342(2): 244–53.

65. Rabjohn, P., E. M. Helm, J. S. Stanley, C. M. West, H. A. Sampson, A. W. Burks et al. (1999). Molecular cloning and epitope analysis of the peanut allergen Ara h 3. The Journal of Clinical Investigation. 103(4): 535–42.

66. Kleber-Janke, T., R. Crameri, U. Appenzeller, M. Schlaak and W. M. Becker (1999). Selective cloning of peanut allergens, including profilin and 2S albumins, by phage display technology. International Archives of Allergy and Immunology. 119(4): 265–74.

67. Mittag, D., J. Akkerdaas, B. K. Ballmer-Weber, L. Vogel, M. Wensing, W. M. Becker et al. (2004). Ara h 8, a Bet v 1-homologous allergen from peanut, is a major allergen in patients with combined birch pollen and peanut allergy. The Journal of Allergy and Clinical Immunology. 114(6): 1410–7.

68. Lauer, I., N. Dueringer, S. Pokoj, S. Rehm, G. Zoccatelli, G. Reese et al. (2009). The non-specific lipid transfer protein, Ara h 9, is an important allergen in peanut. Clinical and Experimental Allergy: Journal of the British Society for Allergy and Clinical Immunology. 39(9): 1427–37.

69. Schwager, C., S. Kull, S. Krause, F. Schocker, A. Petersen, W. M. Becker et al. (2015). Development of a novel strategy to isolate lipophilic allergens (oleosins) from peanuts. PloS one. 10(4): e0123419.

70. Petersen, A., S. Kull, S. Rennert, W. M. Becker, S. Krause, M. Ernst et al. (2015). Peanut defensins: Novel allergens isolated from lipophilic peanut extract. The Journal of Allergy and Clinical Immunology.

71. Rihs, H. P., Z. Chen, F. Rueff, A. Petersen, P. Rozynek, H. Heimann et al. (1999). IgE binding of the recombinant allergen soybean profilin (rGly m 3) is mediated by conformational epitopes. The Journal of Allergy and Clinical Immunology. 104(6): 1293–301.

72. Kleine-Tebbe, J., L. Vogel, D. N. Crowell, U. F. Haustein and S. Vieths (2002). Severe oral allergy syndrome and anaphylactic reactions caused by a Bet v 1-related PR-10 protein in soybean, SAM22. The Journal of Allergy and Clinical Immunology. 110(5): 797–804.

73. Holzhauser, T., O. Wackermann, B. K. Ballmer-Weber, C. Bindslev-Jensen, J. Scibilia, L. Perono-Garoffo et al. (2009). Soybean (Glycine max) allergy in Europe: Gly m 5 (beta-conglycinin) and Gly m 6 (glycinin) are potential diagnostic markers for severe allergic reactions to soy. The Journal of Allergy and Clinical Immunology. 123(2): 452–8.

Table 11.1 contd. ...

...Table 11.1 contd.

74. Joshi, S. and T. Satyanarayana (2014). Optimization of heterologous expression of the phytase (PPHY) of Pichia anomala in P. pastoris and its applicability in fractionating allergenic glycinin from soy protein. Journal of Industrial Microbiology & Biotechnology. 41(6): 977–87.

75. Lin, J., R. Fido, P. Shewry, D. B. Archer and M. J. Alcocer (2004). The expression and processing of two recombinant 2S albumins from soybean (Glycine max) in the yeast Pichia pastoris. Biochimica et Biophysica Acta. 1698(2): 203–12.

76. Lopez-Torrejon, G., G. Salcedo, M. Martin-Esteban, A. Diaz-Perales, C. Y. Pascual and R. Sanchez-Monge (2003). Len c 1, a major allergen and vicilin from lentil seeds: protein isolation and cDNA cloning. The Journal of Allergy and Clinical Immunology. 112(6): 1208–15.

77. Sanchez-Monge, R., C. Y. Pascual, A. Diaz-Perales, J. Fernandez-Crespo, M. Martin-Esteban and G. Salcedo (2000). Isolation and characterization of relevant allergens from boiled lentils. The Journal of Allergy and Clinical Immunology. 106(5): 955–61.

78. Akkerdaas, J., E. I. Finkina, S. V. Balandin, S. Santos Magadan, A. Knulst, M. Fernandez-Rivas et al. (2012). Lentil (Lens culinaris) lipid transfer protein Len c 3: a novel legume allergen. International Archives of Allergy and Immunology. 157(1): 51–7.

79. Nadal, P., A. Pinto, M. Svobodova, N. Canela and C. K. O'Sullivan (2012). DNA aptamers against the Lup an 1 food allergen. PloS one. 7(4): e35253.

80. Zoccatelli, G., S. Pokoj, K. Foetisch, J. Bartra, A. Valero, M. Del Mar San Miguel-Moncin et al. (2010). Identification and characterization of the major allergen of green bean (Phaseolus vulgaris) as a non-specific lipid transfer protein (Pha v 3). Molecular Immunology. 47(7-8): 1561–8.

81. Breiteneder, H., K. Hoffmann-Sommergruber, G. O'Riordain, M. Susani, H. Ahorn, C. Ebner et al. (1995). Molecular characterization of Api g 1, the major allergen of celery (Apium graveolens), and its immunological and structural relationships to a group of 17-kDa tree pollen allergens. European Journal of Biochemistry/FEBS. 233(2): 484–9.

82. Gadermaier, G., M. Egger, T. Girbl, A. Erler, A. Harrer, E. Vejvar et al. (2011). Molecular characterization of Api g 2, a novel allergenic member of the lipid-transfer protein 1 family from celery stalks. Molecular Nutrition & Food Research. 55(4): 568–77.

83. Scheurer, S., A. Wangorsch, D. Haustein and S. Vieths (2000). Cloning of the minor allergen Api g 4 profilin from celery (Apium graveolens) and its cross-reactivity with birch pollen profilin Bet v 2. Clinical and Experimental Allergy: Journal of the British Society for Allergy and Clinical Immunology. 30(7): 962–71.

84. Bublin, M., I. Lauer, C. Oberhuber, S. Alessandri, P. Briza, C. Radauer et al. (2008). Production and characterization of an allergen panel for component-resolved diagnosis of celery allergy. Molecular Nutrition & Food Research. 52 Suppl 2: S241–50.

85. Vejvar, E., M. Himly, P. Briza, S. Eichhorn, C. Ebner, W. Hemmer et al. (2013). Allergenic relevance of nonspecific lipid transfer proteins 2: Identification and characterization of Api g 6 from celery tuber as representative of a novel IgE-binding protein family. Molecular Nutrition & Food Research. 57(11): 2061–70.

86. Diaz-Perales, A., A. I. Tabar, R. Sanchez-Monge, B. E. Garcia, B. Gomez, D. Barber et al. (2002). Characterization of asparagus allergens: a relevant role of lipid transfer proteins. The Journal of Allergy and Clinical Immunology. 110(5): 790–6.

87. Palacin, A., J. Cumplido, J. Figueroa, O. Ahrazem, R. Sanchez-Monge, T. Carrillo et al. (2006). Cabbage lipid transfer protein Bra o 3 is a major allergen responsible for cross-reactivity between plant foods and pollens. The Journal of Allergy and Clinical Immunology. 117(6): 1423–9.

88. Puumalainen, T. J., S. Poikonen, A. Kotovuori, K. Vaali, N. Kalkkinen, T. Reunala et al. (2006). Napins, 2S albumins, are major allergens in oilseed rape and turnip rape. The Journal of Allergy and Clinical Immunology. 117(2): 426–32.

89. Hanninen, A. R., J. H. Mikkola, N. Kalkkinen, K. Turjanmaa, L. Ylitalo, T. Reunala et al. (1999). Increased allergen production in turnip (Brassica rapa) by treatments activating defense mechanisms. The Journal of Allergy and Clinical Immunology. 104(1): 194–201.

90. Leitner, A., E. Jensen-Jarolim, R. Grimm, B. Wuthrich, H. Ebner, O. Scheiner et al. (1998). Allergens in pepper and paprika. Immunologic investigation of the celery-birch-mugwort-spice syndrome. Allergy. 53(1): 36–41.

91. Willerroider, M., H. Fuchs, B. K. Ballmer-Weber, M. Focke, M. Susani, J. Thalhamer et al. (2003). Cloning and molecular and immunological characterisation of two new food allergens, Cap a 2 and Lyc e 1, profilins from bell pepper (Capsicum annuum) and Tomato (Lycopersicon esculentum). International Archives of Allergy and Immunology. 131(4): 245–55.

92. Hoffmann-Sommergruber, K., G. O'Riordain, H. Ahorn, C. Ebner, M. Laimer Da Camara Machado, H. Puhringer et al. (1999). Molecular characterization of Dau c 1, the Bet v 1 homologous protein from carrot and its cross-reactivity with Bet v 1 and Api g 1. Clinical and Experimental Allergy: Journal of the British Society for Allergy and Clinical Immunology. 29(6): 840–7.

93. Ballmer-Weber, B. K., K. Skamstrup Hansen, J. Sastre, K. Andersson, I. Batscher, J. Ostling et al. (2012). Component-resolved in vitro diagnosis of carrot allergy in three different regions of Europe. Allergy. 67(6): 758–66.

94. Hartz, C., M. San Miguel-Moncin Mdel, A. Cistero-Bahima, K. Fotisch, K. J. Metzner, D. Fortunato et al. (2007). Molecular characterisation of Lac s 1, the major allergen from lettuce (Lactuca sativa). Molecular Immunology. 44(11): 2820–30.

95. Santos, K. S., G. Gadermaier, E. Vejvar, H. A. Arcuri, C. E. Galvao, A. C. Yang et al. (2013). Novel allergens from ancient foods: Man e 5 from manioc (Manihot esculenta Crantz) cross reacts with Hev b 5 from latex. Molecular Nutrition & Food Research. 57(6): 1100–9.

96. Foetisch, K., S. Westphal, I. Lauer, M. Retzek, F. Altmann, D. Kolarich et al. (2003). Biological activity of IgE specific for cross-reactive carbohydrate determinants. The Journal of Allergy and Clinical Immunology. 111(4): 889–96.

97. Volpicella, M., C. Leoni, I. Fanizza, S. Rinalducci, A. Placido and L. R. Ceci (2015). Expression and characterization of a new isoform of the 9 kDa allergenic lipid transfer protein from tomato (variety San Marzano). Plant Physiology and Biochemistry:PPB/Societe Francaise De Physiologie Vegetale. 96: 64–71.

98. Wangorsch, A., A. Jamin, K. Foetisch, A. Malczyk, A. Reuter, S. Vierecke et al. (2015). Identification of Sola l 4 as Bet v 1 homologous pathogenesis related-10 allergen in tomato fruits. Molecular Nutrition & Food Research. 59(3): 582–92.

99. Welter, S., S. Dolle, K. Lehmann, D. Schwarz, W. Weckwerth, M. Worm et al. (2013). Pepino mosaic virus infection of tomato affects allergen expression, but not the allergenic potential of fruits. PloS One. 8(6): e65116.

100. Giangrieco, I., C. Alessandri, C. Rafaiani, M. Santoro, S. Zuzzi, L. Tuppo et al. (2015). Structural features, IgE binding and preliminary clinical findings of the 7kDa lipid transfer protein from tomato seeds. Molecular Immunology. 66(2): 154–63.

101. Seppala, U., H. Majamaa, K. Turjanmaa, J. Helin, T. Reunala, N. Kalkkinen et al. (2001). Identification of four novel potato (Solanum tuberosum) allergens belonging to the family of soybean trypsin inhibitors. Allergy. 56(7): 619–26.

102. Seppala, U., H. Alenius, K. Turjanmaa, T. Reunala, T. Palosuo and N. Kalkkinen (1999). Identification of patatin as a novel allergen for children with positive skin prick test responses to raw potato. The Journal of Allergy and Clinical Immunology. 103(1 Pt 1): 165–71.

103. Tuppo, L., I. Giangrieco, P. Palazzo, M. I. Bernardi, E. Scala, V. Carratore et al. (2008). Kiwellin, a modular protein from green and gold kiwi fruits: evidence of in vivo and in vitro processing and IgE binding. Journal of Agricultural and Food Chemistry. 56(10): 3812–7.

Table 11.1 contd. ...

104. Oberhuber, C., Y. Ma, J. Marsh, N. Rigby, U. Smole, C. Radauer et al. (2008). Purification and characterisation of relevant natural and recombinant apple allergens. Molecular Nutrition & Food Research. 52 Suppl 2: S208–19.

105. Bernardi, M. L., I. Giangrieco, L. Camardella, R. Ferrara, P. Palazzo, M. R. Panico et al. (2011). Allergenic lipid transfer proteins from plant-derived foods do not immunologically and clinically behave homogeneously: the kiwifruit LTP as a model. PloS One. 6(11): e27856.

106. Bublin, M., M. Pfister, C. Radauer, C. Oberhuber, S. Bulley, A. M. Dewitt et al. (2010). Component-resolved diagnosis of kiwifruit allergy with purified natural and recombinant kiwifruit allergens. The Journal of Allergy and Clinical Immunology. 125(3): 687–94, 94.e1.

107. Offermann, L. R., I. Giangrieco, M. L. Perdue, S. Zuzzi, M. Santoro, M. Tamburrini et al. (2015). Elusive structural, functional, and immunological features of Act d 5, the green kiwifruit Kiwellin. Journal of Agricultural and Food Chemistry. 63(29): 6567–76.

108. Palazzo, P., L. Tuppo, I. Giangrieco, M. L. Bernardi, C. Rafaiani, R. Crescenzo et al. (2013), Prevalence and peculiarities of IgE reactivity to kiwifruit pectin methylesterase and its inhibitor, Act d 7 and Act d 6, in subjects allergic to kiwifruit. Food Research International. 53(1): 24–30.

109. Chruszcz, M., M. A. Ciardiello, T. Osinski, K. A. Majorek, I. Giangrieco, J. Font et al. (2013). Structural and bioinformatic analysis of the kiwifruit allergen Act d 11, a member of the family of ripening-related proteins. Molecular Immunology. 56(4): 794–803.

110. Sirvent, S., B. Canto, J. Cuesta-Herranz, F. Gomez, N. Blanca, G. Canto et al. (2014). Act d 12 and Act d 13: two novel, masked, relevant allergens in kiwifruit seeds. The Journal of Allergy and Clinical Immunology. 133(6): 1765–7.e4.

111. Reindl, J., H. P. Rihs, S. Scheurer, A. Wangorsch, D. Haustein and S. Vieths IgE reactivity to profilin in pollen-sensitized subjects with adverse reactions to banana and pineapple. International Archives of Allergy and Immunology. 128(2): 105–14.

112. Baur, X. and G. Fruhmann (1979). Allergic reactions, including asthma, to the pineapple protease bromelain following occupational exposure. Clinical Allergy. 9(5): 443–50.

113. Ahrazem, O., M. D. Ibanez, G. Lopez-Torrejon, R. Sanchez-Monge, J. Sastre M. Lombardero et al. (2005). Lipid transfer proteins and allergy to oranges. International Archives of Allergy and Immunology. 137(3): 201–10.

114. Ebo, D. G., O. Ahrazem, G. Lopez-Torrejon, C. H. Bridts, G. Salcedo and W. J. Stevens (2007). Anaphylaxis from mandarin (Citrus reticulata): identification of potential responsible allergens. International Archives of Allergy and Immunology. 144(1): 39–43.

115. Poltl, G., O. Ahrazem, K. Paschinger, M. D. Ibanez, G. Salcedo and I. B. Wilson (2007). Molecular and immunological characterization of the glycosylated orange allergen Cit s 1. Glycobiology. 17(2): 220–30.

116. Lopez-Torrejon, G., J. F. Crespo, R. Sanchez-Monge, M. Sanchez-Jimenez, J. Alvarez, J. Rodriguez et al. (2005). Allergenic reactivity of the melon profilin Cuc m 2 and its identification as major allergen. Clinical and Experimental Allergy: Journal of the British Society for Allergy and Clinical Immunology. 35(8): 1065–72.

117. Sankian, M., F. Talebi, M. Moghadam, F. Vahedi, F. J. Azad and A. R. Varasteh (2011). Molecular cloning and expression of Cucumisin (Cuc m 1), a subtilisin-like protease of Cucumis melo in Escherichia coli. Allergology International: Official Journal of the Japanese Society of Allergology. 60(1): 61–7.

118. Lopez-Torrejon, G., M. D. Ibanez, O. Ahrazem, R. Sanchez-Monge, J. Sastre, M. Lombardero et al. (2005). Isolation, cloning and allergenic reactivity of natural profilin Cit s 2, a major orange allergen. Allergy. 60(11): 1424–9.

119. Asensio, T., J. F. Crespo, R. Sanchez-Monge, G. Lopez-Torrejon, M. L. Somoza, J. Rodriguez et al. (2004). Novel plant pathogenesis-related protein family involved in food allergy. The Journal of Allergy and Clinical Immunology. 114(4): 896–9.

120. Musidlowska-Persson, A., R. Alm and C. Emanuelsson (2007). Cloning and sequencing of the Bet v 1-homologous allergen Fra a 1 in strawberry (*Fragaria ananassa*) shows the presence of an intron and little variability in amino acid sequence. Molecular Immunology. 44(6): 1245–52.

121. Zuidmeer, L., E. Salentijn, M. F. Rivas, E. G. Mancebo, R. Asero, C. I. Matos et al. (2006). The role of profilin and lipid transfer protein in strawberry allergy in the Mediterranean area. Clinical and Experimental Allergy: Journal of the British Society for Allergy and Clinical Immunology. 36(5): 666–75.

122. Fah, J., B. Wuthrich and S. Vieths (1995). Anaphylactic reaction to lychee fruit: evidence for sensitization to profilin. Clinical and Experimental Allergy: Journal of the British Society for Allergy and Clinical Immunology. 25(10): 1018–23.

123. Vanek-Krebitz, M., K. Hoffmann-Sommergruber, M. Laimer da Camara Machado, M. Susani, C. Ebner, D. Kraft et al. (1995). Cloning and sequencing of Mal d 1, the major allergen from apple (*Malus domestica*), and its immunological relationship to Bet v 1, the major birch pollen allergen. Biochemical and Biophysical Research Communications. 214(2): 538–51.

124. Krebitz, M., B. Wagner, F. Ferreira, C. Peterbauer, N. Campillo, M. Witty et al. (2003). Plant-based heterologous expression of Mal d 2, a thaumatin-like protein and allergen of apple (*Malus domestica*), and its characterization as an antifungal protein. Journal of Molecular Biology. 329(4): 721–30.

125. Borges, J.-P., R. Culerrier, D. Aldon, A. Barre, H. Benoist, O. Saurel et al. (2010). GATEWAY technology and *E. coli* recombinant system produce a properly folded and functional recombinant allergen of the lipid transfer protein of apple (Mal d 3). Protein Expression and Purification. 70(2): 277–82.

126. Ciardiello, M. A., P. Palazzo, M. L. Bernardi, V. Carratore, I. Giangrieco, V. Longo et al. (2010). Biochemical, immunological and clinical characterization of a cross-reactive nonspecific lipid transfer protein 1 from mulberry. Allergy. 65(5): 597–605.

127. Sanchez-Monge, R., C. Blanco, A. Diaz-Perales, C. Collada, T. Carrillo, C. Aragoncillo et al. (1999). Isolation and characterization of major banana allergens: identification as fruit class I chitinases. Clinical and Experimental Allergy: Journal of the British Society for Allergy and Clinical Immunology. 29(5): 673–80.

128. Palacin, A., S. Quirce, R. Sanchez-Monge, I. Bobolea, A. Diaz-Perales, F. Martin-Munoz et al. (2011). Sensitization profiles to purified plant food allergens among pediatric patients with allergy to banana. Pediatric Allergy and Immunology: Official Publication of the European Society of Pediatric Allergy and Immunology. 22(2): 186–95.

129. Mrkic, I., M. Abughren, J. Nikolic, U. Andjelkovic, E. Vassilopoulou, A. Sinaniotis et al. (2014). Molecular characterization of recombinant mus a 5 allergen from banana fruit. Molecular Biotechnology. 56(6): 498–506.

130. Sowka, S., L. S. Hsieh, M. Krebitz, A. Akasawa, B. M. Martin, D. Starrett et al. (1998). Identification and cloning of prs a 1, a 32-kDa endochitinase and major allergen of avocado, and its expression in the yeast Pichia pastoris. The Journal of Biological Chemistry. 273(43): 28091–7.

131. Rodriguez, J., J. F. Crespo, A. Lopez-Rubio, J. De La Cruz-Bertolo, P. Ferrando-Vivas, R. Vives et al. (2000). Clinical cross-reactivity among foods of the Rosaceae family. The Journal of Allergy and Clinical Immunology. 106(1 Pt 1): 183–9.

132. Pastorello, E. A., F. P. D'Ambrosio, V. Pravettoni, L. Farioli, G. Giuffrida, M. Monza et al. (2000). Evidence for a lipid transfer protein as the major allergen of apricot. The Journal of Allergy and Clinical Immunology. 105(2 Pt 1): 371–7.

Table 11.1 contd. ...

...Table 11.1 contd.

133. Scheurer, S., A. Wangorsch, J. Nerkamp, P. S. Skov, B. Ballmer-Weber, B. Wuthrich et al. (2001). Cross-reactivity within the profilin panallergen family investigated by comparison of recombinant profilins from pear (Pyr c 4), cherry (Pru av 4) and celery (Api g 4) with birch pollen profilin Bet v 2. Journal of Chromatography B, Biomedical Sciences and Applications. 756(1-2): 315–25.

134. Fuchs, H. C., B. Bohle, Y. Dall'Antonia, C. Radauer, K. Hoffmann-Sommergruber, A. Mari et al. (2006). Natural and recombinant molecules of the cherry allergen Pru av 2 show diverse structural and B cell characteristics but similar T cell reactivity. Clinical and Experimental Allergy: Journal of the British Society for Allergy and Clinical Immunology. 36(3): 359–68.

135. Scheurer, S., I. Lauer, K. Foetisch, M. San Miguel Moncin, M. Retzek, C. Hartz et al. (2004). Strong allergenicity of Pru av 3, the lipid transfer protein from cherry, is related to high stability against thermal processing and digestion. The Journal of Allergy and Clinical Immunology. 114(4): 900–7.

136. Pastorello, E. A., L. Farioli, V. Pravettoni, M. G. Giuffrida, C. Ortolani, D. Fortunato et al. (2001). Characterization of the major allergen of plum as a lipid transfer protein. Journal of Chromatography B, Biomedical Sciences and Applications. 756(1-2): 95–103.

137. Mascheri, A., L. Farioli, V. Pravettoni, M. Piantanida, C. Stafylaraki, J. Scibilia et al. (2015). Hypersensitivity to Tomato (*Lycopersicon esculentum*) in Peach-Allergic Patients: rPru p 3 and rPru p 1 Are Predictive of Symptom Severity. Journal of Investigational Allergology & Clinical Immunology. 25(3): 183–9.

138. Palacin, A., L. Tordesillas, P. Gamboa, R. Sanchez-Monge, J. Cuesta-Herranz, M. L. Sanz et al. (2010). Characterization of peach thaumatin-like proteins and their identification as major peach allergens. Clinical and Experimental Allergy: Journal of the British Society for Allergy and Clinical Immunology. 40(9): 1422–30.

139. Diaz-Perales, A., G. Garcia-Casado, R. Sanchez-Monge, F. J. Garcia-Selles, D. Barber and G. Salcedo (2002). cDNA cloning and heterologous expression of the major allergens from peach and apple belonging to the lipid-transfer protein family. Clinical and Experimental Allergy: Journal of the British Society for Allergy and Clinical Immunology. 32(1): 87–92.

140. Rodriguez-Perez, R., M. Fernandez-Rivas, E. Gonzalez-Mancebo, R. Sanchez-Monge, A. Diaz-Perales and G. Salcedo (2003). Peach profilin: cloning, heterologous expression and cross-reactivity with Bet v 2. Allergy. 58(7): 635–40.

141. Tuppo, L., R. Spadaccini, C. Alessandri, H. Wienk, R. Boelens, I. Giangrieco et al. (2014). Structure, stability, and IgE binding of the peach allergen Peamaclein (Pru p 7). Biopolymers. 102(5): 416–25.

142. Bolla, M., S. Zenoni, S. Scheurer, S. Vieths, M. San Miguel Moncin Mdel, M. Olivieri et al. (2014). Pomegranate (*Punica granatum* L.) expresses several nsLTP isoforms characterized by different immunoglobulin E-binding properties. International Archives of Allergy and Immunology. 164(2): 112–21.

143. Karamloo, F., S. Scheurer, A. Wangorsch, S. May, D. Haustein and S. Vieths (2001). Pyr c 1, the major allergen from pear (*Pyrus communis*), is a new member of the Bet v 1 allergen family. Journal of Chromatography B, Biomedical Sciences and Applications. 756(1-2): 281–93.

144. Ramazzina, I., S. Amato, E. Passera, S. Sforza, G. Mistrello, R. Berni et al. (2012). Isoform identification, recombinant production and characterization of the allergen lipid transfer protein 1 from pear (Pyr c 3). Gene. 491(2): 173–81.

145. Karamloo, F., A. Wangorsch, H. Kasahara, L. B. Davin, D. Haustein, N. G. Lewis et al. (2001). Phenylcoumaran benzylic ether and isoflavonoid reductases are a new class of cross-reactive allergens in birch pollen, fruits and vegetables. European Journal of Biochemistry / FEBS. 268(20): 5310–20.

146. Marzban, G., A. Herndl, D. Kolarich, F. Maghuly, A. Mansfeld, W. Hemmer et al. (2008). Identification of four IgE-reactive proteins in raspberry (*Rubus idaeus* L.). Molecular Nutrition & Food Research. 52(12): 1497–506.

147. Pastorello, E. A., L. Farioli, V. Pravettoni, C. Ortolani, D. Fortunato, M. G. Giuffrida et al. (2003). Identification of grape and wine allergens as an endochitinase 4, a lipid-transfer protein, and a thaumatin. The Journal of Allergy and Clinical Immunology. 111(2): 350–9.

148. Lee, M. F., G. Y. Hwang, Y. H. Chen, H. C. Lin and C. H. Wu (2006). Molecular cloning of Indian jujube (*Zizyphus mauritiana*) allergen Ziz m 1 with sequence similarity to plant class III chitinases. Molecular Immunology. 43(8): 1144–51.

149. Gonzalez de la Pena, M. A., L. Menendez-Arias, R. I. Monsalve and R. Rodriguez (1991). Isolation and characterization of a major allergen from oriental mustard seeds, BraJI. International Archives of Allergy and Applied Immunology. 96(3): 263–70.

150. Regente, M. and L. de la Canal (2003). A cDNA encoding a putative lipid transfer protein expressed in sunflower seeds. Journal of Plant Physiology. 160(2): 201–3.

151. Tai, S. S., L. S. Wu, E. C. Chen and J. T. Tzen (1999). Molecular cloning of 11S globulin and 2S albumin, the two major seed storage proteins in sesame. Journal of Agricultural and Food Chemistry. 47(12): 4932–8.

152. Navuluri, L., S. Parvataneni, H. Hassan, N. P. Birmingham, C. Kelly and V. Gangur (2006). Allergic and anaphylactic response to sesame seeds in mice: identification of Ses i 3 and basic subunit of 11s globulins as allergens. International Archives of Allergy and Immunology. 140(3): 270–6.

153. Leduc, V., D. A. Moneret-Vautrin, J. T. Tzen, M. Morisset, L. Guerin and G. Kanny (2006). Identification of oleosins as major allergens in sesame seed allergic patients. Allergy. 61(3): 349–56.

154. Beyer, K., G. Grishina, L. Bardina and H. A. Sampson (2007). Identification of 2 new sesame seed allergens: Ses i 6 and Ses i 7. The Journal of Allergy and Clinical Immunology. 119(6): 1554–6.

155. Gonzalez De La Pena, M. A., R. I. Monsalve, E. Batanero, M. Villalba and R. Rodriguez (1996). Expression in *Escherichia coli* of Sin a 1, the major allergen from mustard. European Journal of Biochemistry / FEBS. 237(3): 827–32.

156. Palomares, O., A. Vereda, J. Cuesta-Herranz, M. Villalba and R. Rodriguez (2007). Cloning, sequencing, and recombinant production of Sin a 2, an allergenic 11S globulin from yellow mustard seeds. The Journal of Allergy and Clinical Immunology. 119(5): 1189–96.

157. Sirvent, S., O. Palomares, A. Vereda, M. Villalba, J. Cuesta-Herranz and R. Rodriguez (2009). nsLTP and profilin are allergens in mustard seeds: cloning, sequencing and recombinant production of Sin a 3 and Sin a 4. Clinical and Experimental Allergy: Journal of the British Society for Allergy and Clinical Immunology. 39(12): 1929–36.

158. Cabanillas, B., H. Cheng, C. C. Grimm, B. K. Hurlburt, J. Rodriguez, J. F. Crespo et al. (2012). Pine nut allergy: clinical features and major allergens characterization. Molecular Nutrition & Food Research. 56(12): 1884–93.

159. Bauermeister, K., A. Wangorsch, L. P. Garoffo, A. Reuter, A. Conti, S. L. Taylor et al. (2011). Generation of a comprehensive panel of crustacean allergens from the North Sea Shrimp Crangon crangon. Molecular Immunology. 48(15-16): 1983–92.

160. Liu, G. M., M. J. Cao, Y. Y. Huang, Q. F. Cai, W. Y. Weng and W. J. Su (2010). Comparative study of *in vitro* digestibility of major allergen tropomyosin and other food proteins of Chinese mitten crab (Eriocher sinensis). Journal of the Science of Food and Agriculture. 90(10): 1614–20.

161. Garcia-Orozco, K. D., E. Aispuro-Hernandez, G. Yepiz-Plascencia, A. M. Calderon-de-la-Barca and R. R. Sotelo-Mundo (2007). Molecular characterization of arginine kinase, an allergen from the shrimp *Litopenaeus vannamei*. International Archives of Allergy and Immunology. 144(1): 23–8.

Table 11.1 contd. ...

162. Ayuso, R., G. Grishina, L. Bardina, T. Carrillo, C. Blanco, M. D. Ibanez et al. (2008). Myosin light chain is a novel shrimp allergen, Lit v 3. The Journal of Allergy and Clinical Immunology. 122(4): 795–802.

163. Ayuso, R., G. Grishina, M. D. Ibanez, C. Blanco, T. Carrillo, R. Bencharitiwong et al. (2009). Sarcoplasmic calcium-binding protein is an EF-hand-type protein identified as a new shrimp allergen. The Journal of Allergy and Clinical Immunology. 124(1): 114–20.

164. Yadzir, Z. H., R. Misnan, N. Abdullah, F. Bakhtiar, M. Arip and S. Murad (2012). Identification of the major allergen of Macrobrachium rosenbergii (giant freshwater prawn). Asian Pacific Journal of Tropical Biomedicine. 2(1): 50-4.

165. Effect of heat treatment on the major allergen tropomyosin from King prawn [Internet]. 2012.

166. Leung, P. S., K. H. Chu, W. K. Chow, A. Ansari, C. I. Bandea, H. S. Kwan et al. (1994). Cloning, expression, and primary structure of Metapenaeus ensis tropomyosin, the major heat-stable shrimp allergen. The Journal of Allergy and Clinical Immunology. 94(5): 882–90.

167. Myrset, H. R., B. Barletta, G. Di Felice, E. Egaas and M. M. Dooper (2013). Structural and immunological characterization of recombinant Pan b 1, a major allergen of northern shrimp, Pandalus borealis. International Archives of Allergy and Immunology. 160(3): 221–32.

168. Reese, G., B. J. Jeoung, C. B. Daul and S. B. Lehrer (1997). Characterization of recombinant shrimp allergen Pen a 1 (tropomyosin). International Archives of Allergy and Immunology. 113(1-3): 240–2.

169. Shanti, K. N., B. M. Martin, S. Nagpal, D. D. Metcalfe and P. V. Rao (1993). Identification of tropomyosin as the major shrimp allergen and characterization of its IgE-binding epitopes. Journal of Immunology (Baltimore, Md: 1950). 151(10): 5354–63.

170. Kamath, S. D., A. M. Abdel Rahman, T. Komoda and A. L. Lopata (2013). Impact of heat processing on the detection of the major shellfish allergen tropomyosin in crustaceans and molluscs using specific monoclonal antibodies. Food Chemistry. 141(4): 4031–9.

171. Yu, C. J., Y. F. Lin, B. L. Chiang and L. P. Chow (2003). Proteomics and immunological analysis of a novel shrimp allergen, Pen m 2. Journal of Immunology (Baltimore, Md: 1950). 170(1): 445–53.

172. Abdel Rahman, A. M., S. Kamath, A. L. Lopata and R. J. Helleur (2010). Analysis of the allergenic proteins in black tiger prawn (Penaeus monodon) and characterization of the major allergen tropomyosin using mass spectrometry. Rapid Communications in Mass Spectrometry: RCM. 24(16): 2462–70.

173. Mita, H., A. Koketsu, S. Ishizaki and K. Shiomi (2013). Molecular cloning and functional expression of allergenic sarcoplasmic calcium-binding proteins from Penaeus shrimps. Journal of the Science of Food and Agriculture. 93(7): 1737–42.

174. Cloning and expression of P. mondon allergic proteins and their molecular immunological characterization [Internet]. 2010 [cited 2015-10-29].

175. Leung, P. S., Y. C. Chen, M. E. Gershwin, S. H. Wong, H. S. Kwan and K. H. Chu (1998). Identification and molecular characterization of Charybdis feriatus tropomyosin, the major crab allergen. The Journal of Allergy and Clinical Immunology. 102(5): 847–52.

176. Abramovitch, J. B., S. Kamath, N. Varese, C. Zubrinich, A. L. Lopata, R. E. O'Hehir et al. (2013). Ige reactivity of blue swimmer crab tropomyosin, por p 1, and other allergens; cross-reactivity with black tiger prawn and effects of heating. PloS One. 8(6): e67487.

177. Leung, P. S., Y. C. Chen, D. L. Mykles, W. K. Chow, C. P. Li and K. H. Chu (1998). Molecular identification of the lobster muscle protein tropomyosin as a seafood allergen. Molecular Marine Biology and Biotechnology. 7(1): 12–20.

178. Garone, L., J. L. Theibert, A. Miegel, Y. Maeda, C. Murphy and J. H. Collins (1991). Lobster troponin C: amino acid sequences of three isoforms. Archives of Biochemistry and Biophysics. 291(1): 89–91.

179. Chen, H. L., M. J. Cao, Q. F. Cai, W. J. Su, H. Y. Mao and G. M. Liu (2013). Purification and characterisation of sarcoplasmic calcium-binding protein, a novel allergen of red swamp crayfish (*Procambarus clarkii*). Food Chemistry. 139(1–4): 213–23.

180. Lopata, A. L., C. Zinn and P. C. Potter (1997). Characteristics of hypersensitivity reactions and identification of a unique 49 kd IgE-binding protein (Hal-m-1) in abalone (Haliotis midae). The Journal of Allergy and Clinical Immunology. 100(5): 642–8.

181. Asturias, J. A., E. Eraso, M. C. Arilla, N. Gomez-Bayon, F. Inacio and A. Martinez (2002). Cloning, isolation, and IgE-binding properties of Helix aspersa (brown garden snail) tropomyosin. International Archives of Allergy and Immunology. 128(2): 90–6.

182. Miyazawa, H., H. Fukamachi, Y. Inagaki, G. Reese, C. B. Daul S. B. Lehrer et al. (1996). Identification of the first major allergen of a squid (Todarodes pacificus). The Journal of Allergy and Clinical Immunology. 98(5 Pt 1): 948–53.

183. Swoboda, I., N. Balic, C. Klug, M. Focke, M. Weber, S. Spitzauer et al. A general strategy for the generation of hypoallergenic molecules for the immunotherapy of fish allergy. The Journal of Allergy and Clinical Immunology. 132(4): 979–81.e1.

184. Van Do, T., I. Hordvik, C. Endresen and S. Elsayed (2003). The major allergen (parvalbumin) of codfish is encoded by at least two isotypic genes: cDNA cloning, expression and antibody binding of the recombinant allergens. Molecular Immunology. 39(10): 595–602.

185. Kuehn, A., C. Hilger, C. Lehrers-Weber, F. Codreanu-Morel, M. Morisset, C. Metz-Favre et al. (2013). Identification of enolases and aldolases as important fish allergens in cod, salmon and tuna: component resolved diagnosis using parvalbumin and the new allergens. Clinical and Experimental Allergy: Journal of the British Society for Allergy and Clinical Immunology. 43(7): 811–22.

186. Sharp, M. F., S. D. Kamath, M. Koeberl, D. R. Jerry, R. E. O'Hehir, D. E. Campbell et al. (2014). Differential IgE binding to isoallergens from Asian seabass (*Lates calcarifer*) in children and adults. Molecular Immunology. 62(1): 77–85.

187. Griesmeier, U., M. Bublin, C. Radauer, S. Vazquez-Cortes, Y. Ma, M. Ferrandez-Rivas et al. (2010). Physicochemical properties and thermal stability of Lep w 1, the major allergen of whiff. Molecular Nutrition & Food Research. 54(6): 861–9.

188. Shimizu, Y., H. Kishimura, G. Kanno, A. Nakamura, R. Adachi, H. Akiyama et al. Molecular and immunological characterization of beta'-component (Onc k 5), a major IgE-binding protein in chum salmon roe. International Immunology. 26(3): 139–47.

189. Liu, R., A. L. Holck, E. Yang, C. Liu and W. Xue (2013). Tropomyosin from tilapia (*Oreochromis mossambicus*) as an allergen. Clinical and Experimental Allergy: Journal of the British Society for Allergy and Clinical Immunology. 43(3): 365–77.

190. Beale, J. E., M. F. Jeebhay and A. L. Lopata (2009). Characterisation of purified parvalbumin from five fish species and nucleotide sequencing of this major allergen from Pacific pilchard, Sardinops sagax. Molecular Immunology. 46(15): 2985–93.

191. Kuehn, A., T. Scheuermann, C. Hilger and F. Hentges (2010). Important variations in parvalbumin content in common fish species: a factor possibly contributing to variable allergenicity. International Archives of Allergy and Immunology. 153(4): 359–66.

192. Griesmeier, U., S. Vazquez-Cortes, M. Bublin, C. Radauer, Y. Ma, P. Briza et al. (2010). Expression levels of parvalbumins determine allergenicity of fish species. Allergy. 65(2): 191–8.

193. Arrieta, I., M. del Barrio, L. Vidarte, V. del Pozo, C. Pastor, J. Gonzalez-Cabrero et al. (2000). Molecular cloning and characterization of an IgE-reactive protein from Anisakis simplex: Ani s 1. Molecular and Biochemical Parasitology. 107(2): 263–8.

Table 11.1 contd. ...

...*Table 11.1 contd.*

194. Cuellar, C., A. Daschner, A. Valls, C. De Frutos, V. Fernandez-Figares, A. M. Anadon et al. (2012). Ani s 1 and Ani s 7 recombinant allergens are able to differentiate distinct Anisakis simplex-associated allergic clinical disorders. Archives of Dermatological Research. 304(4): 283–8.

195. Perez-Perez, J., E. Fernandez-Caldas, F. Maranon, J. Sastre, M. L. Bernal, J. Rodriguez et al. (2000). Molecular cloning of paramyosin, a new allergen of Anisakis simplex. International Archives of Allergy and Immunology. 123(2): 120–9.

196. Caballero, M. L., A. Umpierrez, T. Perez-Pinar, I. Moneo, C. de Burgos, J. A. Asturias et al. (2012). Anisakis simplex recombinant allergens increase diagnosis specificity preserving high sensitivity. International Archives of Allergy and Immunology. 158(3): 232–40.

197. Rodriguez-Mahillo, A. I., M. Gonzalez-Munoz, F. Gomez-Aguado, R. Rodriguez-Perez, M. T. Corcuera, M. L. Caballero et al. (2007). Cloning and characterisation of the Anisakis simplex allergen Ani s 4 as a cysteine-protease inhibitor. International Journal for Parasitology. 37(8-9): 907–17.

198. Kobayashi, Y., S. Ishizaki, K. Shimakura, Y. Nagashima and K. Shiomi (2007). Molecular cloning and expression of two new allergens from Anisakis simplex. Parasitology Research. 100(6): 1233–41.

199. Garcia-Mayoral, M. F., M. A. Trevino, T. Perez-Pinar, M. L. Caballero, A. Umpierrez et al. (2014). Relationships between IgE/IgG4 epitopes, structure and function in Anisakis simplex Ani s 5, a member of the SXP/RAL-2 protein family. PLoS Neglected Tropical Diseases. 8(3): e2735.

200. Kobayashi, Y., K. Shimakura, S. Ishizaki, Y. Nagashima and K. Shiomi (2007). Purification and cDNA cloning of a new heat-stable allergen from Anisakis simplex. Molecular and Biochemical Parasitology. 155(2): 138–45.

201. Cho, M. K., M. K. Park, S. A. Kang, M. L. Caballero, T. Perez-Pinar, R. Rodriguez-Perez et al. (2014). Allergenicity of two Anisakis simplex allergens evaluated *in vivo* using an experimental mouse model. Experimental Parasitology. 146: 71–7.

202. Kobayashi, Y., K. Ohsaki, K. Ikeda, S. Kakemoto, S. Ishizaki, K. Shimakura et al. (2011). Identification of novel three allergens from Anisakis simplex by chemiluminescent immunoscreening of an expression cDNA library. Parasitology International. 60(2): 144–50.

203. Gonzalez-Fernandez, J., A. Daschner, N. E. Nieuwenhuizen, A. L. Lopata, C. D. Frutos, A. Valls et al. (2015). Haemoglobin, a new major allergen of Anisakis simplex. International Journal for Parasitology. 45(6): 399–407.

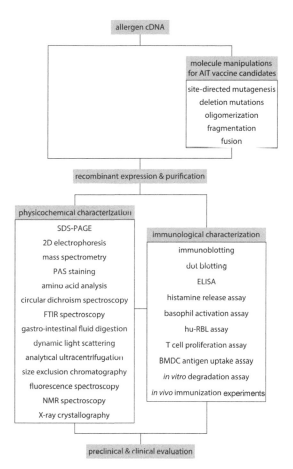

Figure 11.1 Schematic overview of the process towards the production of recombinant food allergens.

It is also possible to obtain sequence information through the analysis of peptides derived from proteolytic digestion with various proteases (Siuti and Kelleher 2007). **Amino acid analysis** is routinely used to check the amino acid content of single amino acid-, peptide- or polypeptide-containing samples. Unlike MS, amino acid analysis gives rather vague information of the primary structure of a protein but offers a very powerful technique to determine the quantity of purified proteins. Upon hydrolysis of all peptide bonds present in the sample by hydrogen chloride, the liberated amino acids are separated, detected and usually quantified by high-performance liquid chromatography. The methodology of amino acid analysis has not changed much since it was invented in the 1950s by Moore and Stein (Stein and Moore 1950, Rutherfurd and Gilani 2009). **Periodic acid-Schiff** (PAS) **stain** is probably the most common procedure to verify glycosylation of proteins. The success of PAS staining is based on the reactivity of free aldehyde groups of monosaccharides with the Schiff reagent, forming a bright red magenta product (Pastorello et al. 1999). After recombinant production verification of protein structure is crucial and several different methods are available to evaluate intact conformations. To identify certain functional groups in purified proteins **Fourier transform infrared** (FTIR) **spectroscopy** is widely used. Thereby, infrared radiation passes through a sample and depending on the unique molecular properties and structure, part of the radiation is absorbed whereas some transmits through. Notably, FTIR can be carried out in liquid and solid samples, thus also formulated allergen preparations or lyophilized allergens can be analyzed by FTIR. The use of an interferometer allows the simultaneous measurement of all infrared frequencies. The resulting interferograms cannot be directly interpreted and need to be Fourier transformed, making data evaluation not trivial. The calculated spectra are used to determine quality and composition of secondary structures (e.g., α-helix, β-sheet) of protein samples and the size of the peaks can give direct information about the amount of protein in the sample (Griffiths and de Haseth 2007). Determination of secondary structure elements can also be achieved by far-UV **circular dichroism** (CD), a widely used spectroscopic method. When a molecule contains one or more light absorbing groups close to a chiral center (in proteins represented by the C_α atoms of amino acids except for glycine), the absorption differences of left- and right-handed circular polarized light can be measured over a range of wavelengths, usually from 190 to 260 nm. CD spectroscopy can be used to monitor the change of secondary structure elements of proteins as a result of changes in environmental factors such as temperature or pH (Han et al. 2015).

This capability of CD is important as stability of food allergens against thermal and/or gastrointestinal degradation and against other harsh environmental conditions is crucial for the classification of class I and class II food allergens. *Per* definition class I food allergens are stable enough to withstand any digestive reaction until they can function as sensitizers, whereas class II food allergens show only cross-reactivity against IgE antibodies produced against other usually inhaled allergens (Harrer et al. 2010). **Gastric and intestinal fluid digestion assays** are valuable tools to simulate food processing in the digestive tract and monitor the stability against proteolytic denaturation. Food allergens, which sensitize through the gastro-intestinal route, show usually extraordinary high resistance towards the harsh conditions of the gastro-intestinal milieu. Moreover, it has been demonstrated that certain class I food allergens can adsorb to the intestinal epithelium triggering the uptake of immunologically intact protein moieties into the organism. A careful evaluation of food allergen processing will therefore provide a deeper understanding of the mechanisms eventually triggering the sensitization process to food allergens (Moreno 2007). The well-known physical technique of **dynamic light scattering** (DLS) is not only used to define the hydrodynamic size distribution profile of proteins but, also gives insight in the aggregation state of a sample. Thereby, the sample is illuminated through a polarizer by a monochromatic light source, usually a laser. As long as the particles are smaller than the wavelength, the light hitting these particles is scattered in all directions. Due to Brownian motions of the proteins, the scattered light fluctuates and forms destructive as well as constructive interference. The time-dependent fluctuations are detected at a known scattering angle by a detector and are used to calculate the hydrodynamic radius of the analyte (Berne and Pecora 1976). Like DLS, **analytical ultracentrifugation** (AU) is applied to get information about homogeneity and aggregation state of protein samples. The sedimentation behavior of macromolecules in a centrifugal field is used to determine their hydrodynamic and thermodynamic characteristics in solution. An advantage of this quantitative analysis is the characterization of a variety of macromolecules in a wide spectrum of solvents allowing the protein to be in its native state and under physiologically relevant conditions. The two main principles of AU, sedimentation velocity and equilibrium, allow the characterization of hydrodynamic properties of proteins and even protein complexes as well as the determination of molar-mass and binding-constants (Lebowitz et al. 2002). Another widely used method to get information about the integrity of a protein and quantity of aggregates is **size exclusion chromatography** (SEC). The chromatographic technique separates molecules by their size and is known as gel-filtration chromatography in case that aqueous solvents transport the sample through a column. However, the molecular weight of detected peaks can often not simply be deduced by comparing the retention times, as aggregates can form irregular shapes and carbohydrates influence the elution position. An elegant way to bypass this problem is the combination of SEC with an on-line light-scattering system to determine the molecular weight of an eluting fraction (Wen et al. 1996). **Fluorescence spectroscopy** is a technique with a broad area of applications but can also be used to detect aggregate formation and conformation of proteins. Electrons of aromatic residues in a sample being in their ground electron state are excited with a beam of light, usually ultraviolet light. Following collisions with other molecules cause them to return to a ground energy level by losing vibrational energy and emitting fluorescence. In case of tryptophan the fluorescence maximum is shifted if exposed to aqueous solvents compared to a hydrophobic protein interior, allowing an assessment about the aggregate state (Caputo and London 2003). Fluorescent dyes can be used for proteins lacking aromatic residues. 1-anilinonaphtalene-8-sulfonic acid (ANS), for example, binds to hydrophobic patches of proteins and is used to detect folded protein conformations (Kane and Bernlohr 1996). Last but not least, **nuclear magnetic resonance spectroscopy** and **X-ray crystallography** are both very sophisticated techniques to unravel the 3-D structure of proteins and therefore not routinely used for allergen characterization. Nevertheless, they are the ultimate approaches of choice for a deep look into the in-solution and crystal structures of proteins.

11.4 IMMUNOLOGICAL ANALYSES OF RECOMBINANT FOOD ALLERGENS

For a comprehensive characterization of recombinant allergens, apart from a detailed physicochemical analysis, the immunological properties of these proteins have to be thoroughly investigated. Such data should confirm that the recombinant molecules exhibit identical immunological behavior as their natural counterparts, allowing their use in either diagnosis or treatment. In this section we provide an overview on routinely applied methods of immunological characterization used to provide a variety of data ranging from IgE-binding and reactivity to cell processing and activation, as well as immunogenic potential.

One of the very first aspects of a recombinant allergen to be tested is the ability to bind allergen-specific antibodies. One way to achieve this is via **immunoblotting** (or **western blotting**). In this technique the protein of interest (or the mixture containing it) is transferred onto a nitrocellulose or PVDF membrane after performing an SDS-PAGE. In a simplified version of the immunoblot, termed **dot blot**, the protein is directly applied on the membrane as a dot, skipping the electrophoresis step. The membrane is then sequentially incubated with a protein-specific primary antibody and a reporter enzyme- or radioactive label-linked secondary antibody directed against the species-specific Fc region of the primary antibody (Renart et al. 1979, Burnette 1981). This two-step (indirect) incubation setup offers great practical flexibility and amplifies the specificity of the assay, while in the less common one-step (direct) setup the protein-specific antibody is the one bearing the detectable label. Primary antibodies can be allergen-specific purified mono- or polyclonal antibodies as well as antibodies in the sera from patients allergic to the respective protein. Depending on the substrate used and its reaction with the antibody-bound enzyme, detection can be based on colorimetric, chemiluminescent or fluorescent imaging methods (Roda and Guardigli 2012, Nishi et al. 2015).

Another method of assessing the antibody binding capacity of proteins is the enzyme-linked immunosorbent assay (**ELISA**).

Although ELISA and immunoblot are methods based on rather similar principles, ELISA is a simpler, cheaper, faster and more quantifiable assay. An ELISA begins with the passive adsorption (coating) of the investigated protein to a solid phase (i.e., polystyrene microtiter plate). The subsequent incubation and detection steps are based on the same principle as described above for immunoblotting; however with ELISA the readout is attained via spectrophotometry. When the target protein is part of a mixture, an ELISA variation is used. In this method, termed **sandwich ELISA**, an antibody is adsorbed on the solid phase subsequently capturing the respective antigen, while the rest of the procedure remains the same (Crowther 2001).

Besides their ability to bind antibodies, the biological activity of recombinant allergens is a crucial attribute to be assessed during immunological characterization. This is determined by the concentration of an allergen needed to induce an *in vitro* activation of allergen-specific IgE-sensitized effector cells by cross-linking of cell-bound IgE. Depending on the cell type used and the way activation is determined, different assays have been developed to evaluate the biological activity of allergenic proteins (Sancho et al. 2010). In the **histamine release** assay, allergen-specific IgE-sensitized basophils from allergic individuals or passively sensitized donor basophils are activated by the allergen and the histamine release is fluorometrically detected. The basophil activation test (**BAT**) uses the same sources of cells but determines activation by flow cytometric detection of activation markers CD203c or CD63 with fluorochrome-tagged antibodies. Both assays are highly dependent on frequent blood donations and are also complex and inflexible. The humanized rat basophilic leukemia cell (**hu-RBL**) mediator release assay is a simpler, quick and very sensitive model to assess biological activity of allergens. A rat basophil cell line (RBL-2H3) transfected with the human high affinity IgE receptor (FcεRI) is passively sensitized and, upon activation by the allergen, IgE-mediated degranulation is determined by fluorometric or colorimetric detection of ß-hexosaminidase release (Vogel et al. 2005).

An essential step for the characterization of recombinant proteins, especially immunotherapy vaccine candidates, is the analysis of their immunogenicity or in other words the ability to activate T cells. A well-established and routine method for that purpose is a **T cell proliferation assay**. This technique requires isolating peripheral blood mononuclear cells (PBMCs) from allergic patients' blood or splenocytes from immunized mice and culturing them in the presence of the relevant protein. After appropriate incubation, proliferation is determined by adding radioactive [^{3}H]thymidine. The incorporated radioactivity in the newly synthesized DNA of the dividing cells, which is proportional to the number of proliferating cells, is measured by scintillation counting. **Proliferation assays** can also be performed with allergen-specific T cell lines (**TCLs**) and clones (**TCCs**), but are much more tedious, complex, and usually not performed in the context of routine immunological characterization (Jahn-Schmid et al. 2002).

Another way to assess the immunogenicity of a protein is to analyze the process of its uptake by murine bone marrow derived dendritic cells (BMDCs). In a **BMDC antigen uptake assay** the DCs are incubated with the respective allergen conjugated with different fluorescent probes (e.g., Alexa Fluor 488, FITC or pHrodo) and subsequently the surface capture, internalization and intracellular degradation of the protein is measured via flow cytometry (Kitzmuller et al. 2015).

Since a link between certain resistance to proteolytic degradation and enhanced immunogenic potential has been established, the immunogenicity of an allergen can also be evaluated by determining its sensitivity to lysosomal proteolysis *in vitro*. The *in vitro* **degradation assay** uses endo-/lysosomes isolated from human monocyte-derived DCs (moDCs), murine BMDCs or a DC line (JAWS II). The tested proteins are incubated with the endo-/lysosomes over a certain time window and quantitative evaluation (amounts of protein degraded) of the assay is achieved via SDS-PAGE densitometry, while qualitative analysis (peptides generated) is performed by mass spectrometry (Egger et al. 2011).

Although the aforementioned techniques provide a thorough and comprehensive overview of a protein's immunological properties, they lack the mechanistic background of a biological organism. This issue is addressed with *in vivo* **immunization experiments** using inbred mouse strains. The allergen is usually adsorbed to aluminum-based adjuvants and administered subcutaneously or intraperitoneally to the animal following several booster injections (Wallner et al. 2011). To mimic the sensitization process of food allergies, also sensitization models via the oral route have been established (Bailon et al. 2012). The immunogenicity of the antigen is thereafter evaluated by analyzing humoral and cellular immune responses. This is achieved via assays determining serum IgE and IgG antibody levels, cytokine secretion profiles and splenic lymphocyte proliferation (Wallner et al. 2011).

11.5 RECOMBINANT FOOD ALLERGENS FOR DIAGNOSIS

Double-blind placebo-controlled food challenge is considered the gold standard for diagnosis of food allergies (Sampson et al. 2012). While reliable results are obtained from clinical studies, difficulties may arise in the daily practice due to the need for trained personal, time consumption, and risk of adverse reactions. In hand with thorough anamnesis, skin prick tests provide information on the sensitization status. Commercially available food extracts for allergy diagnosis however, only provide reasonable results if IgE reactive proteins are highly abundant and stable in the extracts, while otherwise false-negative tests could be obtained (Heinzerling et al. 2013). Therefore, prick-to-prick tests using the offending food are considered an alternative in daily routine allowing patient-tailored selection while standardization of the diagnostics might be challenging (Vlieg-Boerstra et al. 2013). Due to potential adverse reactions observed in *in vivo* food allergy diagnosis, first diagnostic strategies typically involve *in vitro* methods, thus circumventing potential fatal side effects upon exposure to minute allergens amounts (i.e., peanut). Currently, 8 different extracts are available for *in vitro* IgE testing: cow's milk, egg white, cod fish, shrimp, hazelnut, peanut, wheat, and soy. Those food sources are partially covering the so called "big 8" food allergens accounting for 90% of food allergies in the westernized countries (www.fda.gov). In addition to extracts, single allergen molecules are used for more detailed IgE diagnosis in the distinct food sources (Hoffmann-Sommergruber et al. 2015). An advantage of allergen components is the option for standardization and thus reproducibility of the test substances (Ferreira et al. 2014). In contrast, the specificity of food extracts can highly vary due to different cultivars or breeds, maturation stage or processing (Matthes and Schmitz-Eiberger 2009). While recombinant allergens produced in prokaryotic systems without glycan structures can be advantageous for undesired false-positive diagnosis due to cross-reactive carbohydrate determinants (CCD), some allergens might require post-translational modification for proper tertiary structure. In addition, molecule-based diagnostics enable discrimination between genuine sensitization and IgE cross-reactivity due to, i.e., panallergens or CCDs. A schematic overview on the diagnostic workup for food allergies is provided in Figure 11.2.

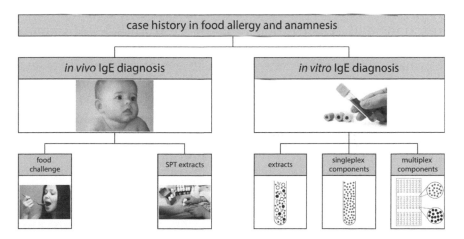

Figure 11.2 Schematic overview of food allergy diagnosis. Photos were obtained from www. fotolia.com.

Currently, 25 recombinant and 10 natural food allergens are commercially available for singleplex IgE allergen diagnosis, while multiplex test systems allow detection of 14 recombinant and 23 natural food components (Table 11.1). In a customized multiplex version (MeDALL allergen-chip), 8 natural and 38 recombinant allergens are additionally spotted on a microarray, initiating molecule-based analysis from pistachio, and further increasing the panel of peanut, walnut, almond, cashew nut, wheat and cow's milk allergens (Lupinek et al. 2014). Microarray test systems offer the possibility of simultaneous detection of numerous allergen molecules using minute amounts of serum. Since allergens are not preselected, valuable data can be obtained for cohort studies and allergen microarrays can aid in refining diagnosis, monitoring disease or therapeutic effects. In contrast to singleplex analysis, multiplex is providing semi-quantitative IgE results. In the subsequent paragraph, relevant food allergen sources and their diagnostic options focusing on commercially available recombinant allergens are explored.

11.5.1 Peanut

Peanut is considered a common trigger of food-induced anaphylactic reactions and sensitization rates range from 0.5–7.2% among European adults (Burney et al. 2014). Values as high as 10.9% among German children of the general population were observed, however this is due to cross-reactivity with pollen and therefore does not reflect primary peanut allergy (Niggemann et al. 2011). A panel of relevant peanut components is available for single component analysis while Ara h 6 and Ara h 8 are only found on multiplex platforms (Table 11.1). Notably, peanut allergy evokes different clinical and immunological pattern in different geographic areas. While American patients frequently recognize Ara h 1, 2, 3, high sensitization rates for Ara h 8, a Bet v 1 homologue, were found in Swedish patients. In contrast, subjects from Spain were more often reactive to the non-specific lipid transfer protein (nsLTP) Ara h 9 (Vereda et al. 2011). Molecule-based diagnosis can thus help to understand different clinical patterns of peanut allergy, which are: (i) sensitization to stable seed storage proteins (e.g., Ara h 2), which is linked with severe systemic reactions, (ii) reactivity to the heat-labile Ara h 8 predominately translating into mild oral allergy syndromes (OAS), and (iii) sensitization to Ara h 9, which is considered a secondary food allergy linked to non-specific lipid transfer protein (nsLTP) allergy (i.e., peach Pru p 3) that might also cause systemic reactions (Vereda et al. 2011, Ballmer-Weber et al. 2015). The seed storage proteins Ara h 2 and Ara h 6 are considered the most reliable marker for peanut allergy with severe reactions (Lopes de Oliveira et al. 2013, Kukkonen et al. 2015). More recently, predictive IgE values for clinically relevant peanut allergy were suggested with 14.4 kU/l (90% probability) and 42.2kU/l (95% probability) of Ara h 2 specific IgE (Beyer et al. 2015).

11.5.2 Tree Nuts and Seeds

The frequency of nut allergy ranges from 0.05 to 4.9%, while data including oral allergy syndrome were significantly higher and predominately observed in Central and Northern Europe (McWilliam et al. 2015). A high but varying degree of IgE cross-reactivity can be observed between tree nuts and seeds which is generally more pronounced in botanically close related sources (Maloney et al. 2008). Hazelnut represents one of the major causes of tree nut allergy, and relevant allergens are available for molecule-based analysis (Table 11.1) (Costa et al. 2015). The EuroPrevall study showed varying sensitization pattern among European hazelnut allergic patients from different regions. Sensitization to seed storage proteins (Cor a 9 and Cor a 14) was observed in less than 10% of patients, which were typically children, and was found to correlate with IgE from other nuts, seeds, and legumes. In Mediterranean countries, hazelnut allergy is driven by the nsLTP Cor a 8, while in birch pollen endemic areas, predominately adult patients are tested positive for Cor a 1 (Datema et al. 2015). Using components is thus clearly advantageous over extract based diagnosis for tree nut and seed allergy as primary sensitization can be distinguished from secondary food (nsLTP) or pollen-food (Fagales) allergy. This is of particular importance considering that clinical symptoms of nut and seed allergies as PR-10 mediated adverse reactions are frequently milder and present themselves as OAS, whereas genuine sensitization to the highly stable seed storage proteins is associated with more severe reactions. For hazelnut allergy, a 90% probability for a positive hazelnut challenge was estimated with Cor a 14-specific IgE of 47.8 kU/l (Beyer et al. 2015). It was recently reported that Ana o 3, the 2S albumin of cashew nut, is superior over extract based diagnosis and predictive for symptomatic versus tolerant patients in cashew but also pistachio sensitized individuals (Savvatianos et al. 2015).

11.5.3 Fruits and Vegetables

Adverse reactions to plant food allergens are observed in 0.1–4.3% of the population with allergens from the Rosaceae family (e.g., peach, apple) are the most frequent in triggering allergic symptoms (Zuidmeer et al. 2008, Burney et al. 2010). Currently, molecule-based IgE diagnosis is available for fruits (apple, peach, and kiwi), while purified allergens from vegetables are restricted to Api g 1 from celery (Table 11.1). Commercially available components mainly belong to the PR-10 (Bet v 1 homologs) and PR-14 (nsLTP) family, with the only exception being the profilin Pru p 4 from peach. In comparison to extract-based diagnosis of fruits and vegetables with low sensitivity, component diagnosis also allows determination of clinical pattern (Kollmann et al. 2013). This was, for example, demonstrated in a study of apple allergy in different patients' cohorts, where sensitization to Mal d 1 is typically linked to mild symptoms mainly restricted to the oral cavity. OAS to plant food in birch endemic areas is frequently caused by primary sensitization to Bet v 1 and is observed in 73% of birch pollen allergic patients (Geroldinger-Simic et al. 2011). While Mal d 1 was considered a reliable tool for diagnosis of birch pollen-associated apple allergy (Kollmann et al. 2013), another study came to the conclusion that IgE levels to PR-10 proteins are not predictive for the respective food allergy with the exception of Api g 1, which correlates to adverse reactions to celeriac (Guhsl et al. 2015). Apple allergy in Spain is in contrast driven by Mal d 3 and associated with more severe reactions as a consequence of a primary peach allergy (Fernandez-Rivas et al. 2006). In nsLTP allergy, Pru p 3 is regarded the marker molecule for genuine peach sensitization and may be used to explain allergic reactions to other nsLTP-containing plant food sources (Egger et al. 2010, Hartz et al. 2010, Scala et al. 2015).

11.5.4 Wheat

Wheat is considered an important source of vegetable proteins in most food cultures and IgE-mediated reactions to wheat can present themselves as (i) IgE-mediated food allergy, (ii) wheat-dependent exercise induced anaphylaxis (WDEIA) or (iii) baker's asthma. Depending on the age and region of study cohorts, the sensitization frequency for wheat allergy has been reported to be 0.4–1% (Ostblom et al. 2008, Zuidmeer et al. 2008, Longo et al. 2013). However, extract-based diagnosis demonstrates rather low specificity as test results are frequently biased by IgE cross-reactivity with grass pollen (van Kampen et al. 2009, Sander et al. 2015). Although a large panel of wheat allergens is identified, only Tri a 19 (omega-5-gliadin) and Tri a 14 (nsLTP) are commercially available for molecule-based diagnosis (Table 11.1). Tri a 19 represents the most important allergen in WDEIA (82% sensitization frequency), a rare but potentially severe allergy typically observed after wheat ingestion followed by physical exercise (Matsuo et al. 2005, Hofmann et al. 2012). In patients suffering from baker's asthma, the heat and proteolytically stable Tri a 14 showed a sensitization prevalence of 60% (Palacin et al. 2007).

11.5.5 Soy

Soybeans can trigger allergic reactions after consumption of unprocessed but also highly processed soy and sensitization rates of 0.4–3.1% to soy extract have been reported (Katz et al. 2014). The PR-10 protein Gly m 4 plays a relevant role in pollen-food allergies leading mostly to oral allergy syndromes after soy consumption but can occasionally also trigger severe generalized symptoms (Kleine-Tebbe et al. 2002, Kosma et al. 2011, Berneder et al. 2013). It is noteworthy to mention, that Gly m 4 is underrepresented in diagnostic soy extracts, often leading to false-negative results in patients with type II soy allergy (Berneder et al. 2013). Especially in Fagales pollen allergic patients, molecule-based diagnosis of recombinant Gly m 4 therefore represents a valuable tool for birch-pollen induced soy allergy (De Swert et al. 2012).

11.5.6 Fish

Sensitization frequency to fish is below 1% within the general population, but increased prevalence is found in regions with high fish consumption. Allergic patients demonstrate mild to severe reactions upon ingestion, contact or inhalation of fish allergens (Burney et al. 2010, Sharp and Lopata 2014). Parvalbumins, the major fish allergens (sensitization frequency 70%–90% among fish allergic patients) (Kuehn et al. 2013), are commercially available for component resolved diagnosis (Table 11.1). It is suggested to group fish allergic patients into three diagnostic clusters, which are (i) patients with reactions to several fish parvalbumins, (ii) patients reacting to parvalbumins from one (or few) fish species, and (iii) patients sensitized to other fish allergens (enolase and aldolase) (Kuehn et al. 2013). While for cluster (i) the commercially available Cyp c 1 and Gad c 1 are sufficient to diagnose fish allergy, for cluster (ii) and (iii) more components would be needed. Especially those patients could benefit from detailed component resolved diagnosis, because it could provide them with information about which fish species they could tolerate and which they should avoid (Mourad and Bahna 2015). In addition to allergens from fish meat, proteins with allergenic potential from fish blood, gelatin, caviar and the fish parasite *Anisakis simplex* (Table 11.1) are also known (Nieuwenhuizen and Lopata 2014).

11.5.7 Shellfish

Shellfish allergy has a prevalence of 2% in the general population, but in contrast to many other food allergies, children are less frequently affected than adults (Sicherer et al. 2004). Amongst food allergens, shellfish has one of the highest rates of anaphylaxis (Ross et al.

2008). Currently, 4 shrimp allergens are commercially available for molecule base diagnosis (Table 11.1). A reliable marker predicting shrimp allergy is IgE against the major shrimp allergen tropomyosin (Pen a 1, Pen m 1) (Pascal et al. 2015). The majority (93%) of patients with anaphylaxis caused by crustacean ingestion with IgE to shrimp tropomyosin, were also allergic to mollusks' (Vidal et al. 2015), suggesting cross-reactivity between tropomyosins from a broad variety of shellfish. Among house-dust mite allergic patients, 5–10% are sensitized to the tropomyosin Der p 10. Due to IgE cross-reactivity between house-dust mite and shellfish tropomyosin, those patients are at risk to develop allergic symptoms upon the consumption of shellfish (Lopata et al. 2010, Barber et al. 2012). Therefore, it might be worthwhile to test Der p 10 positive patients for Pen a 1 and Pen m 1 to predict possible food allergy.

Diagnosis of food allergy is predominately based on clinical symptoms and *in vitro* or *in vivo* tests of allergen extracts. Purified molecules and in particular CCD-free recombinant components have the advantage of further refining diagnosis and allowing prediction of potential symptom severity. The constant increase in molecular allergen research, in hand with microarray technologies, allows determination of refined IgE profiles and thus supports clinicians providing dietary interventions without limiting the diversity of healthy food intake.

11.6 RECOMBINANT FOOD ALLERGENS FOR ALLERGY THERAPY

Allergen immunotherapy (AIT) generally aims to restore or induce immunologic tolerance to disease-eliciting allergens. The efficacy of the treatment strategy for inhalant and venom allergies is well accepted, and recent data support the idea that AIT is also beneficial for food allergies and atopic dermatitis (Jutel et al. 2015). A serious problem for the treatment of food allergies is the fact that fatal reactions to food allergens may occur. In 1992, a small study of subcutaneous immunotherapy to treat peanut allergy was performed using unmodified peanut extracts as active vaccine component.

The study was placebo controlled and showed clinical efficacy in the actively treated group; however, the rate of systemic side effects was >13%. Moreover, due to a formulation error in the pharmacy, a patient within the placebo group received the maintenance dose of peanut extract and died of anaphylaxis—a fact that led to the termination of the study (Oppenheimer et al. 1992). In a small study with 12 peanut allergic patients published five years later, the aim was again the treatment of peanut allergies using a rush protocol based on peanut extracts. The study was clinically efficacious in some patients, whereas half of the actively treated patients did not show much of improvement in a double-blind placebo-controlled food challenge after one year of treatment. Of note, the patients who could not tolerate the full maintenance dose had a high rate of systemic reactions during updosing (average 9.8 epinephrine injections per subject were necessary), but also side effects during the maintenance phase (12.6 epinephrine injections per subject) were high (Nelson et al. 1997, Nowak-Wegrzyn and Sampson 2011). This indicates a general problem of extract-based AIT of food allergies. Allergen extracts are hardly standardizable, heterogeneous mixtures of easily extractable compounds from a food source. The cocktail ideally contains sufficient amounts of the disease-eliciting allergens to induce tolerance. However, unmodified allergen extracts bear the risk of the induction of IgE-mediated side effects, which are difficult to control and may end fatal. To avoid such problems, AIT concepts based on recombinant food allergens have been developed. Therefore, the disease eliciting allergens need to be identified within the allergen source and selected for the use in AIT. Most AIT trials were performed for inhalant allergies based on formulations with a single allergen (Ferreira et al. 2014), though, for grass pollen allergies, a cocktail consisting of four major allergens has been successfully tested (Jutel et al. 2005). In peanut allergy, Ara h 1, 2, and 3 represent major allergenic components. The IgE-binding epitopes of the three allergens have been identified and amino acids critical for the binding sites have been replaced resulting in proteins with reduced IgE-binding properties (Bannon et al. 2001). In a new vaccination strategy, the recombinant allergens were independently expressed in

E. coli, the host cells were killed by treatment with heat and phenol, and thereafter the *E. coli*-encapsulated allergens were formulated as vaccine for rectal application (EMP-123). In a phase I trial, EMP-123 was tested in five healthy individuals in a four week schedule without inducing adverse reactions. Thereafter, ten peanut-allergic patients were treated in an updosing regimen for ten weeks followed by three biweekly administered maintenance doses. Due to side-effects five patients could not complete the trial, whereas four had no and one patient only mild symptoms (Wood et al. 2013).

Similar to allergies to peanuts, tree nuts, and shellfish, as well as egg and milk in infants, allergic reactions to fish can be severe to life threatening. Parvalbumin has been identified as major IgE binding component in various allergenic fish species (Table 11.1). The small (10–13 kDa) proteins encode three EF-hand domains and are extraordinarily stable to heat treatment, denaturation and proteolysis. Under the umbrella of an EU-funded study for the development of food allergy-specific immunotherapy (FAST), a modified version of the carp parvalbum Cyp c 1 was selected as AIT candidate. The allergen was shown to cover most cross-reactive IgE epitopes of homologous fish parvalbumins (Zuidmeer-Jongejan et al. 2015). The mutant Cyp c 1 was generated by introducing point mutations in the two EF-hand domains which actually bind calcium ions, resulting in drastic reduction of IgE binding (Swoboda et al. 2007). Within the FAST project a GMP-batch of the mutant Cyp c 1 was produced and is currently evaluated in a clinical trial (ClinicalTrials.gov, Valenta et al. 2015, Zuidmeer-Jongejan et al. 2015).

Many allergic reactions towards food allergens are triggered by cross-reactive IgE antibodies primarily directed against inhalant allergens. Thus, attempts were made to treat pollen induced cross-reactive food allergies by cross-immunotherapy. Thereby, primarily the pollen allergy (i.e., birch pollen allergy) is treated with subcutaneous or sublingual AIT, whereas concomitant food allergies (i.e., allergies to apple or hazelnut) can be modified simultaneously. Although it is still unclear to what extent birch pollen AIT can actually improve related food allergies, studies indicate that the treatment can induce cross-reactive blocking antibodies against the major birch pollen allergen Bet v 1, which would eventually recognize the food homologues from apple and hazelnut (Subbarayal et al. 2013). The clinical effects of such interventions appear to be long-lasting (Asero 2003). Nevertheless, the treatment is based on allergen extracts (Nowak-Wegrzyn and Sampson 2011). For birch pollen allergy many trials have been successfully completed using recombinant Bet v 1 or derivatives thereof (Ferreira et al. 2014), thus one can expect that recombinant allergens will also be used to study the effects of cross-immunotherapy in more detail.

As mentioned, immunotherapy strategies based on recombinant allergens or derivatives thereof are most advanced for inhalant allergies. However, many of the concepts have been adapted for food allergens including site-directed mutagenesis as shown, for instance, for the peach lipid transfer protein Pru p 3, the apple allergen Mal d 1, or the peanut allergens Ara h 1, 2, and 3 (Bannon et al. 2001, Bolhaar et al. 2005, Gomez-Casado et al. 2013). For shellfish allergy, two hypoallergenic variants were designed, one by replacing 49 residues of shellfish Met e 1 with homologous residues from fish tropomyosin, the other by the deletion of nine IgE epitopes resulting in a truncated version of the allergen. Both recombinant proteins showed markedly reduced IgE-binding properties, but were still able to induce Met e 1 reactive antibodies in a mouse model (Wai et al. 2014). Reese et al. studied the effects of oligomerization on carrot Dau c 1 wildtype, as well as mutant allergens, and found that the low IgE-binding mutants were more immunogenic as dimers compared to their monomeric forms (Reese et al. 2007). The fusion of T cell epitopes from carrot Dau c 1 and Api g 1 from celery to the homologous pollen allergen Bet v 1 was explored in a prophylactic mouse model. In this setup nasal pretreatment with the recombinant chimer could effectively suppress IgE-mediated immune responses to the parental allergens in poly-sensitized animals, which was associated with the induction of regulatory mechanisms (Hoflehner et al. 2012). The hen egg white ovomucoid Gal d 1 consists of three tandem domains. To generate a vaccine candidate, the third domain of Gal d 1 was mutated and produced as recombinant protein in *E. coli*. The hypoallergen was used to desensitize mice primarily sensitized with the unmodified

Gal d 1 third domain. The treatment could suppress allergen-specific IgE levels and in parallel induce the production of regulatory cytokines (Rupa and Mine 2006). The encouraging preclinical data of these examples show that the development of safer food allergy therapeutics is possible. Moreover, naturally occurring low IgE-binding proteins may provide an alternative to engineered allergens to be used for the rational design of food allergy vaccines (Ramos et al. 2009). Thus, recombinant allergens promise significant advances in the therapy of food allergies.

11.7 CONCLUSIONS

The use of recombinant allergens has brought tremendous progress in the fields of food allergy diagnosis and therapy. A large number of important food allergens are available in recombinant form. Nevertheless, strict quality controls of the recombinant products are mandatory to guarantee highest standards. Especially in allergy diagnosis the use of recombinant allergens has revolutionized the process, enabling clinicians to identify disease eliciting allergens as well as cross-reactivity pattern by analyzing patients' individual IgE recognition profiles. Moreover, first attempts to use recombinant allergens for therapeutic setups are on the way and we can be quite curious about the outcome of the trials. In principle recombinant technology would provide us with the tools necessary for personalized allergy medicine. However, whether this can be realized in the future not only depends on the recombinant allergen products but also on the regulatory guidelines from the medical agencies (Cromwell et al. 2011).

ACKNOWLEDGEMENTS

This work was supported by the Priority Program "Allergy-Cancer-BioNano Research Centre" of the University of Salzburg and the EU-funded project FAST (201871).

Keywords: Recombinant allergens; recombinant allergens and diagnosis; recombinant allergen production; recombinant allergens and therapy; food allergy

REFERENCES

Asero, R. (2003). How long does the effect of birch pollen injection SIT on apple allergy last? Allergy. 58(5): 435–438.

Bailon, E., M. Cueto-Sola, P. Utrilla, J. Rodriguez-Ruiz, N. Garrido-Mesa, A. Zarzuelo, J. Xaus, J. Galvez and M. Comalada (2012). A shorter and more specific oral sensitization-based experimental model of food allergy in mice. J Immunol Methods. 381(1-2): 41–49.

Ballmer-Weber, B. K., J. Lidholm, M. Fernandez-Rivas, S. Seneviratne, K. M. Hanschmann, L. Vogel, P. Bures, P. Fritsche, C. Summers, A. C. Knulst, T. M. Le, I. Reig, N. G. Papadopoulos, A. Sinaniotis, S. Belohlavkova, T. Popov, T. Kralimarkova, F. de Blay, A. Purohit, M. Clausen, M. Jedrzejczak-Czechowcz, M. L. Kowalski, R. Asero, R. Dubakiene, L. Barreales, E. N. Clare Mills, R. van Ree and S. Vieths (2015). IgE recognition patterns in peanut allergy are age dependent: perspectives of the EuroPrevall study. Allergy. 70(4): 391–407.

Bannon, G. A., G. Cockrell, C. Connaughton, C. M. West, R. Helm, J. S. Stanley, N. King, P. Rabjohn, H. A. Sampson and A. W. Burks (2001). Engineering, characterization and *in vitro* efficacy of the major peanut allergens for use in immunotherapy. Int Arch Allergy Immunol. 124(1-3): 70–72.

Barber, D., J. Arias, M. Boquete, V. Cardona, T. Carrillo, G. Gala, P. Gamboa, J. C. Garcia-Robaina, D. Hernandez, M. L. Sanz, A. I. Tabar, C. Vidal, H. Ipsen, F. de la Torre and M. Lombardero (2012). Analysis of mite allergic patients in a diverse territory by improved diagnostic tools. Clin Exp Allergy. 42(7): 1129–1138.

Berne, B. J. and R. Pecora (1976). Dynamic Light Scattering With Applications to Chemistry, Biology , and Physics, Dover Publications, Inc.

Berneder, M., M. Bublin, K. Hoffmann-Sommergruber, T. Hawranek and R. Lang (2013). Allergen chip diagnosis for soy-allergic patients: Gly m 4 as a marker for severe food-allergic reactions to soy. Int Arch Allergy Immunol. 161(3): 229–233.

Beyer, K., L. Grabenhenrich, M. Hartl, A. Beder, B. Kalb, M. Ziegert, A. Finger, N. Harandi, R. Schlags, M. Gappa, L. Puzzo, H. Roblitz, M. Millner-Uhlemann, S. Busing, H. Ott, L. Lange and B. Niggemann (2015). Predictive values of component-specific IgE for the outcome of peanut and hazelnut food challenges in children. Allergy. 70(1): 90–98.

Bolhaar, S. T., L. Zuidmeer, Y. Ma, F. Ferreira, C. A. Bruijnzeel-Koomen, K. Hoffmann-Sommergruber, R. van Ree and A. C. Knulst (2005). A mutant of the major apple allergen, Mal d 1, demonstrating hypo-allergenicity in the target organ by double-blind placebo-controlled food challenge. Clin Exp Allergy. 35(12): 1638–1644.

Burnette, W. N. (1981). Western blotting: electrophoretic transfer of proteins from sodium dodecyl sulfate—polyacrylamide gels to unmodified nitrocellulose and radiographic detection with antibody and radioiodinated protein A. Anal Biochem. 112(2): 195–203.

Burney, P., C. Summers, S. Chinn, R. Hooper, R. van Ree and J. Lidholm (2010). Prevalence and distribution of sensitization to foods in the European Community Respiratory Health Survey: a EuroPrevall analysis. Allergy. 65(9): 1182–1188.

Burney, P. G., J. Potts, I. Kummeling, E. N. Mills, M. Clausen, R. Dubakiene, L. Barreales, C. Fernandez-Perez, M. Fernandez-Rivas, T. M. Le, A. C. Knulst, M. L. Kowalski, J. Lidholm, B. K. Ballmer-Weber, C. Braun-Fahlander, T. Mustakov, T. Kralimarkova, T. Popov, A. Sakellariou, N. G. Papadopoulos, S. A. Versteeg, L. Zuidmeer, J. H. Akkerdaas, K. Hoffmann-Sommergruber and R. van Ree (2014). The prevalence and distribution of food sensitization in European adults. Allergy. 69(3): 365–371.

Caputo, G. A. and E. London (2003). Cumulative effects of amino acid substitutions and hydrophobic mismatch upon the transmembrane stability and conformation of hydrophobic alpha-helices. Biochemistry. 42(11): 3275–3285.

ClinicalTrials.gov. Identifier: NCT02017626.

Costa, J., I. Mafra, I. Carrapatoso and M. B. Oliveira (2015). Hazelnut allergens: Molecular characterisation, detection and clinical relevance. Crit Rev Food Sci Nutr. 0.

Cromwell, O., D. Hafner and A. Nandy (2011). Recombinant allergens for specific immunotherapy. J Allergy Clin Immunol. 127(4): 865–872.

Crowther, J. R. (2001). The ELISA Guidebook, Humana Press.

Datema, M. R., L. Zuidmeer-Jongejan, R. Asero, L. Barreales, S. Belohlavkova, F. de Blay, P. Bures, M. Clausen, R. Dubakiene, D. Gislason, M. Jedrzejczak-Czechowicz, M. L. Kowalski, A. C. Knulst, T. Kralimarkova, T. M. Le, A. Lovegrove, J. Marsh, N. G. Papadopoulos, T. Popov, N. Del Prado, A. Purohit, G. Reese, I. Reig, S. L. Seneviratne, A. Sinaniotis, S. A. Versteeg, S. Vieths, A. H. Zwinderman, C. Mills, J. Lidholm, K. Hoffmann-Sommergruber, M. Fernandez-Rivas, B. Ballmer-Weber and R. van Ree (2015). Hazelnut allergy across Europe dissected molecularly: A EuroPrevall outpatient clinic survey. J Allergy Clin Immunol. 136(2): 382–391.

De Swert, L. F., R. Gadisseur, S. Sjolander, M. Raes, J. Leus and E. Van Hoeyveld (2012). Secondary soy allergy in children with birch pollen allergy may cause both chronic and acute symptoms. Pediatr Allergy Immunol. 23(2): 117–123.

Egger, M., M. Hauser, A. Mari, F. Ferreira and G. Gadermaier (2010). The role of lipid transfer proteins in allergic diseases. Curr Allergy Asthma Rep. 10(5): 326–335.

Egger, M., A. Jurets, M. Wallner, P. Briza, S. Ruzek, S. Hainzl, U. Pichler, C. Kitzmuller, B. Bohle, C. G. Huber and F. Ferreira (2011). Assessing protein immunogenicity with a dendritic cell line-derived endolysosomal degradome. PLoS One. 6(2): e17278.

Fernandez-Rivas, M., S. Bolhaar, E. Gonzalez-Mancebo, R. Asero, A. van Leeuwen, B. Bohle, Y. Ma, C. Ebner, N. Rigby, A. I. Sancho, S. Miles, L. Zuidmeer, A. Knulst, H. Breiteneder, C. Mills, K. Hoffmann-Sommergruber and R. van Ree (2006). Apple allergy across Europe: how allergen sensitization profiles determine the clinical expression of allergies to plant foods. J Allergy Clin Immunol 118(2): 481–488.

Ferreira, F., M. Wolf and M. Wallner (2014). Molecular approach to allergy diagnosis and therapy. Yonsei Med J. 55(4): 839–852.

Geroldinger-Simic, M., T. Zelniker, W. Aberer, C. Ebner, C. Egger, A. Greiderer, N. Prem, J. Lidholm, B. K. Ballmer-Weber, S. Vieths and B. Bohle (2011). Birch pollen-related food allergy: clinical aspects and the role of allergen-specific IgE and IgG4 antibodies. J Allergy Clin Immunol. 127(3): 616–622. e611.

Gomez-Casado, C., M. Garrido-Arandia, P. Gamboa, N. Blanca-Lopez, G. Canto, J. Varela, J. Cuesta-Herranz, L. F. Pacios, A. Diaz-Perales and L. Tordesillas (2013). Allergenic characterization of new mutant forms of Pru p 3 as new immunotherapy vaccines. Clin Dev Immunol. 2013: 385615.

Griffiths, P. R. and J. A. de Haseth (2007). Fourier Transform Infared Spectroscopy, Wiley-Interscience A John Wiley & Sons, Inc.

Guhsl, E. E., G. Hofstetter, N. Lengger, W. Hemmer, C. Ebner, R. Froschl, M. Bublin, C. Lupinek, H. Breiteneder and C. Radauer (2015). IgE, IgG4 and IgA specific to Bet v 1-related food allergens do not predict oral allergy syndrome. Allergy. 70(1): 59–66.

Han, Y., J. Wang, Y. Li, Y. Hang, X. Yin and Q. Li (2015). Circular dichroism and infrared spectroscopic characterization of secondary structure components of protein Z during mashing and boiling processes. Food Chem. 188: 201–209.

Harrer, A., M. Egger, G. Gadermaier, A. Erler, M. Hauser, F. Ferreira and M. Himly (2010). Characterization of plant food allergens: an overview on physicochemical and immunological techniques. Mol Nutr Food Res. 54(1): 93–112.

Hartz, C., I. Lauer, M. del Mar San Miguel Moncin, A. Cistero-Bahima, K. Foetisch, J. Lidholm, S. Vieths and S. Scheurer (2010). Comparison of IgE-binding capacity, cross-reactivity and biological potency of allergenic non-specific lipid transfer proteins from peach, cherry and hazelnut. Int Arch Allergy Immunol. 153(4): 335–346.

Heinzerling, L., A. Mari, K. C. Bergmann, M. Bresciani, G. Burbach, U. Darsow, S. Durham, W. Fokkens, M. Gjomarkaj, T. Haahtela, A. T. Bom, S. Wohrl, H. Maibach and R. Lockey (2013). The skin prick test—European standards. Clin Transl Allergy. 3(1): 3.

Hoffmann-Sommergruber, K., S. Pfeifer and M. Bublin (2015). Applications of molecular diagnostic testing in food allergy. Curr Allergy Asthma Rep. 15(9): 56.

Hoflehner, E., K. Hufnagl, I. Schabussova, J. Jasinska, K. Hoffmann-Sommergruber, B. Bohle, R. M. Maizels and U. Wiedermann (2012). Prevention of birch pollen-related food allergy by mucosal treatment with multi-allergen-chimers in mice. PLoS One. 7(6): e39409.

Hofmann, S. C., J. Fischer, C. Eriksson, O. Bengtsson Gref, T. Biedermann and T. Jakob (2012). IgE detection to alpha/beta/gamma-gliadin and its clinical relevance in wheat-dependent exercise-induced anaphylaxis. Allergy. 67(11): 1457–1460.

Jahn-Schmid, B., P. Kelemen, M. Himly, B. Bohle, G. Fischer, F. Ferreira and C. Ebner (2002). The T cell response to Art v 1, the major mugwort pollen allergen, is dominated by one epitope. J Immunol. 169(10): 6005–6011.

Jutel, M., I. Agache, S. Bonini, A. W. Burks, M. Calderon, W. Canonica, L. Cox, P. Demoly, A. J. Frew, R. O'Hehir, J. Kleine-Tebbe, A. Muraro, G. Lack, D. Larenas, M. Levin, H. Nelson, R. Pawankar, O. Pfaar, R. van Ree, H. Sampson, A. F. Santos, G. Du Toit, T. Werfel, R. Gerth van Wijk, L. Zhang and C. A. Akdis (2015). International consensus on allergy immunotherapy. J Allergy Clin Immunol. 136(3): 556–568.

Jutel, M., L. Jaeger, R. Suck, H. Meyer, H. Fiebig and O. Cromwell (2005). Allergen-specific immunotherapy with recombinant grass pollen allergens. J Allergy Clin Immunol. 116(3): 608–613.

Kane, C. D. and D. A. Bernlohr (1996). A simple assay for intracellular lipid-binding proteins using displacement of 1-anilinonaphthalene 8-sulfonic acid. Anal Biochem. 233(2): 197–204.

Katz, Y., P. Gutierrez-Castrellon, M. G. Gonzalez, R. Rivas, B. W. Lee and P. Alarcon (2014). A comprehensive review of sensitization and allergy to soy-based products. Clin Rev Allergy Immunol. 46(3): 272–281.

Kitzmuller, C., N. Zulehner, A. Roulias, P. Briza, F. Ferreira, I. Fae, G. F. Fischer and B. Bohle (2015). Correlation of sensitizing capacity and T-cell recognition within the Bet v 1 family. J Allergy Clin Immunol. 136(1): 151–158.

Kleine-Tebbe, J., L. Vogel, D. N. Crowell, U. F. Haustein and S. Vieths (2002). Severe oral allergy syndrome and anaphylactic reactions caused by a Bet v 1- related PR-10 protein in soybean, SAM22. J Allergy Clin Immunol. 110(5): 797–804.

Kollmann, D., M. Geroldinger-Simic, T. Kinaciyan, H. Huber, C. Ebner, J. Lidholm and B. Bohle (2013). Recombinant Mal d 1 is a reliable diagnostic tool for birch pollen allergen-associated apple allergy. J Allergy Clin Immunol. 132(4): 1008–1010.

Kosma, P., S. Sjolander, E. Landgren, M. P. Borres and G. Hedlin (2011). Severe reactions after the intake of soy drink in birch pollen-allergic children sensitized to Gly m 4. Acta Paediatr. 100(2): 305–306.

Kuehn, A., C. Hilger, C. Lehners-Weber, F. Codreanu-Morel, M. Morisset, C. Metz-Favre, G. Pauli, F. de Blay, D. Revets, C. P. Muller, L. Vogel, S. Vieths and F. Hentges (2013). Identification of enolases and aldolases as important fish allergens in cod, salmon and tuna: component resolved diagnosis using parvalbumin and the new allergens. Clin Exp Allergy. 43(7): 811–822.

Kukkonen, A. K., A. S. Pelkonen, S. Makinen-Kiljunen, H. Voutilainen and M. J. Makela (2015). Ara h 2 and Ara 6 are the best predictors of severe peanut allergy: A double-blind placebo-controlled study. Allergy. 70(10): 1239–1245.

Laemmli, U. K. (1970). Cleavage of structural proteins during the assembly of the head of bacteriophage T4. Nature. 227(5259): 680–685.

Lebowitz, J., M. S. Lewis and P. Schuck (2002). Modern analytical ultracentrifugation in protein science: A tutorial review. Protein Sci. 11(9): 2067–2079.

Longo, G., I. Berti, A. W. Burks, B. Krauss and E. Barbi (2013). IgE-mediated food allergy in children. Lancet. 382(9905): 1656–1664.

Lopata, A. L., R. E. O'Hehir and S. B. Lehrer (2010). Shellfish allergy. Clin Exp Allergy. 40(6): 850–858.

Lopes de Oliveira, L. C., M. Aderhold, M. Brill, G. Schulz, C. Rolinck-Werninghaus, E. N. Clare Mills, B. Niggemann, C. K. Naspitz, U. Wahn and K. Beyer (2013). The value of specific IgE to peanut and its component Ara h 2 in the diagnosis of peanut allergy. J Allergy Clin Immunol Pract. 1(4): 394–398.

Lupinek, C., E. Wollmann, A. Baar, S. Banerjee, H. Breiteneder, B. M. Broecker, M. Bublin, M. Curin, S. Flicker, T. Garmatiuk, H. Hochwallner, I. Mittermann, S. Pahr, Y. Resch, K. H. Roux, B. Srinivasan, S. Stentzel, S. Vrtala, L. N. Willison, M. Wickman, K. C. Lodrup-Carlsen, J. M. Anto, J. Bousquet, C. Bachert, D. Ebner, T. Schlederer, C. Harwanegg and R. Valenta (2014). Advances in allergen-microarray technology for diagnosis and monitoring of allergy: the MeDALL allergen-chip. Methods. 66(1): 106–119.

Maloney, J. M., M. Rudengren, S. Ahlstedt, S. A. Bock and H. A. Sampson (2008). The use of serum-specific IgE measurements for the diagnosis of peanut, tree nut, and seed allergy. J Allergy Clin Immunol. 122(1): 145–151.

Matsuo, H., K. Kohno and E. Morita (2005). Molecular cloning, recombinant expression and IgE-binding epitope of omega-5 gliadin, a major allergen in wheat-dependent exercise-induced anaphylaxis. FEBS J. 272(17): 4431–4438.

Matthes, A. and M. Schmitz-Eiberger (2009). Apple (Malus domestica L. Borkh.) allergen Mal d 1: effect of cultivar, cultivation system, and storage conditions. J Agric Food Chem. 57(22): 10548–10553.

McWilliam, V., J. Koplin, C. Lodge, M. Tang, S. Dharmage and K. Allen (2015). The prevalence of tree nut allergy: A systematic review. Curr Allergy Asthma Rep. 15(9): 54.

Moreno, F. J. (2007). Gastrointestinal digestion of food allergens: effect on their allergenicity. Biomed Pharmacother. 61(1): 50–60.

Mourad, A. A. and S. L. Bahna (2015). Fish-allergic patients may be able to eat fish. Expert Rev Clin Immunol. 11(3): 419–430.

Nelson, H. S., J. Lahr, R. Rule, A. Bock and D. Leung (1997). Treatment of anaphylactic sensitivity to peanuts by immunotherapy with injections of aqueous peanut extract. J Allergy Clin Immunol. 99(6 Pt 1): 744–751.

Nieuwenhuizen, N. E. and A. L. Lopata (2014). Allergic reactions to Anisakis found in fish. Curr Allergy Asthma Rep. 14(8): 455.

Niggemann, B., R. Schmitz and M. Schlaud (2011). The high prevalence of peanut sensitization in childhood is due to cross-reactivity to pollen. Allergy. 66(7): 980–981.

Nishi, K., S. Isobe, Y. Zhu and R. Kiyama (2015). Fluorescence-based bioassays for the detection and evaluation of food materials. Sensors (Basel). 15(10): 25831–25867.

Nowak-Wegrzyn, A. and H. A. Sampson (2011). Future therapies for food allergies. J Allergy Clin Immunol. 127(3): 558–573; quiz 574-555.

Oppenheimer, J. J., H. S. Nelson, S. A. Bock, F. Christensen and D. Y. Leung (1992). Treatment of peanut allergy with rush immunotherapy. J Allergy Clin Immunol. 90(2): 256–262.

Ostblom, E., G. Lilja, G. Pershagen, M. van Hage and M. Wickman (2008). Phenotypes of food hypersensitivity and development of allergic diseases during the first 8 years of life. Clin Exp Allergy. 38(8): 1325–1332.

Palacin, A., S. Quirce, A. Armentia, M. Fernandez-Nieto, L. F. Pacios, T. Asensio, J. Sastre, A. Diaz-Perales and G. Salcedo (2007). Wheat lipid transfer protein is a major allergen associated with baker's asthma. J Allergy Clin Immunol. 120(5): 1132–1138.

Pascal, M., G. Grishina, A. C. Yang, S. Sanchez-Garcia, J. Lin, D. Towle, M. D. Ibanez, J. Sastre, H. A. Sampson and R. Ayuso (2015). Molecular diagnosis of shrimp allergy: Efficiency of several allergens to predict clinical reactivity. J Allergy Clin Immunol Pract. 3(4): 521–529 e510.

Pastorello, E. A., L. Farioli, V. Pravettoni, C. Ortolani, M. Ispano, M. Monza, C. Baroglio, E. Scibola, R. Ansaloni, C. Incorvaia and A. Conti (1999). The major allergen of peach (Prunus persica) is a lipid transfer protein. J Allergy Clin Immunol. 103(3 Pt 1): 520–526.

Ramazzina, I., S. Amato, E. Passera, S. Sforza, G. Mistrello, R. Berni and C. Folli (2012). Isoform identification, recombinant production and characterization of the allergen lipid transfer protein 1 from pear (Pyr c 3). Gene. 491(2): 173–181.

Ramos, M. L., J. J. Huntley, S. J. Maleki and P. Ozias-Akins (2009). Identification and characterization of a hypoallergenic ortholog of Ara h 2.01. Plant Mol Biol. 69(3): 325–335.

Reese, G., B. K. Ballmer-Weber, A. Wangorsch, S. Randow and S. Vieths (2007). Allergenicity and antigenicity of wild-type and mutant, monomeric, and dimeric carrot major allergen Dau c 1: destruction of conformation, not oligomerization, is the roadmap to save allergen vaccines. J Allergy Clin Immunol. 119(4): 944–951.

Renart, J., J. Reiser and G. R. Stark (1979). Transfer of proteins from gels to diazobenzyloxymethyl-paper and detection with antisera: a method for studying antibody specificity and antigen structure. Proc Natl Acad Sci USA. 76(7): 3116–3120.

Roda, A. and M. Guardigli (2012). Analytical chemiluminescence and bioluminescence: latest achievements and new horizons. Anal Bioanal Chem. 402(1): 69–76.

Ross, M. P., M. Ferguson, D. Street, K. Klontz, T. Schroeder and S. Luccioli (2008). Analysis of food-allergic and anaphylactic events in the National Electronic Injury Surveillance System. J Allergy Clin Immunol. 121(1): 166–171.

Rupa, P. and Y. Mine (2006). Engineered recombinant ovomucoid third domain can desensitize Balb/c mice of egg allergy. Allergy. 61(7): 836–842.

Rutherfurd, S. M. and G. S. Gilani (2009). Amino acid analysis. Curr Protoc Protein Sci. Chapter 11: Unit 11.19.

Sampson, H. A., R. Gerth van Wijk, C. Bindslev-Jensen, S. Sicherer, S. S. Teuber, A. W. Burks, A. E. Dubois, K. Beyer, P. A. Eigenmann, J. M. Spergel, T. Werfel and V. M. Chinchilli (2012). Standardizing double-blind, placebo-controlled oral food challenges: American Academy of Allergy, Asthma & Immunology-European Academy of Allergy and Clinical Immunology PRACTALL consensus report. J Allergy Clin Immunol. 130(6): 1260–1274.

Sancho, A. I., K. Hoffmann-Sommergruber, S. Alessandri, A. Conti, M. G. Giuffrida, P. Shewry, B. M. Jensen, P. Skov and S. Vieths (2010). Authentication of food allergen quality by physicochemical and immunological methods. Clin Exp Allergy. 40(7): 973–986.

Sander, I., H. P. Rihs, G. Doekes, S. Quirce, E. Krop, P. Rozynek, V. van Kampen, R. Merget, U. Meurer, T. Bruning and M. Raulf (2015). Component-resolved diagnosis of baker's allergy based on specific IgE to recombinant wheat flour proteins. J Allergy Clin Immunol. 135(6): 1529–1537.

Savvatianos, S., A. P. Konstantinopoulos, A. Borga, G. Stavroulakis, J. Lidholm, M. P. Borres, E. Manousakis and N. G. Papadopoulos (2015). Sensitization to cashew nut 2S albumin, Ana o 3, is highly predictive of cashew and pistachio allergy in Greek children. J Allergy Clin Immunol. 136(1): 192–194.

Scala, E., S. J. Till, R. Asero, D. Abeni, E. C. Guerra, L. Pirrotta, R. Paganelli, D. Pomponi, M. Giani, O. De Pita and L. Cecchi (2015). Lipid transfer protein sensitization: Reactivity profiles and clinical risk assessment in an Italian cohort. Allergy. 70(8): 933–943.

Sharp, M. F. and A. L. Lopata (2014). Fish allergy: in review. Clin Rev Allergy Immunol. 46(3): 258–271.

Sicherer, S. H., A. Munoz-Furlong and H. A. Sampson (2004). Prevalence of seafood allergy in the United States determined by a random telephone survey. J Allergy Clin Immunol. 114(1): 159–165.

Siuti, N. and N. L. Kelleher (2007). Decoding protein modifications using top-down mass spectrometry. Nat Methods. 4(10): 817–821.

Son, D. Y., S. Scheurer, A. Hoffmann, D. Haustein and S. Vieths (1999). Pollen-related food allergy: cloning and immunological analysis of isoforms and mutants of Mal d 1, the major apple allergen, and Bet v 1, the major birch pollen allergen. Eur J Nutr. 38(4): 201–215.

Stein, W. H. and S. Moore (1950). Chromatographic determination of the amino acid composition of proteins. Cold Spring Harb Symp Quant Biol. 14: 179–190.

Subbarayal, B., D. Schiller, C. Mobs, N. W. de Jong, C. Ebner, N. Reider, D. Bartel, J. Lidholm, W. Pfutzner, R. Gerth van Wijk, S. Vieths and B. Bohle (2013). Kinetics, cross-reactivity, and specificity of Bet v 1-specific IgG4 antibodies induced by immunotherapy with birch pollen. Allergy. 68(11): 1377–1386.

Swoboda, I., A. Bugajska-Schretter, B. Linhart, P. Verdino, W. Keller, U. Schulmeister, W. R. Sperr, P. Valent, G. Peltre, S. Quirce, N. Douladiris, N. G. Papadopoulos, R. Valenta and S. Spitzauer (2007). A recombinant hypoallergenic parvalbumin mutant for immunotherapy of IgE-mediated fish allergy. J Immunol. 178(10): 6290–6296.

Swoboda, I., A. Bugajska-Schretter, P. Verdino, W. Keller, W. R. Sperr, P. Valent, R. Valenta and S. Spitzauer (2002). Recombinant carp parvalbumin, the major cross-reactive fish allergen: a tool for diagnosis and therapy of fish allergy. J Immunol. 168(9): 4576–4584.

Valenta, R., H. Hochwallner, B. Linhart and S. Pahr (2015). Food allergies: the basics. Gastroenterology. 148(6): 1120–1131 e1124.

van Kampen, V., R. Merget, S. Rabstein, I. Sander, T. Bruening, H. C. Broding, C. Keller, H. Muesken, A. Overlack, G. Schultze-Werninghaus, J. Walusiak and M. Raulf-Heimsoth (2009). Comparison of wheat and rye flour solutions for skin prick testing: a multi-centre study (Stad 1). Clin Exp Allergy. 39(12): 1896–1902.

Vereda, A., M. van Hage, S. Ahlstedt, M. D. Ibanez, J. Cuesta-Herranz, J. van Odijk, M. Wickman and H. A. Sampson (2011). Peanut allergy: Clinical and immunologic differences among patients from 3 different geographic regions. J Allergy Clin Immunol. 127(3): 603–607.

Vidal, C., B. Bartolome, V. Rodriguez, M. Armisen, A. Linneberg and A. Gonzalez-Quintela (2015). Sensitization pattern of crustacean-allergic individuals can indicate allergy to molluscs. Allergy. 70(11): 1493–1496.

Vlieg-Boerstra, B. J., W. E. van de Weg, S. van der Heide and A. E. Dubois (2013). Where to prick the apple for skin testing? Allergy. 68(9): 1196–1198.

Vogel, L., D. Luttkopf, L. Hatahet, D. Haustein and S. Vieths (2005). Development of a functional *in vitro* assay as a novel tool for the standardization of allergen extracts in the human system. Allergy. 60(8): 1021–1028.

Wai, C. Y., N. Y. Leung, M. H. Ho, L. J. Gershwin, S. A. Shu, P. S. Leung and K. H. Chu (2014). Immunization with Hypoallergens of shrimp allergen tropomyosin inhibits shrimp tropomyosin specific IgE reactivity. PLoS One. 9(11): e111649.

Wallner, M., M. Hauser, M. Himly, N. Zaborsky, S. Mutschlechner, A. Harrer, C. Asam, U. Pichler, R. van Ree, P. Briza, J. Thalhamer, B. Bohle, G. Achatz and F. Ferreira (2011). Reshaping the Bet v 1 fold modulates T(H) polarization. J Allergy Clin Immunol. 127(6): 1571–1578. e1579.

Wallner, M., M. Himly, A. Neubauer, A. Erler, M. Hauser, C. Asam, S. Mutschlechner, C. Ebner, P. Briza and F. Ferreira (2009). The influence of recombinant production on the immunologic behavior of birch pollen isoallergens. PLoS One. 4(12): e8457.

Wen, J., T. Arakawa and J. S. Philo (1996). Size-exclusion chromatography with on-line light-scattering, absorbance, and refractive index detectors for studying proteins and their interactions. Anal Biochem. 240(2): 155–166.

Werfel, T., R. Asero, B. K. Ballmer-Weber, K. Beyer, E. Enrique, A. C. Knulst, A. Mari, A. Muraro, M. Ollert, L. K. Poulsen, S. Vieths, M. Worm and K. Hoffmann-Sommergruber (2015). Position paper of the EAACI: food allergy due to immunological cross-reactions with common inhalant allergens. Allergy. 70(9): 1079–1090.

Wood, R. A., S. H. Sicherer, A. W. Burks, A. Grishin, A. K. Henning, R. Lindblad, D. Stablein and H. A. Sampson (2013). A phase 1 study of heat/phenol-killed, E. coli-encapsulated, recombinant modified peanut proteins Ara h 1, Ara h 2, and Ara h 3 (EMP-123) for the treatment of peanut allergy. Allergy. 68(6): 803–808.

Zuidmeer-Jongejan, L., H. Huber, I. Swoboda, N. Rigby, S. A. Versteeg, B. M. Jensen, S. Quaak, J. H. Akkerdaas, L. Blom, J. Asturias, C. Bindslev-Jensen, M. L. Bernardi, M. Clausen, R. Ferrara, M. Hauer, J. Heyse, S. Kopp, M. L. Kowalski, A. Lewandowska-Polak, B. Linhart, B. Maderegger, B. Maillere, A. Mari, A. Martinez, E. N. Mills, A. Neubauer, C. Nicoletti, N. G. Papadopoulos, A. Portoles, V. Ranta-Panula, S. Santos-Magadan, H. J. Schnoor, S. T. Sigurdardottir, P. Stahl-Skov, G. Stavroulakis, G. Stegfellner, S. Vazquez-Cortes, M. Witten, F. Stolz, L. K. Poulsen, M. Fernandez-Rivas, R. Valenta and R. van Ree (2015). Development of a hypoallergenic recombinant parvalbumin for first-in-man subcutaneous immunotherapy of fish allergy. Int Arch Allergy Immunol. 166(1): 41–51.

Zuidmeer, L., K. Goldhahn, R. J. Rona, D. Gislason, C. Madsen, C. Summers, E. Sodergren, J. Dahlstrom, T. Lindner, S. T. Sigurdardottir, D. McBride and T. Keil (2008). The prevalence of plant food allergies: a systematic review. J Allergy Clin Immunol. 121(5): 1210–1218 e1214.

12

Peanut Allergy
Biomolecular Characterization for Development of a Peanut T-Cell Epitope Peptide Therapy

Jennifer M. Rolland, Sara R. Prickett and
Robyn E. O'Hehir*

CONTENTS

Department of Allergy, Immunology and Respiratory Medicine, Alfred Hospital and Central
    Clinical School and Department of Immunology and Pathology, Monash University,
    Melbourne, Victoria, Australia.
* Corresponding author: r.ohehir@alfred.org.au

351

12.1 INTRODUCTION

Peanut (*Arachis hypogaea*) allergy is a growing problem globally carrying a huge socioeconomic burden for patients, families and the community. Studies suggest prevalence rates from 1–3% in many westernized countries with peanut allergy typically being a lifelong affliction (Nwaru et al. 2014, Osborne et al. 2011, Sicherer et al. 2003). Although fatalities are fortunately rare, the fear of death is very real for each patient. Currently, there is no cure for peanut allergy with management strategies focusing on complete avoidance and utilization of adrenaline as the emergency antidote for anaphylaxis following inadvertent exposure. There is a very strong imperative for a safe and effective specific therapy for peanut allergy.

Molecular characterization of peanut allergens has facilitated elucidation of the underlying immune response to peanut proteins, which drives the clinical reactivity. This knowledge has paved the way for current active research towards a safe and effective specific peanut allergy immunotherapy, building on a growing body of data and early clinical trials of intradermally administered Synthetic Peptide Immune-Regulatory Epitope (SPIRE) therapies for other allergies (reviewed by Prickett et al. 2015).

This chapter discusses the clinical features of peanut allergy in the context of the mucosal immune response to food allergens and the molecular characterization of peanut allergens. Current knowledge of the cellular immune response to peanut allergens is presented together with how this knowledge is being harnessed for the development of peanut-SPIRE as a potential therapeutic modality for peanut allergy.

12.2 CLINICAL FEATURES OF PEANUT ALLERGY

Peanut allergy generally emerges early in life, typically by the second year after encounter with peanut food products, frequently peanut butter in westernized societies. It is a well-recognized component of the atopic march, occurring usually in individuals with a history of infantile eczema and subsequent emergence of allergic rhinitis with or without asthma. Peanut is the food most commonly responsible for anaphylaxis in children, adolescents and adults (Bock et al. 2007, Burks 2008). Various culprits have been suggested to explain the current epidemic of peanut allergy including reduced microbial encounter through increased hygiene practices ("hygiene hypothesis"), altered digestive tract microbiota (for example from increased antibiotic use), dietary changes towards "fast foods", increased consumption of peanuts in filler foods and inclusion of peanut oils in cosmetics used to manage eczematous skin (Lack 2012, Marrs et al. 2013). *In utero* exposure to peanut proteins has been demonstrated, as has exposure through breast milk but the role of these routes in sensitization remains contentious. However, it is certainly clear that the rise in peanut allergy is too fast for genetics to explain the changing prevalence.

Ingestion of peanuts may trigger allergic symptoms within minutes to several hours with clinical features ranging from mild oropharyngeal irritation through to life threatening anaphylaxis. Symptoms and signs may include nausea, vomiting, mouth itch (pruritus), hives (urticaria), lip, tongue or throat swelling (angioedema), wheezing (bronchospasm), dizziness or collapse due to rapid loss of blood pressure (hypotension) and, infrequently, death. Currently there is no specific therapy with patients advised to practise strict avoidance and, with reaction to inadvertent exposure, to implement their anaphylaxis action plans including prompt self-administration of intramuscular adrenaline with medical follow-up as needed.

12.3 THE MUCOSAL IMMUNE RESPONSE TO PEANUT ALLERGENS

Development of allergen immunotherapy for potent allergens such as peanut requires elucidation of underlying immune events. Encounter of the immune system with peanut allergens typically occurs at the gut mucosal surface and results in sensitization after first exposure in genetically susceptible individuals with symptoms of peanut allergy on re-exposure (de Leon et al. 2007; Figure 12.1). In the gastric mucosa, peanut allergens are engulfed by specialized epithelial M cells and transferred to antigen presenting cells (APC) such as dendritic cells for processing into peptide fragments. As discussed above, genetic and environmental factors play a role in promoting allergen uptake by the gut epithelium and driving the adverse immune response

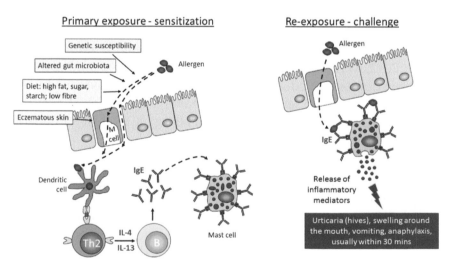

Figure 12.1 Cellular interactions during the mucosal allergic immune response to peanut and factors that promote this response.

in peanut allergic individuals. Peptide-MHC class II complexes are presented to naïve T cells with appropriate T-cell receptor specificity causing their differentiation into T helper 2 (Th2) cells. Activated Th2 cells also recognize peanut allergen peptide-MHC class II complexes on the surface of B cells and through their release of the cytokines interleukin-4 (IL-4) and IL-13, drive B-cell immunoglobulin (Ig) class switching to IgE antibody production. Secreted allergen-specific IgE antibodies bind to high affinity FcεRI receptors on effector cells such as mast cells and basophils (sensitization). Upon re-encountering peanut allergen, cross-linking of cell-bound adjacent IgE molecules activates these cells to release inflammatory mediators such as potent and rapidly acting histamine. These mediators induce the clinical features of peanut allergy. In contrast, non-atopic individuals exhibit a Th1 polarized response to allergen, characterized by predominant secretion of interferon-gamma (IFN-γ) and IL-2 cytokines. An enhanced regulatory T-cell (Treg) response to allergen is also shown by non-allergic subjects.

12.4 ALLERGENIC COMPONENTS OF PEANUT

To date, 17 peanut allergens have been identified (Ara h 1-17; WHO/International Union of Immunological Societies Allergen Nomenclature Subcommittee, http://www.allergen.org/; cited 2015; summarized in Table 12.1). The allergens belong to several different biochemical families, but several are seed storage proteins. Ara h 1 and Ara h 2 are cupin and conglutin family members, respectively and are designated major peanut allergens, each being recognized by serum IgE in > 50% of peanut allergic cohorts tested (Burks et al. 1995, Kleber-Janke 1999). Together these two allergens constitute ~ 25% of total peanut protein. Although Ara h 3 is reported by some as a major allergen, serum IgE recognition has been inconsistent between cohorts with others reporting < 50% responders (e.g., Rabjohn et al. 1999). In contrast, Ara h 6 is classified a minor allergen, but many reports support its reclassification as a major allergen, which may be attributed to the high degree of sequence homology with Ara h 2

Table 12.1 Molecular characteristics of peanut (*Arachis hypogaea*) allergens.*

Peanut Allergen	Biochemical Name/Family	MW (SDS-PAGE)	% IgE Reactivity in Peanut Allergic Subjects
Ara h 1	Cupin (Vicillin-type, 7S globulin)	64 kDa	89% (Burks et al. 1995)
Ara h 2	Conglutin (2S albumin)	17 kDa	85% (Kleber-Janke et al. 1999)
Ara h 3	Cupin (Legumin-type, 11S globulin, Glycinin)	60 kDa, 37 kDa (fragment)	44% (Rabjohn et al. 1999)
Ara h 4	Renamed to Ara h 3.02		53% (Kleber-Janke et al. 1999)
Ara h 5	Profilin	15 kDa	13% (Kleber-Janke et al. 1999)
Ara h 6	Conglutin (2S albumin)	15 kDa	38% (Kleber-Janke et al. 1999)
Ara h 7	Conglutin (2S albumin)	15 kDa	43% (Kleber-Janke et al. 1999)
Ara h 8	Pathogenesis-related protein, PR-10, Bet v 1 family member	17 kDa	85% (Mittag et al. 2004)
Ara h 9	Nonspecific lipid-transfer protein type 1	9.8 kDa	<50% (Krause et al. 2009)
Ara h 10	16 kDa oleosin	16 kDa	n/a
Ara h 11	14 kDa oleosin	14 kDa	n/a
Ara h 12	Defensin	8 kDa (reducing), 12 kDa (non-reducing), 5.184 kDa (mass)	n/a
Ara h 13	Defensin	8 kDa (reducing), 11 kDa (non-reducing), 5.472 kDa (mass)	n/a
Ara h 14	Oleosin	17.5 kDa	46% (www.allergen.org)
Ara h 15	Oleosin	17 kDa	46% (www.allergen.org)
Ara h 16	Non-specific lipid transfer protein 2	8.5 kDa by SDS PAGE reducing	16% (www.allergen.org)
Ara h 17	Non-specific lipid transfer protein 1	11 kDa by SDS-PAGE reducing	16% (www.allergen.org)

*From WHO/IUIS Allergen Nomenclature Sub-Committee data base: www.allergen.org
n/a = not available

(Kleber-Janke et al. 1999). Ara h 8 is a major allergen in patients with combined birch pollen and peanut allergy as it shows homology to Bet v 1; both are members of the pathogenesis-related protein (PR10) family (Mittag et al. 2004). Ara h 1 is the most abundant major peanut allergen comprising 12–16% of total peanut protein (Koppelman et al. 2001, de Leon et al. 2007) while several studies have confirmed that Ara h 2 is the most potent peanut allergen (Blanc et al. 2009, Koppelman et al. 2004, Palmer et al. 2005). There is consensus that both Ara h 1 and Ara h 2 are important in clinical peanut allergy and would need to be considered in a novel therapy. The two allergens have been cloned and sequenced (Burks et al. 1995, Chatel et al. 2003) facilitating molecular characterization and design of a composite therapeutic.

12.5 BIOCHEMICAL PROPERTIES OF PEANUT ALLERGENS

It has been suggested that peanut allergens possess certain physicochemical properties that enhance resistance to acidic digestive enzymes and cooking, thus augmenting their ability to reach intestinal villi and be presented to the immune system. It is known that Ara h 1 can form stable dimers, trimers and larger complexes with heating while retaining IgE reactivity (Koppelman et al. 1999, Maleki et al. 2000). Large proteolytic fragments with retained IgE binding affinity are formed by the action of the gastrointestinal enzymes pepsin, trypsin and chymotrypsin on Ara h 1 (Maleki et al. 2000). The resistance of Ara h 1 to degradation in this manner may be attributable to its stable, homotrimeric structure. The monomer-monomer interaction decreases access to the catalytic sites within the protein, allowing Ara h 1 to survive intact during food processing or passage along the gastrointestinal tract.

Ara h 2 and Ara h 6 have also been shown to retain allergenicity following thermal treatment and proteolytic digestion (Suhr et al. 2004, Lehmann et al. 2006). Following digestion, structural studies have identified immunologically active core structures within both allergens that induced inflammatory cell mediator release despite decreased serum IgE binding (Lehmann et al. 2006). It has been shown that Ara h 2 has sequence homology with trypsin inhibitors and can therefore itself act as a weak trypsin inhibitor. Furthermore, this activity of Ara h 2 is enhanced by roasting, which protects Ara h 2, and also Ara h 1, from trypsin breakdown; roasted peanuts are known to be more allergenic than boiled peanuts (Maleki et al. 2003). In contrast, Ara h 8 has lower resistance to thermal and gastric digestion (Mittag et al. 2004).

Another reported contributing factor to peanut allergenicity is the Maillard reaction (Chung and Champagne 1999, Maleki et al. 2000, Chung et al. 2002). During cooking a non-enzymatic reaction occurs between proteins and reducing sugars (Namiki 1988). Glycosylated amino groups of proteins form Amadori products that degrade into dicarbonyl intermediates and subsequently react with amino groups of other proteins to form stable advanced glycation end products (AGE). It has been shown that AGE formed after heating a previously non-allergenic peanut lectin protein in the presence of sugars, could inhibit binding of serum IgE to untreated peanut allergens (Chung and Champagne 1999) suggesting that the Maillard reaction could change a non-allergenic protein into a potential allergen. Independently, roasted peanut proteins inhibited IgE binding to raw peanut proteins 90-fold more effectively than raw peanut proteins (Maleki et al. 2000). These studies suggest that the formation of AGE after heat-treatment of peanuts could be significant contributors to their enhanced allergenicity.

12.6 SPECIFIC IMMUNOTHERAPY FOR PEANUT ALLERGY

Conventional allergen immunotherapy for peanut allergy using unfractionated peanut extracts is an area of active clinical research but is not currently recommended in clinical practice due to the unacceptably high risk of anaphylaxis (Anagnostou 2015). Several groups are exploring oral (including sublingual) immunotherapy regimens (Jones et al. 2009, 2014, Blumchen et al. 2010, Varshney et al. 2011, Syed et al. 2014, Vickery et al. 2014) but it is generally considered that the high frequency of adverse reactions observed is concerning and unpredictable. More recently epicutaneous peanut patches have been explored to achieve clinical desensitization with less severe side effects (reviewed by Jones et al. 2014). However, in all these trials, despite daily dosing for months or years, efficacy has been variable and unpredictable. Even when effective, protection is only sustained with ongoing daily peanut administration (via patches or consumption) posing a substantial compliance challenge. Insights into the immune response to peanut allergens and factors that influence clinical outcome as well as detailed characterization of the allergenic components of peanut as described above, point to the development of a rational new strategy for a safe and effective specific immunotherapy for peanut allergy. This is based on targeting the allergen specific T cell, which by its cytokine profile, whether polarized towards Th2 or Th1/Treg, plays a pivotal role in determining clinical outcome on peanut encounter (Rolland et al. 2013). For this, short T-cell epitope containing peptides, recently designated as SPIRE therapy, can be used to target these T cells without engaging cell-bound allergen specific IgE (Prickett et al. 2015).

12.7 DEVELOPMENT OF A SPIRE THERAPY

12.7.1 Rationale for SPIRE Therapy

T-cell epitopes are short linear amino acid sequences containing key residues for anchoring the peptide into pockets of the presenting MHC class II molecule as well as residues, which engage with sites on the T-cell receptor $V\alpha$ and $V\beta$ loops. In contrast, B-cell epitopes are usually conformational, requiring native structure to bring discontinuous segments of the allergen molecule together to engage antibody binding sites. In the case of peanut allergens, linear IgE epitopes have been described (Burks et al. 1997, Stanley et al. 1997, Bøgh et al. 2012) but by sequence comparison and careful pretesting, IgE reactivity of a T-cell targeting peptide can be avoided. Regardless, the peptides comprising minimal T-cell epitope sequences are too small to cross-link surface bound IgE on mast cells and basophils, and therefore cannot invoke release of inflammatory mediators as do extracts containing intact allergenic components. This is a seminal requirement for a safe therapeutic for peanut allergy.

Further promoting their efficacy, T-cell epitope peptides can load directly on to HLA class II molecules on the surface of APC and hence be presented at higher frequency than peptides processed from the whole molecule by APC. This also allows peptides to be presented by non-professional or immature APC, including human T cells, without activating pro-inflammatory and co-stimulatory signals (in contrast to whole allergen extracts). The net effect is to promote the induction of immunological tolerance (frequently called anergy), apoptosis and/or suppressive activity in responding T cells, a property believed to be pivotal in achieving successful outcome during SPIRE treatment. Importantly, findings to date indicate that T-cell epitope peptides used in SPIRE therapy do not themselves elicit antibody production with the consequent risk of immune complex formation. In addition to their potential for increased efficacy and safety, peptides are an attractive alternative to whole extracts due to their ease of standardization, cost-effective production in large quantities at high purity and consistency, stability in lyophilized form and ease of modification to achieve desired chemical and biological properties.

It has been demonstrated directly that using peptides to target T cells specific for dominant epitopes of major allergens can alter responses to whole allergen extracts, known as linked epitope suppression. In early *in vitro* and murine studies, O'Hehir and colleagues showed that the dominant T-cell epitope peptide of Der p 1, the major allergen of house dust mite (HDM), could induce tolerance not only to this peptide, but to the whole Der p 1 allergen and HDM extract (Higgins et al. 1992, Hoyne et al. 1993). This phenomenon has been validated subsequently in different murine models of allergy, and also in human studies of SPIRE therapy for cat allergy (Briner et al. 1993, Couroux et al. 2015, Patel et al. 2013, Worm et al. 2013). Clinical administration of dominant Fel d 1 (major cat allergen) T-cell epitope peptides altered T-cell responses to those peptides, other non-related Fel d 1 peptides, and whole cat allergen extract.

12.7.2 Validation of Allergen SPIRE Therapeutics in Clinical Trials

Placebo-controlled phase IIb trials using SPIRE therapy for cat allergy, and more recently, grass, ragweed and HDM allergy (Circassia Ltd; www.circassia.co.uk.) have confirmed safety and efficacy of SPIRE immunotherapy. Each of the vaccines consists of a mixture of seven peptides <20 residues in length which encompass CD4+ T-cell epitopes of clinically relevant major allergens. Pilot studies in cat allergy indicated that four intradermal injections of 6 nmol/peptide (optimal dose dependent on the allergen) given at four week intervals were most effective at reducing symptoms following subsequent allergen challenge in an environmental exposure chamber, with effects lasting at least two years (Patel et al. 2013, Worm et al. 2013). Phase 3 trial of cat peptides (EudraCT Number:2012-001733-13) confirmed safety and tolerability but failed to demonstrate clinical efficacy over placebo. Active treatment improved symptom and medication scores by ~60% but a similar placebo effect was observed. During early studies with higher dose cat peptides, late asthmatic reactions occurred following treatment in some patients. The observed bronchospasm at higher peptide doses is consistent with the early release of IL-4 (a known bronchoconstrictor) from peptide-tolerized clonal T cells first reported by O'Hehir (O'Hehir et al. 1996) and subsequently confirmed by others (Hoyne et al. 1996). Refined protocols with shorter peptides (<20 aa) and lower peptide doses eliminated these reactions in subsequent trials. In summary, SPIRE immunotherapy has improved clinical outcomes in a range of aeroallergies and offers considerable promise for other allergies.

12.7.3 Mechanisms of Action of Allergen SPIRE Therapy

As clinical translation of SPIRE therapy for allergy progresses, underlying immunological mechanisms are becoming evident

(Larche 2014, Larche and Wraith 2005). Distinct differences are apparent from conventional allergen immunotherapy via the subcutaneous or sublingual routes (Bohle et al. 2007, O'Hehir et al. 2007, Sandrini et al. 2015), particularly regarding the contribution of IgG antibodies. In contrast to conventional immunotherapy, evidence for a role for specific IgG_4 blocking antibody in successful peptide immunotherapy is not yet forthcoming. As the peptides used for SPIRE therapy are short and screened for lack of inflammatory cell activating potential, they are unlikely to induce antibody production. Further studies of SPIRE therapy are needed to clarify whether allergen-specific IgG_4 antibodies are induced and play a role in long-term clinical efficacy. High-level changes in allergen-specific T cells appear more consistent between conventional and peptide immunotherapy approaches, although the underlying mechanisms and/or their kinetics are thought to vary. Clinical success with peptide therapy seems to be associated with the induction of anergy, and/or apoptosis of allergen-specific Th2-type CD4$^+$ T cells and induction of IL-10-mediated suppressor mechanisms.

Decreased T-cell proliferative and cytokine responses to allergen are consistent observations following peptide therapy (e.g., Oldfield et al. 2002, Smith et al. 2004). In a T-cell peptide clinical study using bee venom, decreased PLA2-induced T-cell proliferation and IL-2, IL-4, IL-5, IL-13 and IFN-γ production were reversed by IL-2 and IL-15, consistent with anergy as the underlying mechanism (Muller et al. 1998). Deletion of allergen-specific Th2 cells has also been suggested from murine models of peptide-induced tolerance. Clinical HLA class II tetrameric studies are required to quantify allergen-specific clonal T-cell populations. Using this technology for conventional subcutaneous immunotherapy for Timothy grass pollen allergy, preferential loss of clonal Th2-type T cells specific for dominant epitopes of major Timothy grass pollen allergens over T cells specific for less-dominant epitopes with a Th1 or Tr1-phenotype was observed (Wambre et al. 2014). However, an important potential confounder of tetramer-based approaches is the reliance on TCR expression on the cell surface, making it difficult to distinguish deletion from anergy given that anergic T cells have decreased TCR expression. In this study the pathogenic Th2 cells were further identified by lack of CD27 expression providing another marker to support selective loss of these cells. There was also associated decreased expression of the apoptotic inhibitor Bcl-2 over the cells that escaped deletion. These data support the view that dominant T-cell epitope-based peptides of major allergens could cause targeted inactivation or deletion of the most pathogenic T cells in allergic subjects.

Pilot studies of cat and bee venom peptide immunotherapy showed increased IL-10 production during therapy suggesting a role for Treg (Oldfield et al. 2002, Tarzi et al. 2006). Further investigations indicated a requirement for IL-10 for cat peptide-induced suppression of allergen-specific immune responses and linked epitope suppression (Campbell et al. 2009), and induction of an antigen-specific CD4$^+$ T-cell population with regulatory function was evident (Verhoef et al. 2005). Skin biopsies from allergen challenge sites showed an increased number of CD4$^+$IFN-γ^+ and CD25$^+$ cells after peptide therapy consistent with immune deviation and Treg induction (Alexander et al. 2005). Further functional analyses and phenotyping of T cells from peripheral blood and tissues are required to delineate activated CD4$^+$ T cells from natural or induced Treg (Rolland et al. 2010).

12.8 DESIGN OF A SPIRE THERAPEUTIC FOR PEANUT ALLERGY

The above described promising clinical trials of SPIRE therapies for aeroallergens and the molecular and immunochemical characterization of clinically significant peanut allergens, pave the way for utilisation of this new therapeutic class for potent allergens such as peanut. The following sections describe the strategy for identifying T-cell reactive but non-IgE reactive short peptides suitable for a safe and effective peanut allergy therapy. An overview of key steps in this strategy is given in Figure 12.2.

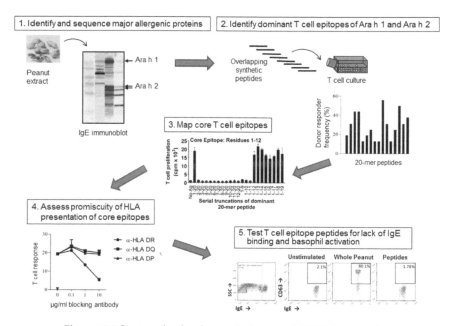

Figure 12.2 Strategy for development of a peanut SPIRE therapeutic.

12.8.1 Mapping T-Cell Epitopes of Major Peanut Allergens

Detailed mapping of the dominant T-cell epitopes of the major peanut allergens Ara h 1 and Ara h 2 has been reported (Glaspole 2005, Prickett et al. 2011, 2013). For this, Ara h 1- or Ara h 2-specific T cells were selected by stimulating PBMC from different peanut allergic donors with the respective major allergen for seven days then isolating the reactive (dividing) CD4+ T cells. The Ara h 1- or 2-reactive T cells were expanded, then tested for specificity to overlapping 20-mer peptides (11 aa overlap) spanning the entire allergen sequences. T-cell reactive 20-mers were identified and the dominant 20-mers selected based on donor and T-cell line responder frequency, magnitude of T-cell response, and patterns of donor recognition. Subsequently, the minimum peptide sequence(s) required to stimulate T cells (core T-cell epitope) was determined within each dominant 20-mer by stimulating 20-mer specific T cells with sets of peptides progressively truncated from the N- or C-terminus of the 20-mer.

12.8.2 Determination of HLA-II Molecules which Present Peptides to T Cells

There is no known HLA-association with peanut allergy. An important consideration when selecting peptides for immunotherapy is whether they can be presented by different HLA class II molecules and therefore be suitable for treating a genetically diverse human population. Consistent with T-cell epitopes of other major allergens, the dominant T-cell epitopes of Ara h 1 and 2 demonstrate strong and degenerate HLA-binding (Prickett et al. 2011, 2013). The HLA-restriction of T-cell recognition of each epitope was assessed in different donors using HLA blocking antibodies and HLA-genotyping and showed that each epitope could be presented on two or more different HLA-molecules. The epitopes were collectively presented on a combination of HLA-DR, HLA-DQ and HLA-DP molecules. Inclusion of HLA-DQ and/or HLA-DP restricted T-cell epitopes is particularly advantageous for a T-cell targeted therapeutic since these HLA-molecules tend to be more conserved in mixed populations than HLA-DR molecules, enabling broader population coverage with fewer T-cell epitope sequences. DeLong et al. (2011) used a tetramer guided epitope mapping approach to identify a panel of Ara h 1 T-cell epitopes and similarly demonstrated presentation of the HLA-DR-restricted epitopes by multiple HLA-DR molecules.

12.8.3 Refinement of Peptides for Ease of Production and Solubility, Confirmation of T-Cell Reactivity and Lack of IgE-mediated Basophil Activation

Having identified the dominant T-cell epitopes of the major peanut allergens and demonstrated broad presentation by HLA class II molecules, a candidate peptide set can be selected for a peanut SPIRE therapeutic. Where possible, epitopes with sequence overlap should be combined into a single peptide no more than 20 residues long to minimize the number of peptides in the final therapeutic set. Since cysteine residues favour disulphide bridging, peptides containing cysteines can be problematic for peptide stability and biological reactivity, so cysteine residues can be substituted with structurally conserved but non-reactive serine residues (Prickett et al. 2011). Further amino acid modifications can be performed to enhance solubility and stability (Prickett et al. 2015) but all such peptide variants necd to be retested for T-cell reactivity. Once a candidate peptide panel is identified, this is validated for PBMC T-cell reactivity using a new cohort of peanut-allergic subjects to confirm comparable response with whole peanut extract. Importantly before clinical trial, the candidate peptides must be tested singly and in combination over a dose range for basophil reactivity in order to confirm lack of potential to elicit an acute IgE-mediated reaction. The basophil activation test by flow cytometry or histamine release using blood samples from peanut allergic donors is a convenient and reliable assay for clinically relevant, functional IgE reactivity (Prickett et al. 2013, Worm et al. 2011, Santos et al. 2014). Notably, all potential vaccine recipients should be screened with a basophil activation test and/or skin prick test prior to clinical testing, and any responders excluded from receiving the vaccine.

12.9 CONCLUSIONS

The molecular characterization of allergens has greatly facilitated the development of specific therapeutics for allergy. However, an optimal preparation suitable for the prevention or treatment of peanut allergy is currently lacking due to high risk of adverse reactions. Harnessing vital data from the molecular characterization of peanut allergens and following the success in early phase clinical trials of novel SPIRE therapies in cat, HDM and grass pollen allergy, the scene is set to progress a peanut-SPIRE therapeutic into first-in-human testing. Detailed selection and immunological characterization of dominant peanut T-cell epitope peptides together with the ability to check these for lack of functional IgE reactivity before administering to patients provides encouragement that a safe and effective therapy for peanut allergy is imminent.

ACKNOWLEDGEMENTS

Research by the authors described in this chapter was funded by the National Health and Medical Research Council Australia, the Ilhan Food Allergy Foundation and the Alfred Trusts, Australia. The authors declare that they are founders and shareholders of Aravax Pty. Ltd., a company established for development and commercialization of a peanut-SPIRE therapy.

Keywords: Peanut allergy; T-cell epitopes; specific immunotherapy; peptide immuno therapy; SPIRE therapy; food allergy

REFERENCES

Anagnostou, K. (2015). Recent advances in immunotherapy and vaccine development for peanut allergy. Ther Adv Vaccines. 3: 55–65.

Alexander, C., M. Tarzi, M. Larche and A.B. Kay (2005). The effect of Fel d 1-derived T-cell peptides on upper and lower airway outcome measurements in cat-allergic subjects. Allergy. 60: 1269–1274.

Blanc, F., K. Adel-Patient, M. F. Drumare, E. Paty, J. M. Wal and H. Bernard (2009). Capacity of purified peanut allergens to induce degranulation in a functional *in vitro* assay: Ara h 2 and Ara h 6 are the most efficient elicitors. Clin Exp Allergy. 39: 1277–1285.

Blumchen, K., H. Ulbricht, U. Staden, K. Dobberstein, J. Beschorner, L. C. de Oliveira et al. (2010). Oral peanut immunotherapy in children with peanut anaphylaxis. J Allergy Clin Immunol. 126: 83–91.

Bock, S. A., A. Munoz-Furlong and H. A. Sampson (2007). Further fatalities caused by anaphylactic reactions to food, 2001–2006. J Allergy Clin Immunol. 119: 1016–1018.

Bøgh, K. L., H. Nielsen, C. B. Madsen, E. N. Mills, N. M. Rigby, T. Eiwegger et al. (2012). IgE epitopes of intact and digested Ara h 1: A comparative study in humans and rats. Mol Immunol. 51: 337–346.

Bohle, B., T. Kinaciyan, M. Gerstmayr, A. Radakovics, B. Jahn-Schmid and C. Ebner (2007). Sublingual immunotherapy induces IL-10-producing T regulatory cells, allergen-specific T-cell tolerance, and immune deviation. J Allergy Clin Immunol. 120: 707–713.

Briner, T. J., M. C. Kuo, K. M. Keating, B. L. Rogers and J. L. Greenstein (1993). Peripheral T-cell tolerance induced in naïve and primed mice by subcutaneous injection of peptides from the major cat allergen Fel d 1. Proc Natl Acad Sci USA. 90: 7608–7612.

Burks, A. W. (2008). Peanut allergy. Lancet. 371: 1538–1546.

Burks, A. W., G. Cockrell, J. S. Stanley, R. M. Helm and G. A. Bannon (1995). Recombinant peanut allergen Ara h I expression and IgE binding in patients with peanut hypersensitivity. J Clin Invest. 96: 1715–1721.

Burks, A. W., D. Shin, G. Cockrell, J. S. Stanley, R. M. Helm and G. A. Bannon (1997). Mapping and mutational analysis of the IgE-binding epitopes on Ara h 1, a legume vicilin protein and a major allergen in peanut hypersensitivity. Eur J Biochem. 245: 334–339.

Burks, A. W., L. W. Williams, C. Connaughton, G. Cockrell, T. J. O'Brien and R. M. Helm (1992). Identification and characterization of a second peanut major peanut allergen Ara h II, with use of the sera of patients with atopic dermatitis and positive peanut challenge. J Allergy Clin Immunol. 90: 962–969.

Campbell, J. D., K. F. Buckland, S. J. McMillan, J. Kearley, W. L. Oldfield, L. J. Stern et al. (2009). Peptide immunotherapy in allergic asthma generates IL-10-dependent immunological tolerance associated with linked epitope-suppression. J Exp Med. 206: 1535–1547.

Chatel, J. M., H. Bernard and F. M. Orson (2003). Isolation and characterization of two complete Ara h 2 isoforms cDNA. Int Arch Allergy Immunol. 131: 14–18.

Chung, S. Y. and E. T. Champagne (1999). Allergenicity of Maillard reaction products from peanut proteins. J Agric Food Chem. 47: 5227–5231.

Chung, S. Y., S. Maleki, E. T. Champagne, K. L. Buhr and D. W. Gorbet (2002). High-oleic peanuts are not different from normal peanuts in allergenic properties. J Agric Food Chem. 50: 878–882.

Couroux, P., D. Patel, K. Armstrong, M. Larche and R. P. Hafner (2015). Fel d 1-derived synthetic peptide immune-regulatory epitopes show a long-term treatment effect in cat allergic subjects. Clin Exp Allergy. 45: 974–981.

de Leon, M. P., J. M. Rolland and R. E. O'Hehir (2007). The peanut allergy epidemic: allergen molecular characterisation and prospects for therapy. Expert Rev Mol Med. 9: 1–18.

DeLong, J. H., K. H. Simpson , E. Wambre, E. A. James, D. Robinson and W. W. Kwok (2011). Ara h 1-reactive T cells in individuals with peanut allergy. J Allergy Clin Immunol. 127: 1211–1218.

Glaspole, I. N., M. P. de Leon, J. M. Rolland and R. E. O'Hehir (2005). Characterization of the T-cell epitopes of a major peanut allergen, Ara h 2. Allergy. 60: 35–40.

Higgins, J. A., J. R. Lamb, S. G. E. Marsh, S. Tonks, J. Hayball, S. Rosen-Bronson et al. (1992). Peptide induced non-responsiveness to HLA-DP restricted human T cells reactive with *Dermatophagoides* spp. (house dust mite). J Allergy Clin Immunol. 90: 749–756.

Hoyne, G. F., B. A. Askonas, C. Hetzel, W. R. Thomas and J. R. Lamb (1996). Regulation of house dust mite responses by intranasally administered peptide: transient activation of CD4+ T cells precedes the development of tolerance *in vivo*. Int Immunol. 8: 335–342.

Hoyne, G. F., R. E. O'Hehir, D. C. Wraith, W. R. Thomas and J. R. Lamb (1993). Inhibition of T cell and antibody responses to house dust mite allergen by inhalation of the dominant T cell epitope in naive and sensitised mice. J Exp Med. 178: 1783–1788.

Jones, S. M., A. W. Burks and C. Dupont (2014). State of the art on food allergen immunotherapy: oral, sublingual, and epicutaneous. J Allergy Clin Immunol. 133: 318–323.

Jones, S. M., L. Pons, J. L. Roberts, A. M. Scurlock, T. T. Perry, M. Kulis et al. (2009). Clinical efficacy and immune regulation with peanut and oral immunotherapy. J Allergy Clin Immunol. 24: 292–300.

Kleber-Janke, T., R. Crameri, U. Appenzeller, M. Schlaak and W. M. Becker (1999). Selective cloning of peanut allergens, including profilin and 2S albumins, by phage display technology. Int Arch Allergy Immunol. 119: 265–274.

Koppelman, S. J., C. A. Bruijnzeel-Koomen, M. Hessing and H. H. de Jongh (1999). Heat-induced conformational changes of Ara h 1, a major peanut allergen, do not affect its allergenic properties. J Biol Chem. 274: 4770–4777.

Koppelman, S. J., R. A. Vlooswijk, L. M. Knippels, M. Hessing, E. F. Knol, F. C. van Reijsen et al. (2001). Quantification of major peanut allergens Ara h 1 and Ara h 2 in the peanut varieties Runner, Spanish, Virginia, and Valencia, bred in different parts of the world. Allergy. 56: 132–137.

Koppelman, S. J., M. Wensing, M. Ertmann, A. C. Knulst and E. F. Knol (2004). Relevance of Ara h 1, Ara h 2 and Ara h 3 in peanut-allergic patients, as determined by immunoglobulin E Western blotting, basophil-histamine release and intracutaneous testing; Ara h 2 is the most important peanut allergen. Clin Exp Allergy. 34: 583–590.

Krause, S., G. Reese, S. Randow, D. Zennaro, D. Quaratino, P. Palazzo et al. (2009). Lipid transfer protein (Ara h 9) as a new peanut allergen relevant for a Mediterranean allergic population. J Allergy Clin Immunol. 124: 771–778.

Lack, G. (2012). Update on risk factors for food allergy J Allergy Clin Immunol. 129: 1187–1197.

Larche, M. (2014). Mechanisms of peptide immunotherapy in allergic airways disease. Ann Am Thorac Soc 11(Suppl. 5): S292–296.

Larche, M. and D. C. Wraith (2005). Peptide-based therapeutic vaccines for allergic and autoimmune diseases. Nat Med. 11: S69–76.

Lehmann, K., K. Schweimer, G. Reese, S. Randow, M. Suhr, W. M. Becker et al. (2006). Structure and stability of 2S albumin-type peanut allergens: implications for the severity of peanut allergic reactions. Biochem J. 395: 463–472.

Maleki, S. J., S. Y. Chung, E. T. Champagne and J. P. Raufman (2000). The effects of roasting on the allergenic properties of peanut proteins. J Allergy Clin Immunol. 106: 763–768.

Maleki, S. J., R. A. Kopper, D. S. Shin, C. W. Park, C. M. Compadre, H. Sampson et al. (2000). Structure of the major peanut allergen Ara h 1 may protect IgE-binding epitopes from degradation. J Immunol. 164: 5844–5849.

Maleki, S. J., O. Viquez, T. Jacks, H. Dodo, E. T. Champagne, S. Y. Chung et al. (2003). The major peanut allergen, Ara h 2, functions as a trypsin inhibitor, and roasting enhances this function. J Allergy Clin Immunol. 112: 190–195.

Marrs, T., K. D. Bruce, K. Logan, D. W. Rivett, M. R. Perkin, G. Lack et al. (2013). Is there an association between microbial exposure and food allergy? A systematic review. Pediatr Allergy Immunol. 24: 311–320.

Mittag, D., J. Akkerdaas, B. K. Ballmer-Weber, L. Vogel, M. Wensing, W. M. Becker et al. (2004). Ara h 8, a Bet v 1-homologous allergen from peanut, is a major allergen in patients with combined birch pollen and peanut allergy. J Allergy Clin Immunol. 114: 1410–1417.

Muller, U., C. A. Akdis, M. Fricker, M. Akdis, T. Blesken, F. Bettens et al. (1998). Successful immunotherapy with T-cell epitope peptides of bee venom phospholipase A2 induces specific T-cell anergy in patients allergic to bee venom. J Allergy Clin Immunol. 101: 747–754.

Namiki, M. (1988). Chemistry of Maillard reactions: recent studies on the browning reaction mechanism and the development of antioxidants and mutagens. Adv Food Res. 32: 115–184.

Nwaru, B. I., L. Hickstein, S. S. Panesar, A. Murano, T. Werfel, V. Cardona et al. (2014). The epidemiology of food allergy in Europe: a systematic review and meta-analysis. Allergy. 69: 62–75.

O'Hehir, R. E., R. A. Lake, T. J. Schall, H. Yssel, E. Panagiotopoulou and J. R. Lamb (1996). Regulation of cytokine and chemokine transcription in a human TH2 type T cell clone during the induction phase of anergy. Clin Exp Allergy. 26: 20–27.

O'Hehir, R. E., A. Sandrini, G. P. Anderson and J. M. Rolland (2007). Sublingual allergen immunotherapy: Immunological mechanisms and prospects for refined vaccine preparations. Curr Med Chem. 14: 2235–2244.

Oldfield, W. L., M. Larche and A. B. Kay (2002). Effect of T-cell peptides derived from Fel d 1 on allergic reactions and cytokine production in patients sensitive to cats: a randomised controlled trial. Lancet. 360: 47–53.

Osborne, N. J., J. J. Koplin, P. E. Martin, L. C. Gurrin, A. J. Lowe, M. C. Matheson et al. (2011). Prevalence of challenge-proven IgE-mediated food allergy using population based sampling and predetermined challenge criteria in infants. J Allergy Clin Immunol. 127: 668–676.

Palmer, G. W., D. A. Dibbern Jr, A. W. Burks, G. A. Bannon, S. A. Bock, H. S. Porterfield et al. (2005). Comparative potency of Ara h 1 and Ara h 2 in immunochemical and functional assays of allergenicity. Clin Immunol. 115: 302–312.

Patel, D., P. Couroux, P. Hickey, A. M. Salapatek, P. Laidler et al. (2013). Fel d 1-derived peptide antigen desensitization shows a persistent treatment effect 1 year after the start of dosing: a randomized, placebo-controlled study. J Allergy Clin Immunol. 131: 103–109.

Prickett, S. R., J. M. Rolland and R. E. O'Hehir (2015). Immunoregulatory T cell epitope peptides: the new frontier in allergy therapy. Clin Exp Allergy. 45: 1015–1026.

Prickett, S. R., A. L. Voskamp, A. Dacumos, K. Symons, J. M. Rolland and R. E. O'Hehir (2011). Ara h 2 peptides comprising dominant CD4⁺ T-cell epitopes: candidates for a peanut allergy therapeutic. J Allergy Clin Immunol. 127: 608–615.

Prickett, S., A. Voskamp, T. Phan, A. Dacumos-Hill, S. Mannering, J. M. Rolland et al. (2013). Ara h 1 CD4+ T-cell epitope-based peptides: candidates for a peanut allergy therapeutic. Clin Exp Allergy. 43: 684–697.

Rabjohn, P., E. M. Helm, J. S. Stanley, C. M. West, H. A. Sampson, A. W. Burks et al. (1999). Molecular cloning and epitope analysis of the peanut allergen Ara h 3. J Clin Invest. 103: 535–542.

Rolland, J. M., L. M. Gardner and R. E. O'Hehir (2010). Functional regulatory T cells and allergen immunotherapy. Curr Opin Allergy Clin Immunol. 10: 559–566.

Rolland, J. M., S. Prickett, L. M. Gardner and R. E. O'Hehir (2013). T cell targeted strategies for improved efficacy and safety of specific immunotherapy for allergic disease. Antiinflamm. Antiallergy Agents Med Chem. 12: 201–22.

Sandrini, A., J. M. Rolland and R. E. O'Hehir (2015). Current developments for improving efficacy of allergy vaccines. Expert Rev Vaccines. 14: 1073–1087.

Santos, A. F., A. Douiri, N. Becares, S. Y. Wu, A. Stephens, S. Radulovic et al. (2014). Basophil activation test discriminates between allergy and tolerance in peanut-sensitized children. J Allergy Clin Immunol. 134: 645–652.

Sicherer, S. H., A. Munoz-Furlong, J. H. Godbold and H. A. Sampson (2003). US prevalence of self-reported peanut, tree nut and sesame allergy: 11-year follow-up. J Allergy Clin Immunol. 125: 1322–1326.

Smith, T. R., C. Alexander, A. B. Kay, M. Larche and D. S. Robinson (2004). Cat allergen peptide immunotherapy reduces CD4(+) T cell responses to cat allergen but does not alter suppression by CD4(+) CD25(+) T cells: a double-blind placebo-controlled study. Allergy. 59: 1097–1101.

Stanley, J. S., N. King, A. W. Burks, S. K. Huang, H. Sampson, G. Cockrell et al. (1997). Identification and mutational analysis of the immunodominant IgE binding epitopes of the major peanut allergen Ara h 2. Arch Biochem Biophys. 342: 244–253.

Suhr, M., D. Wicklein, U. Lepp and W. M. Becker (2004). Isolation and characterization of natural Ara h 6: evidence for a further peanut allergen with putative clinical relevance based on resistance to pepsin digestion and heat. Mol Nutr Food Res. 48: 390–399.

Syed, A., M. A. Garcia, S. C. Lyu, R. Bucayu, A. Kohli, S. Ishida et al. (2014). Peanut oral immunotherapy results in increased antigen-induced regulatory T-cell function and hypomethylation of Forkhead box protein 3 (Foxp3). J Allergy Clin Immunol. 133: 500–510.

Tarzi, M., S. Klunker, C. Texier, A. Verhoef, S. O. Stapel, C. A. Akdis et al. (2006). Induction of interleukin-10 and suppressor of cytokine signalling-3 gene expression following peptide immunotherapy. Clin Exp Allergy. 36: 465–474.

Varshney, P., S. M. Jones, A. M. Scurlock, T. T. Perry, A. Kemper, P. Steele et al. (2011). A randomized controlled study of peanut oral immunotherapy: clinical desensitization and modulation of the allergic response. J Allergy Clin Immunol. 127: 654–660.

Verhoef, A., C. Alexander, A. B. Kay and M. Larche (2005). T cell epitope immunotherapy induces a CD4+ T cell population with regulatory activity. PLoS Med. 2: e78.

Vickery, B. P., A. M. Scurlock, M. Kulis, P. H. Steele, J. Kamilaris, J. P. Berglund et al. (2014). Sustained unresponsiveness to peanut in subjects who have completed peanut oral immunotherapy. J Allergy Clin Immunol. 133: 468–475.

Viquez, O. M., C. G. Summer and H. W. Dodo (2001). Isolation and molecular characterization of the first genomic clone of a major peanut allergen, Ara h 2. J Allergy Clin Immunol. 107: 713–717.

Wambre, E., J. H. DeLon, E. A. James, N. Torres-Chin, W. Pfutzner, C. Mobs et al. (2014). Specific immunotherapy modifies allergen-specific CD4(+) T-cell responses in an epitope-dependent manner. J Allergy Clin Immunol. 133: 872–879.

Worm, M., H. H. Lee, J. Kleine-Tebbe, R. P. Hafner, P. Laidler, D. Healey et al. (2011). Development and preliminary clinical evaluation of a peptide immunotherapy vaccine for cat allergy. J Allergy Clin Immunol. 127: 89–97.

Worm, M., D. Patel and P. S. Creticos (2013). Cat peptide antigen desensitization for treating cat allergic rhinoconjunctivitis. Expert Opin Investig Drugs. 22: 1347–1357.

Index

T - #0356 - 071024 - C392 - 234/156/17 - PB - 9780367781996 - Gloss Lamination